TECHNICAL WRITING
A Practical Approach

Second Edition

TECHNICAL WRITING
A Practical Approach

William S. Pfeiffer
Southern College of Technology

Merrill, an imprint of
Macmillan Publishing Company
New York

Maxwell Macmillan Canada
Toronto

Maxwell Macmillan International
New York Oxford Singapore Sydney

Editor: Steven Helba
Developmental Editor: Monica Ohlinger
Production Editor: Mary Irvin
Art Coordinator: Ruth A. Kimpel
Photo Editor: Anne Vega
Text Designer: Anne Flanagan
Cover Designer: Russ Maselli
Production Buyer: Pamela Bennett

This book was set in Meridian by Macmillan Publishing Company and was printed and bound by R.R. Donnelley & Sons. The cover was printed by Phoenix Color Corp.

The Publisher offers discounts on this book when ordered in bulk quantities. For more information, write to: Special Sales Department, Macmillan Publishing Company, 445 Hutchinson Avenue, Columbus, OH 43235, or call 1-800-228-7854

Macmillan Publishing Company
866 Third Avenue
New York, NY 10022

Macmillan Publishing Company is part of the
Maxwell Communication Group of Companies.

Maxwell Macmillan Canada, Inc.
1200 Eglinton Avenue East, Suite 200
Don Mills, Ontario M3C 3N1

Library of Congress Cataloging-in Publication Data
Pfeiffer, William S.
    Technical writing: a practical approach/William S. Pfeiffer. —
2nd ed.
      p. cm.
    Includes index.
    ISBN 0-02-395111-7
    1. English language—Technical English. 2. English language—
Rhetoric. I. Title.
PE1475.P47 1994   93-792
808'.0666—dc20   CIP

Printing: 1 2 3 4 5 6 7 8 9   Year: 4 5 6 7

Photo credits: All chapter-opener photos: Todd Yarrington/Macmillan
"McDuff" color insert: Page 2: Photo courtesy of Ohio Department of Natural Resources; Page 3: Photo courtesy of Schlumberger/Sedco Forex; Page 4: Photo courtesy of Grant Medical Center, Columbus, Ohio; Page 5: Photo property of the Houston Oilers; Page 6: © 1993 Terence E. Seidel; Page 7: Photo courtesy of U.S. Council for Energy Awareness; Page 8: © 1993 Todd Yarrington.

Preface

Be always employed in something
useful. Cut off all unnecessary
actions.
Ben Franklin, *Autobiography*

Either write things worth reading,
or do things worth writing.
Ben Franklin, *Poor Richard's Almanac*

Ben Franklin followed his own advice and always preferred the practical approach. His life reflected this virtue, as did his writing. Works like the *Autobiography* and *Poor Richard's Almanac* seek to instruct *and* entertain, without the "unnecessary actions" that might detract from their usefulness. Franklin's preference for practical writing has meaning even today, over 200 years later, as we search for alternatives to the lifeless, confusing prose so often produced by many organizations.

A "no-frills" approach to writing also pertains to books *about* writing, like this one. Such books should use every page to offer something practical to the student. With limited time both inside and outside of class, students need a text that provides the main tools for setting the writing process in motion. *Technical Writing: A Practical Approach* meets this need.

The first edition of the book created McDuff, Inc., a simulated yet realistic company used in examples and assignments throughout the book. This second edition expands the world of McDuff to include many more professions and case studies. Further benefits of the new edition are highlighted in the next section.

FEATURES OF THE SECOND EDITION

Besides introducing the textbook firm of McDuff, the first edition offered clear guidelines for all documents, annotated writing models, realistic assignments, and a writing handbook. While preserving these features of the first edition, the second edition adds the following changes in format or content:

v

■ *Feature #1: Four New Chapters*

Four topics deserving further emphasis earned their own chapters in this new edition:

- **Organizing Information** (Chapter 3): Presents the simple "ABC" pattern of organization used for all documents covered in the text.
- **Page Design** (Chapter 4): Explains and displays techniques for creating visually interesting documents, such as lists, headings, white space, and changes in font type and size.
- **The Job Search** (Chapter 14): Brings together and expands upon topics included in several different chapters in the first edition, including job letters, resumes, interviews, and negotiating.
- **Style in Technical Writing** (Chapter 15): Gives students the elements of good writing style, while reserving matters of grammar and mechanics for the Handbook at the end of the book.

■ *Feature #2: New, Conveniently Located Models*

Annotated models helped users of the first edition incorporate the book's guidelines into their own work. This edition adds more models from diverse professions—health care, corporate training, engineering, computer science, and more. For easy reference the text groups models for writing are at the end of each chapter. This placement keeps chapter discussion from being interrupted by document models, especially longer ones. With color coding on the paper edges and page references, models are easy for students to find.

■ *Feature #3: Many New Assignments, in New Formats*

Engaging, realistic assignments form the core of this book's practical approach to writing. Many chapters now include two assignment sections, one for short assignments and the other for longer assignments. Thus, instructors can select projects to be completed in class or those that require more preparation time outside of class.

■ *Feature #4: Color Insert*

A new color insert makes McDuff, Inc., even more accessible to students. That is, the text now contains seven McDuff "Project Sheets" in a color insert. The sheets describe specific projects completed in seven diverse project areas—such as environmental work, construction, and corporate training. They provide data for assignments used throughout the book and help make McDuff, Inc., seem like an actual potential employer of college graduates. Students can place themselves into professional roles and respond to real-world technical problems, even if their own experience is limited.

■ *Feature #5: More Emphasis on Collaborative Writing*

Chapter 1 offers a step-by-step procedure for helping students learn to write in groups, which is common practice in many organizations. Then the following chapters include assignments that require group efforts within the McDuff, Inc., context. The text also gives suggestions for using computer technology to foster group communication. Electronic mail is one such medium discussed.

■ *Feature #6: An Expanded Ancillary Package*

The ancillary package now includes the following:

- 50 two-color transparencies
- A thorough Instructor's Manual, which includes over 200 transparency masters, sample syllabi, editing revisions, a bibliography for the teaching of technical writing, and master copies of the Planning Form and the McDuff letterhead for photocopying and distribution to students
- A copy of RightWriter Release 6.0 (ask your Macmillan sales representative for a request form)

In addition, we have included at the end of the text a coupon with which students can purchase RightWriter 6.0 for $29.95.

Besides the major changes described above, the second edition includes many smaller but important improvements in the book. Here are a few:

- The inside cover lists common proofreading marks
- The text gives simple instructions for using the Planning Form during the early stages of the writing process
- The research chapter covers questionnaires and surveys
- The book places even more emphasis on writing all types of summaries

ACKNOWLEDGEMENTS

I relied on the help of many generous people and companies in writing the second edition of this book. First, these four technical firms allowed me to use some written material gathered during my consulting work for them: Fugro-McClelland, Inc., Law Environmental, Inc., McBride-Ratliff and Associates, Inc., and Westinghouse Environmental and Geotechnical Services, Inc. My hands-on experience with these companies shaped my approach to teaching technical writing and writing about it.

Second, many colleagues at Southern College of Technology and in companies gave me excellent suggestions used in this book. I especially want to thank George Ferguson, Bob Harbort, Dory Ingram, Becky Kelly, Randy Nipp, Jo Pevey, Ken Rainey, Hattie Schumaker, Herb Smith, James Stephens, and Tom Wiseman. In addition, these reviewers provided useful advice: Linda Brasher, Mississippi State University; Sandra Christianson, National College; Betty Cramton, Kansas State University—Salina; Kyle Anne Gearhart, DeVRY Institute of Technology; Chandice Johnson, North Dakota State University; Rebecca Kelly, Southern College of Technology; Marcia Noe, University of Tennessee at Chattanooga; Sondra Saunders, DeVRY Institute of Technology; Eileen M. Schwartz, Purdue University Calumet; Kathy G. Titus, University of Detroit Mercy; and James L. Zachary, University of Tampa.

Third, I appreciate the help of the following students for allowing me to adapt their written work for use in this text: Michael Alban, Becky Austin, Natalie Birnbaum, Corey Baird, Cedric Bowden, Gregory Braxton, Ishmael Chigumira, Bill Darden, Jeffrey Daxon, Rob Duggan, William English, Joseph Fritz, Jon Guffey, Gary Harvey, Sam Harkness, Lee Harvey, Hammond Hill, Sudhir Kapoor, Steven Knapp, Wes Matthews, Kim Meyer, James Moore, James Porter, James Roberts, Mort Rolleston, Chris Ruda, Barbara Serkedakis, Tom Skywark, Tom Smith, DaTonja Stanley, Chris Swift, James Stephens, and Jeff Woodward.

Finally, it was my very good fortune to have the same first-rate developmental editor, Monica Ohlinger, for both the first and second editions of this book. Monica always kept me focused on just what this second edition could become. I also want to thank these other Macmillan people for their hard work on the project: Stephen Helba, Executive Editor; Mary Irvin, Production Editor; Ruth Kimpel, Art Coordinator; Anne Flanagan, Text Designer; and Russ Maselli, Cover and Color Insert Designer.

Again, I owe a huge debt to my family for seeing me through another textbook project. My son Zachary, my daughter Katie, and my wife Evelyn understood the time I had to commit and helped me every step of the way. I especially depended on Ev's editorial eye and good judgment throughout the project.

CLOSING

A final comment to the students about to read this book: At the start of my classes, I sometimes ask students to describe their professional goals for the next ten years. As you might expect, their comments suggest they hope to rise to important positions in the workplace and make genuine contributions to their professions. Such long-term thinking is crucial, keeping you on course in your life.

Yet, ultimately, the way you handle the small, daily details of life most influences the real contribution you make in the long run. If you do good technical work, believe in what you do, and communicate well with others—both interpersonally and in writing—success will come your way. The author Robert Pirsig put it this way in his 1971 novel, *Zen and the Art of Motorcycle Maintenance:* "The place to improve the world is first in one's own heart and head and hands, and then work outward from there." This writer believes—and this book tries to show—that clear, concise, and honest writing is one of the most powerful tools of your heart, head, *and* hands.

MERRILL'S INTERNATIONAL SERIES IN ENGINEERING TECHNOLOGY

Zanger & Zanger, *Fiber Optics: Communication and Other Applications*, 0-675-20944-7

Microcomputer Servicing

Adamson, *Microcomputer Repair*, 0-02-300825-3

Asser, Stigliano, & Bahrenburg, *Microcomputer Servicing: Practical Systems and Troubleshooting, 2nd Edition*, 0-02-304241-9

Asser, Stigliano, & Bahrenburg, *Microcomputer Theory and Servicing, 2nd Edition*, 0-02-304231-1

Programming

Adamson, *Applied Pascal for Technology*, 0-675-20771-1

Adamson, *Structured BASIC Applied to Technology, 2nd Edition*, 0-02-300827-X

Adamson, *Structured C for Technology*, 0-675-20993-5

Adamson, *Structured C for Technology (with disk)*, 0-675-21289-8

Nashelsky & Boylestad, *BASIC Applied to Circuit Analysis*, 0-675-20161-6

Instrumentation and Measurement

Berlin & Getz, *Principles of Electronic Instrumentation and Measurement*, 0-675-20449-6

Buchla & McLachlan, *Applied Electronic Instrumentation and Measurement*, 0-675-21162-X

Gillies, *Instrumentation and Measurements for Electronic Technicians, 2nd Edition*, 0-02-343051-6

Transform Analysis

Kulathinal, *Transform Analysis and Electronic Networks with Applications*, 0-675-20765-7

Biomedical Equipment Technology

Aston, *Principles of Biomedical Instrumentation and Measurement*, 0-675-20943-9

Mathematics

Monaco, *Essential Mathematics for Electronics Technicians*, 0-675-21172-7

Davis, *Technical Mathematics*, 0-675-20338-4

Davis, *Technical Mathematics with Calculus*, 0-675-20965-X

INDUSTRIAL ELECTRONICS/INDUSTRIAL TECHNOLOGY

Bateson, *Introduction to Control System Technology, 4th Edition*, 0-02-306463-3

Fuller, *Robotics: Introduction, Programming, and Projects*, 0-675-21078-X

Goetsch, *Industrial Safety and Health: In the Age of High Technology*, 0-02-344207-7

Goetsch, *Industrial Supervision: In the Age of High Technology*, 0-675-22137-4

Geotsch, *Introduction to Total Quality: Quality, Productivity, and Competitiveness*, 0-02-344221-2

Horath, *Computer Numerical Control Programming of Machines*, 0-02-357201-9

Hubert, *Electric Machines: Theory, Operation, Applications, Adjustment, and Control*, 0-675-20765-7

Humphries, *Motors and Controls*, 0-675-20235-3

Hutchins, *Introduction to Quality: Management, Assurance, and Control*, 0-675-20896-3

Laviana, *Basic Computer Numerical Control Programming*, 0-675-21298-7

Pond, *Fundamentals of Statistical Quality Control*

Reis, *Electronic Project Design and Fabrication, 2nd Edition*, 0-02-399230-1

Rosenblatt & Friedman, *Direct and Alternating Current Machinery, 2nd Edition*, 0-675-20160-8

Smith, *Statistical Process Control and Quality Improvement*, 0-675-21160-3

Webb, *Programmable Logic Controllers: Principles and Applications, 2nd Edition*, 0-02-424970-X

Webb & Greshock, *Industrial Control Electronics, 2nd Edition*, 0-02-424864-9

MECHANICAL/CIVIL TECHNOLOGY

Dalton, *The Technology of Metallurgy*, 0-02-326900-6

Keyser, *Materials Science in Engineering, 4th Edition*, 0-675-20401-1

Kokernak, *Fluid Power Technology*, 0-02-305705-X

Kraut, *Fluid Mechanics for Technicians*, 0-675-21330-4

Mott, *Applied Fluid Mechanics, 4th Edition*, 0-02-384231-8

Mott, *Machine Elements in Mechanical Design, 2nd Edition*, 0-675-22289-3

Rolle, *Thermodynamics and Heat Power, 4th Edition*, 0-02-403201-8

Spiegel & Limbrunner, *Applied Statics and Strength of Materials, 2nd Edition*, 0-02-414961-6

Spiegel & Limbrunner, *Applied Strength of Materials*, 0-02-414970-5

Wolansky & Akers, *Modern Hydraulics: The Basics at Work*, 0-675-20987-0

Wolf, *Statics and Strength of Materials: A Parallel Approach to Understanding Structures*, 0-675-20622-7

DRAFTING TECHNOLOGY

Cooper, *Introduction to VersaCAD*, 0-675-21164-6

Ethier, *AutoCAD in 3 Dimensions*, 0-02-334232-3

Goetsch & Rickman, *Computer-Aided Drafting with AutoCAD*, 0-675-20915-3

Kirkpatrick & Kirkpatrick, *AutoCAD for Interior Design and Space Planning*, 0-02-364455-9

Kirkpatrick, *The AutoCAD Book: Drawing, Modeling, and Applications, 2nd Edition*, 0-675-22288-5

Kirkpatrick, *The AutoCAD Book: Drawing, Modeling, and Applications, Including Release 12, 3rd Edition*, 0-02-364440-0

Lamit & Lloyd, *Drafting for Electronics, 2nd Edition*, 0-02-367342-7

Lamit & Paige, *Computer-Aided Design and Drafting*, 0-675-20475-5

Maruggi, *Technical Graphics: Electronics Worktext, 2nd Edition*, 0-675-21378-9

Maruggi, *The Technology of Drafting*, 0-675-20762-2

Sell, *Basic Technical Drawing*, 0-675-21001-1

TECHNICAL WRITING

Croft, *Getting a Job: Resume Writing, Job Application Letters, and Interview Strategies*, 0-675-20917-X

Panares, *A Handbook of English for Technical Students*, 0-675-20650-2

Pfeiffer, *Proposal Writing: The Art of Friendly Persuausion*, 0-675-20988-9

Pfeiffer, *Technical Writing: A Practical Approach, 2nd Edition*, 0-02-395111-7

Roze, *Technical Communications: The Practical Craft, 2nd Edition*, 0-02-404171-8

Weisman, *Basic Technical Writing, 6th Edition*, 0-675-21256-1

Contents

12 Oral Communication   381

13 Technical Research   407

TECHNICAL WRITING
A Practical Approach

PART I

1 | *Process in Technical Writing*

A McDuff engineer collects data for a technical report. This task is part of the planning stage of writing. The other two stages are drafting and revising.

G ood communication skills are essential in any career you choose. Jobs, promotions, raises, and professional prestige result from your ability to write and speak effectively. With so much at stake, you need a simple road map to direct you toward writing excellence. *Technical Writing: A Practical Approach* is such a map.

The four chapters in Part 1 give you an overview of technical writing and prepare you to complete the assignments in this book:

Chapter 1: Defines technical writing and describes the writing process. As shown in Figure 1–1, the technical writing process has three main parts: planning, drafting, and revising. Careful completion of this *process* is the best guarantee of a successful *product*—the final document.

Chapter 2: Introduces you to life in the corporate world and to the fictional company of McDuff, Inc. This organization will provide the framework for most examples and assignments in this book.

Chapter 3: Focuses on organizing information for diverse readers who are often very busy and unfamiliar with a project.

Chapter 4: Describes techniques for using the best elements of page design and computer technology in the writing process.

FIGURE 1–1
Flowchart for the technical writing process

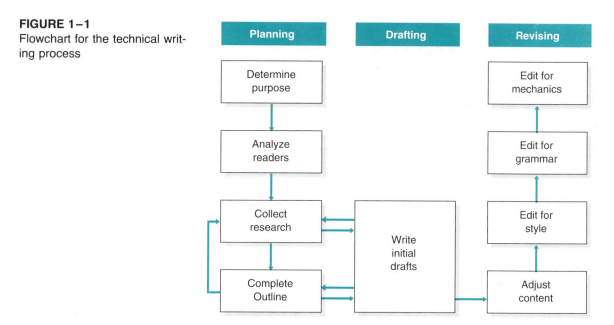

So that you can begin writing early in the course, preliminaries are kept to a minimum. Now let's take a closer look at the nature of technical writing.

GETTING FROM HERE TO THERE

You probably learned how to write short essays in previous writing courses. This book helps you transfer that basic knowledge to the kind of writing done on the job. In the process, you will discover that learning technical writing is a bit like studying a foreign language, with a new set of rules. Yet career writing is so practical, so well grounded in common sense, that it will seem to proceed smoothly from your previous work. This section highlights features of both traditional academic writing and job-related technical writing.

Features of Academic Writing

Writing you have done in school probably has had these characteristics:

- **Purpose:** Demonstrating what you know about the topic, in a way that justifies a high grade
- **Your knowledge of topic:** Less than the teacher who evaluates the writing
- **Audience:** Teacher who requests the assignment and who will read it from beginning to end
- **Criteria for evaluation:** Depth, logic, clarity, unity, and grammar

Academic writing requires that you display your learning to someone who knows more about the subject than you do. Because this person's job is to evaluate your work, you have what might be called a "captive audience." Here are two common situations in which academic writing would be appropriate, along with brief examples:

■ Example 1: Essays for English Class

In a typical high school or college writing class, you write short essays based on your experience or readings. Instructors expect you to (1) state the topic, purpose, and main points in the introduction, (2) develop the supporting points in the body, and (3) wrap up the paper in the conclusion. Your reader wants a coherent, unified, grammatically sound, and interesting piece of writing. What follows is the first paragraph from one such essay. The student was asked to attend and then comment on a local cultural event.

> The recent exhibit at Atlanta's High Museum of Art, "Masterpieces of the American West," gives the viewer a small glimpse of the beautiful land and fascinating peoples that confronted the explorer-artists who ventured west in the 1800s. This paper will describe my response to three works—Catlin's *Mandan Dance*, Bierstadt's *Rocky Mountain Waterfall*, and Remington's *Turn Him Loose, Bill*. The first portrays the Indian culture; the second, Western landscape; and the third, the encroaching civilization from the East.

■ Example 2: Exams and Papers in General Studies Courses

Courses like history and science often require that you write analytical papers and exams. As in essays for a writing class, these papers should start with a topic statement that outlines main points and then proceed with supporting paragraphs. For example, you may be asked to describe both the immediate and long-term reasons for a particular event. Here is the first paragraph of a cause-effect essay in environmental science. The writer must define and give causes for the "greenhouse effect."

> The term *greenhouse effect* refers to the global warming trend of the earth. It occurs when the sun's reflected heat is unable to escape the earth's atmosphere into space, due to a buildup of carbon monoxide and other gases in the atmosphere. This essay will describe these main reasons for the dramatic increase in the greenhouse effect: the burning of rain forests in the Amazon, accelerated emission of auto pollutants around the world, and increases in the use of coal-fired plants.

> In both examples, the *purpose* is to demonstrate knowledge, and the *audience* is someone already familiar with the subject or approach. In this sense, academic writing shows your command of information to someone more knowledgeable about the subject than you are. The next section examines a different kind of writing—the kind you will be doing in your career and in this course. The purpose and audience differ considerably from those of academic writing.

Features of Technical Writing

The ground rules for writing shift somewhat when you begin your career. Those unprepared for this change often flounder for years, never quite understanding the new rules. *Technical writing* is a generic term for all written communications done on the job—whether in business, industry, or other professions. It is particularly identified with jobs in technology, engineering, science, the health professions, and other fields with specialized vocabularies. The terms *technical writing, professional writing, business writing,* and *occupational writing* all mean essentially the same thing—writing done in your career. Here are the main characteristics of technical writing:

- **Purpose:** Getting something done within an organization (completing a project, persuading a customer, pleasing your boss, etc.)
- **Your knowledge of topic:** Usually greater than that of the reader
- **Audience:** Often several people, with differing technical backgrounds
- **Criteria for evaluation:** Clear and simple organization of ideas, in a format that meets the needs of busy readers

In particular, note these differences between technical writing and academic writing:

1. Technical writing has a practical role on the job, whereas academic writing aims only to display your knowledge in school.
2. Technical writing is done by an informed writer conveying needed information to an uninformed reader, whereas academic writing is done by a student as learner for a teacher as source of knowledge.
3. Technical writing often is read by many readers, whereas academic writing aims to satisfy only one person, the teacher.

Finally, technical writing places greater emphasis on techniques of organization and format that help readers find important information as quickly as possible.

Figure 1–2 lists some typical on-the-job writing assignments. While not exhaustive, the list does include many of the writing projects you will encounter. Also, see Figure 1–3 for an example of a short technical document.

Besides projects that involve writing, you will also have speaking responsibilities during your career—for example, formal speeches at conferences and informal presentations at meetings. The term *technical communication* is used to include both the writing and speaking tasks involved with any job.

Although technical communication plays a key role in the success of all technical professionals and managers, the amount of time you devote to it will depend on your job. A 1989 survey of technical managers gives some idea of the time involved. Conducted by the National Aeronautics and Space Administration (NASA), the survey canvassed managers in profit-making and nonprofit organizations in the field of aeronautics. As Figure 1–4 shows, 100% of the profit managers and 98% of the nonprofit managers in the study consider technical communication a "somewhat important" or "very important" part of their jobs.

Even more telling is the amount of time the NASA survey respondents spend communicating. Figure 1–4 indicates that they use (1) over one-third of their

---

Correspondence: In-House or External
- Memos to your boss and to your subordinates
- Routine letters to customers, vendors, etc.
- "Good news" letters to customers
- "Bad news" letters to customers
- Sales letters to potential customers
- Electronic mail (E-mail) messages to coworkers or customers over a computer network

Short Reports: In-House or External
- Analysis of a problem
- Recommendation
- Equipment evaluation
- Progress report on project or routine periodic report
- Report on the results of laboratory or field work
- Description of the results of a company trip

Long Reports: In-House or External
- Complex problem analysis, recommendation, or equipment evaluation
- Project report on field and/or laboratory work
- Feasibility study

Other Documents
- Proposal to boss for new product line
- Proposal to boss for change in procedures
- Proposal to customer to sell a product, service, or idea
- Feasibility study for your firm or for a customer
- Abstract or summary of technical article
- Technical article or presentation
- Operation manuals or other manual

FIGURE 1–2
Examples of technical writing

work time conveying information *to* others and (2) another one-third working with technical information sent to them *by* others. Based on a 40-hour week, therefore, both groups spend roughly *two-thirds* of their working week on job duties associated with technical communication.

Now that you know the nature and importance of technical writing, the next section examines the first part of the planning stage: determining a document's purpose.

DISCOVERING YOUR PURPOSE

Kate Paulsen works as a training supervisor for the Boston office of McDuff, Inc., a firm described in more detail in chapter 2. The company is growing so quickly that hiring, training, and retraining employees have become major goals. Kate

Mc Duff, Inc.

MEMORANDUM

DATE: December 6, 1993
TO: Holly Newsome
FROM: Michael Allen ma
SUBJECT: Printer Recommendation

Introductory Summary

Recently you asked for my evaluation of the Hemphill LaserFast printer currently used in my department. Having analyzed the printer's features, print quality, and cost, I am quite satisfied with its performance.

LaserFast Features

Among the LaserFast's features, I have found these five to be most useful:

1. Portrait and landscape print modes
2. Selectable character sizes (10/12/16.7 characters per inch)
3. 1.5 megabytes of print-buffer memory
4. Selectable paper sizes (letter, legal, half-letter)
5. Print speed of six pages per minute

In addition, the LaserFast printer is equipped with two built-in fonts and accepts several additional fonts from specially purchased cartridges. This combination of features makes the LaserFast a most versatile printer.

LaserFast Quality

The Hemphill LaserFast printer produces laser-sharp clarity that rivals professional typeset quality. The print resolution is an amazing 300 x 300 dots per inch, among the highest attainable in current desktop printers. This memo was printed on LaserFast, and as you can see, the quality speaks for itself.

LaserFast Costs

Considering the features and quality, the LaserFast is an excellent printer for the money. At a retail price of $2500, it is one of the lowest-priced laser printers. In fact, the highest-quality daisy wheel and dot matrix printers cost just $500 less than the LaserFast, yet neither offers the superior features or quality of the laser printer.

Conclusion

On the basis of my observations, I strongly recommend that our firm continue to use and purchase the LaserFast printer. Please call me at ext. 204 if you want further information about this excellent machine.

FIGURE 1–3
Short report

FIGURE 1–4

Data from NASA aeronautics survey

Source: Thomas E. Pinelli et al., *Technical Communications in Aeronautics: Results of an Exploratory Study—An Analysis of Profit Managers' and Nonprofit Managers' Responses* (Washington, D.C.: National Aeronautics and Space Administration, NASA TM-101626, October 1989), 71. (Available from NTIS Springfield, Va.)

TABLE 1. Importance of Technical Communications

How Important	Profit Managers		Nonprofit Managers	
	No.	%	No.	%
Very	86	92.5	43	84.3
Somewhat	7	7.5	7	13.7
Not at all	0	0.0	1	2.0
Total	93	100.0	51	100.0

TABLE 2. Time Spent Communicating Technical Information to Others

Time Spent Per Week, Hour	Profit Managers		Nonprofit Managers	
	No.	%	No.	%
5 or less	13	14.3	9	18.0
6 to 10	33	36.2	16	30.0
11 to 20	37	40.7	21	42.0
21 or more	8	8.8	5	10.0
Total	91	100.0	51	100.0
Mean	13.5		13.9	

TABLE 3. Time Spent Working with Technical Information Received from Others

Time Spent Per Week, Hour	Profit Managers		Nonprofit Managers	
	No.	%	No.	%
5 or less	8	8.7	6	12.0
6 to 10	42	46.2	23	46.0
11 to 20	36	39.6	18	36.0
21 or more	5	5.5	3	6.0
Total	91	100.0	50	100.0
Mean	13.0		13.0	

recently flew to Cleveland to attend a workshop sponsored by a major professional training organization. The workshop emphasized a new in-house procedure for surveying employee training needs. After returning to Boston, Kate must write her manager a trip report that describes the survey technique. She ponders three different approaches to the report:

- **Giving an overview** of the survey procedure she studied during the three-day workshop—stressing a few key points so that her manager could decide whether to inquire further
- **Providing details** of exactly how the survey procedure could be applied to her firm—with enough specifics for her manager to see exactly how the survey could be used at McDuff
- **Proposing** that the procedure for conducting the needs survey be used at McDuff—in language that argues strongly for adoption

For Kate, the first step is to decide what she wants to accomplish. Likewise, every piece of *your* writing should have a specific reason for being. The purpose may be dictated by someone else or selected by you. In either case, it must be firmly understood *before* you start writing. Purpose statements guide every decision you make while you plan, draft, and revise.

Kate Paulsen's three choices indicate some of your options, but there are others. Your choice of purpose will fall somewhere within this continuum:

For example, when reporting to your boss on the feasibility of adding a new wing to your office building, you should be quite objective. You must provide facts that can lead to an informed decision. If you are an outside contractor proposing to construct such a wing, however, your purpose is more persuasive. You will be trying to convince readers that your firm should receive the construction contract.

When preparing to write, therefore, you need to ask yourself two related questions about your purpose.

■ Question 1: Why Am I Writing This Document?

This question should be answered in just one or two sentences, even in complicated projects. Often the resulting purpose statement can be moved "as is" to the beginning of your outline and later to the first draft.

For example, Kate Paulsen finally decides on the following purpose statement, which becomes the first passage in her trip report: "This memo will highlight main features of the training needs survey introduced at the workshop I attended in Cleveland. I will focus on several possible applications you might want to consider for our office training." Note that Kate's purpose would rest about halfway across the persuasive continuum shown earlier. Though she will not be strongly advocating McDuff's use of the survey, she will be giving information that suggests the company might benefit by using it.

■ Question 2: What Response Do I Want from Readers?

The first question about purpose leads inevitably to the second about results. Again, your response should be only one or two sentences long. Though brief, it should pinpoint exactly what you want to happen as a result of your document. Are you just giving data for the file? Will information you provide help others do their jobs? Will your document recommend a major change?

In Kate Paulsen's case, she decides on this results statement: "Though I'm not yet sure if this needs survey is right for McDuff and is worth purchasing, I want my boss to consider it." Unlike the purpose statement, the results statement may not go directly into your document. Kate's statement hints at a "hidden" agenda that may be implicit in her trip report but will not be explicitly stated. This statement, written for her own use, becomes an essential part of her planning. It is a concrete goal to keep in mind as she writes.

The answers to these two questions about purpose and results are included on the Planning Form your instructor may ask you to use for assignments. Figure 1–5 on pp. 14 and 15 includes a copy of the form, along with instructions for using it. The last page of this book contains another copy you can duplicate for use with assignments.

Having established your purpose, you are now ready to consider the next part of the writing process: audience analysis.

ANALYZING YOUR READERS

One cardinal rule governs all on-the-job writing:

> ### WRITE FOR YOUR READER, NOT FOR YOURSELF.

This rule especially applies to science and technology because many readers may know little about your field. In fact, writing experts agree that most technical writing assumes too much knowledge on the part of the reader. The key to avoiding this problem is to examine the main obstacles readers face and adopt a strategy for overcoming them.

This section (1) highlights problems that readers have understanding technical writing, (2) suggests techniques to prevent these problems, and (3) describes some main classifications of technical readers. At first, analyzing your audience might seem awkward and even unproductive. You are forced out of your own world to consider that of your reader, before you even put pen to paper. The payoff, however, will be a document that has clear direction and gives the audience what it wants.

Obstacles for Readers

As purchasing agent for McDuff, Inc., Charles Blair must recommend one automobile sedan for fleet purchase by the firm's sales force and executives. First, he will conduct some research—interviewing car firm representatives, reading car evaluations in consumer magazines, and inquiring about the needs of his firm's salespeople. Then he will submit a recommendation report to the selection committee consisting of the company president, the accounting manager, several salespeople, and the supervisor of company maintenance. As Charles will discover, readers of all backgrounds often have these four problems when reading any technical document:

1. Constant interruptions
2. Impatience finding information they need
3. A different technical background from the writer
4. Shared decision-making authority with others

If you think about these obstacles every time you write, you will be better able to understand and respond to your readers.

■ Obstacle 1: Readers Are Always Interrupted

As a professional, how often will you have the chance to read a report or other document without interruption? Such times are rare. Your reading time will be interrupted by meetings and phone calls, so a report often gets read in several sittings. Aggravating this problem is the fact that readers may have forgotten details of the project.

■ Obstacle 2: Readers Are Impatient

Many readers lose patience with vague or digressive writing. They are thinking "What's the point?" or "So what?" as they plod through memos, letters, reports, and proposals. They want to know the significance of the document right away.

■ Obstacle 3: Readers Lack Your Technical Knowledge

In college courses, the readers of your writing are professors with knowledge of the subject. In your career, however, you will write to readers who lack the information and background that you have. They expect a technically sophisticated response, but in language they can understand. If you write over their heads, you will not accomplish your purpose. Think of yourself as an educator; if readers do not learn from your reports, you have failed in your objective.

■ Obstacle 4: Most Documents Have More Than One Reader

If you always wrote to only one person, technical writing would be easy. Each document could be tailored to the background, interests, and technical education of just that individual. However, this is not so in the actual world of business and industry. Readers usually share decision-making authority with others, who may read all or just part of the text. That reality requires you to respond to the needs of many individuals—most of whom have a hectic schedule, are impatient, and possess a technical background different from yours.

Ways to Understand Readers

Obstacles to communication can be frustrating. Yet there are techniques for overcoming them. First, you must find out exactly what information each reader needs. Think of the problem this way—would you give a speech without learning about the background of your audience? Writing depends just as much, if not more, on such analysis. Follow these four steps to determine your readers' needs:

PLANNING FORM

NAME: _____   ASSIGNMENT: _____

I. Purpose: Answer each question in one or two sentences.

    A. Why are you writing this document?_____

_____

    B. What response do you want from readers? _____

_____

II. Reader Matrix: Fill in names and positions of people who may read the document.

	Decision-Makers	Advisors	Receivers
Managers	_____ _____ _____	_____ _____ _____	_____ _____ _____
Experts	_____ _____ _____	_____ _____ _____	_____ _____ _____
Operators	_____ _____ _____	_____ _____ _____	_____ _____ _____
General Readers	_____ _____ _____	_____ _____ _____	_____ _____ _____

III. Information on Individual Readers: Answer these questions about selected members of your audience. Attach additional sheets as is necessary.

    1. What is this reader's technical or educational background?

_____

_____

    2. What main question does this person need answered?

_____

_____

    3. What main action do you want this person to take?

_____

_____

    4. What features of this person's personality might affect his or her reading?

_____

_____

    5. What features does this person prefer in

        Format?_____

        Style?_____

        Organization? _____

IV. Outline: Attach an outline (topic) to use in drafting the **BODY** of this document.

FIGURE 1–5

Planning form for all technical documents

Instructions for Completing "Planning Form"

The Planning Form is for your use in preparing assignments in your technical writing course. It focuses only on the planning stage of writing. Complete it *before* you begin your first draft.

1. Use the Planning Form to help plan your strategy for all writing assignments. Your instructor may or may not require that it be submitted with assignments.

2. Photocopy the form on the back page of this book or write the aswers to questions on separate sheets of paper, whatever option your instructor prefers. (Your instructor may hand out enlarged, letter-size copies of the form, which are included in the Instructor's Manual.)

3. Answer the two purpose questions in one or two sentences each. Be as specific as possible about the purpose of the document and the response you want—especially from the decision-makers.

4. Note that the reader matrix, which includes 12 categories, classifies each reader by two criteria: (a) technical levels (shown on the vertical axis) and (b) relationship to the decision-making process (shown on the horizontal axis). Some of the boxes will be filled with one or more names, while others may be blank. How you fill out the form depends on the complexity of your audience and, of course, on the directions of your instructor.

5. Refer to Chapter 2 for names and positions of any McDuff employees to place in the reader matrix, if your paper is based on a simulated case from McDuff, Inc.

6. Note that questions in the "Information on Individual Readers" section can be filled out for one or more readers depending on your instructor. But you must answer *all* five questions for *each* reader you choose.

7. Complete the topic outline after you have collected whatever information or research your document requires. The outline should be specific and should include two or three levels (see Figure 1-8).

FIGURE 1–5, *continued*

■ *Audience Analysis Step 1: Write Down What You Know About Your Reader*

To build a framework for analyzing your audience, you need to write down—not just casually think about—the answers to these questions for each reader:

1. What is this reader's technical or educational background?
2. What main question does this person need answered?
3. What main action do you want this person to take?
4. What features of this person's personality might affect his or her reading?
5. What features does this reader prefer in
   Format?
   Style?
   Organization?

The Planning Form in Figure 1–5 includes these five questions.

■ *Audience Analysis Step 2: Talk With Colleagues Who Have Written to the Same Readers*

Often your best source of information about your readers is a colleague where you work. Ask around the office or check company files to discover who else may have written to the same audience. Useful information could be as close as the next office.

■ *Audience Analysis Step 3: Find Out Who Makes Decisions*

Almost every document requires action of some kind. Identify decision-makers ahead of time so that you can design the document with them in mind. Know the needs of your *most important* reader.

■ *Audience Analysis Step 4: Remember That All Readers Prefer Simplicity*

Even if you uncover little specific information about your readers, you can always rely on one basic fact: Readers of all technical backgrounds prefer concise, simple writing. The popular KISS principle (Keep It Short and Simple) is a worthy goal.

Types of Readers

You have learned some typical problems readers face and some general solutions to these problems. To complete the audience-analysis stage, this section shows you how to classify readers by two main criteria: knowledge and influence. Specifically, you need to answer two questions about every potential reader:

1. How much does this reader already know about the subject?
2. What part will this reader play in making decisions?

Then use the answers to these questions to plan your document. Figure 1–6 (adapted from the Planning Form in Figure 1–5) provides a reader matrix by which you can quickly view the technical levels and decision-making roles of all your readers. For complex documents, your audience may include many of the 12 categories shown on the matrix. Also, you may have more than one person in each box—that is, there may be more than one reader with the same background and decision-making role.

Technical Levels.   On-the-job writing requires that you translate technical ideas into language that nontechnical people can understand. This task can be very complicated because you often have several readers, each with different levels of knowledge about the topic. If you are to ''write for your reader, not for yourself,'' you must identify the technical background of *each* reader. Four categories will help you classify each reader's knowledge of the topic.

■   *Reader Group 1: Managers*

Many technical professionals aspire to become managers. Once into management, they are removed from hands-on technical details of their profession. Instead, they manage people, set budgets, and make decisions of all kinds. Thus you should assume that management readers will not be familiar with fine technical points, will have forgotten details of your project, or both. These managers often need:

- Background information
- Definitions of technical terms
- Lists and other format devices that highlight points
- Clear statements about what is supposed to happen next

Technical Level	Decision-Making Level		
	Decision-Makers	Advisers	Receivers
Managers			
Experts			
Operators			
General Readers			

FIGURE 1–6
Reader matrix

In chapter 3, we will discuss an all-purpose "ABC format" for organization that responds to the needs of managers.

■ *Reader Group 2: Experts*

Experts include anyone with a good understanding of your topic. They may be well educated—as with engineers and scientists—but that is not necessarily the case. In the example mentioned earlier, the maintenance supervisor with no college training could be considered an "expert" about selecting a new automobile for fleet purchase. That supervisor will understand any technical information about car models and features. Whatever their educational levels, most experts in your audience need:

- Thorough explanations of technical details
- Data placed in tables and figures
- References to outside sources used in writing the report
- Clearly labeled appendices for supporting information

■ *Reader Group 3: Operators*

Because decision-makers are usually managers or technical experts, they tend to get most of the attention. However, some documents also have readers who are operators. They may be technicians in a field crew, workers on an assembly line, salespeople in a department store, or drivers for a trucking firm—anyone who puts the ideas in your document into practice. These readers expect:

- A clear table of contents for locating sections that relate to them
- Easy-to-read listings for procedures or instructions
- Definitions of technical terms
- A clear statement of exactly how the report affects their jobs

■ *Reader Group 4: General Readers*

General readers, also called "laypersons," compose a group with the least amount of information about your topic or field. For example, a report on the environmental impact of a toxic waste dump might be read by many general readers who are home owners in the surrounding area. Most will have little technical understanding of toxic wastes and associated environmental hazards. These general readers often need:

- Definitions of technical terms
- Frequent use of graphics like charts and photographs
- A clear distinction between facts and opinions

Like managers, general readers need to be assured that (1) all implications of the document have been put down on paper and (2) important information has not been buried in overly technical language.

Decision-Making Levels.   Figure 1–6 shows that your readers, whatever their technical level, also can be classified by the degree to which they make decisions based on your document. Pay special attention to those most likely to use your report to create change. Use three levels to classify your audience during the planning process:

■ *First-Level Audience: Decision-Makers*

The first-level audience must act on the information. If you are proposing a new fax machine for your office, first-level readers will decide whether to accept or reject the idea. If you are comparing two computer systems for storing records at a hospital, the first-level audience will decide which unit to purchase. If you are describing electrical work your firm completed in a new office building, it will decide whether the project has fulfilled agreed-upon guidelines.

In other words, decision-makers translate information into action. They are usually, but not always, managers within the organization. One exception occurs in highly technical companies, wherein decision-makers may be technical experts with advanced degrees in science or engineering. Another exception occurs when decision-making committees consist of a combined audience. For example, the deacons' committee of a church may be charged with the task of choosing a firm to build a new addition to the sanctuary.

■ *Second-Level Audience: Advisers*

This second group could be called "influencers." Although they don't make decisions themselves, they read the document and give advice to those who will decide. Often the second-level audience comprises experts such as engineers and accountants, who are asked to comment on technical matters. After reading the summary, a decision-making manager may refer the rest of the document to advisers for their views.

■ *Third-Level Audience: Receivers*

Some readers do not take part in the decision-making process. They only receive information contained in the document. For example, a report recommending changes in the hiring of fast-food workers may go to the store managers after it has been approved, just so they can put the changes into effect. This third-level audience usually includes readers defined as "operators" in the previous section—that is, those who may be asked to follow guidelines or instructions contained in a report.

Using all this information about technical and decision-making levels, you can analyze each reader's (1) technical background with respect to your document and (2) potential for making decisions after reading what you present. Then you can move on to the research and outline stages of writing.

COLLECTING INFORMATION

Having established a clear sense of purpose and your readers' needs, you're ready to collect information for writing. Although you may want to use a scratch outline to guide the research process, a detailed outline normally gets written *after* you have collected research necessary to support the document.

This section lays out a general strategy for research. Details about research are included in chapter 13 ("Technical Research").

■ *Research Step 1: Decide What Kind of Information You Need*

There are two types of research—primary and secondary. Primary research is that which you collect on your own, whereas secondary information is generated by others and found in books, periodicals, or other sources. Figure 1–7 gives examples of both types. Use the kind of research that will be most helpful in supporting the goals of your project. Here are two examples:

- **Report context for using primary research:** A recommendation report to purchase new light tables for the drafting department is supported by your survey of the office drafters. All of them recommend the brand of table you have selected.
- **Report context for using secondary research:** Your report on light tables depends on data found in several written sources, such as an article in a mechanical engineering journal that contrasts features of three tables. On the basis of this article, you recommend a particular table.

■ *Research Step 2: Devise a Research Strategy*

Before you start searching through libraries or conducting interviews, you need a plan. In its simplest form, this plan may list the questions that you expect to answer in your quest for information. For example, a research strategy for a report on office chairs for word-processing operators might pose these questions:

- What kind of chair design do experts in the field of workplace environment recommend for word processors?
- Are there any data that connect the design of chairs with the efficiency of operators?

Primary	Secondary
1. **Interviews** 2. **Surveys** 3. **Laboratory Work** 4. **Field Work** 5. **Personal Observation**	1. **Bibliographies** (lists of possible sources—in print or on computer data bases) 2. **Periodical Indexes** (lists of journal and magazine articles, by subject) 3. **Newspaper Indexes** 4. **Books** 5. **Journals** 6. **Newspapers** 7. **Reference Books** (encyclopedias, dictionaries, directories, etc.) 8. **Government Reports** 9. **Company Reports**

FIGURE 1–7
Research sources

- Have any specific chair brands been recommended by experts?
- Is there information that suggests a connection between poor chair design and specific health problems?

■ *Research Step 3: Record Notes Carefully*

See chapter 13 for the variety of resources available to you at a well-stocked library. Once you have located the information you need in these sources, you must be very careful incorporating it into your own document. As chapter 13 denotes, you must clearly distinguish direct quotations, paraphrasing, and summaries in your notes. Then, when you are ready to translate these notes into a first draft, you will know exactly how much borrowed information you have used and in what form.

■ *Research Step 4: Acknowledge Your Sources*

The care that you took in step 3 must be accompanied by thorough acknowledgement of the specific sources. Chapter 13 demonstrates one correct citation system.

■ *Research Step 5: Keep a Bibliography for Future Use*

Consider any research you do for a writing project as an investment in later efforts. Even after your research for a project is complete and you have submitted the report, keep active files on any subjects that relate to your work. Update these files every time you complete a research-related project, such as the two mentioned previously on chair design and light tables. If you or a colleague wants to examine the subject later, you will have developed your own data base from which to start.

WRITING AN OUTLINE

After determining purpose and audience and completing research steps, you are ready to write your outline. Outlines provide the best method for planning any piece of writing, especially long documents. They do not have to be pretty; they just have to guide your writing of the draft. If you begin using outlines now, you will find it easier to organize and write documents of all kinds throughout your career. Refer to the following steps in preparing functional outlines. Figures 1–8 and 1–9 show the outline process in action.

■ *Outline Step 1: Record Your Random Ideas Quickly*

At first, ideas need *not* be placed in a pattern. Just jot down as many major and minor points as possible. For this exercise, try to use only one piece of paper, even if it is oversized. Putting points on one page helps prepare the way for the next step, in which you begin to make connections among points.

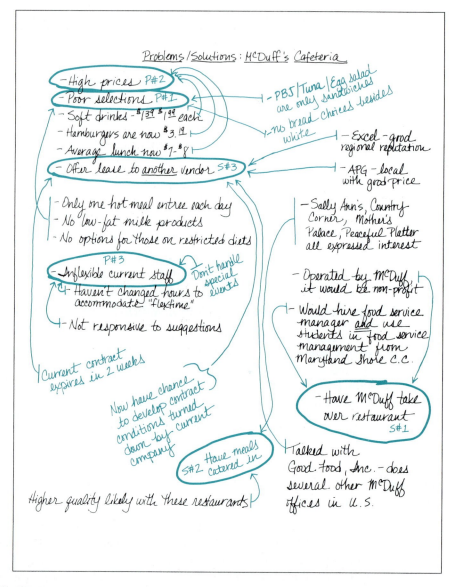

FIGURE 1–8
The outlining process: Early stage

■ *Outline Step 2: Show Relationships*

Next you need to connect related ideas. Using your brainstorming sheet, follow these three steps:

1. Circle or otherwise mark the points that will become main sections.
2. Connect each main point with its supporting ideas, using lines or arrows.
3. Delete material that seems irrelevant to your purpose.

PROBLEMS AND SOLUTIONS: CURRENT CAFETERIA IN BUILDING

I. Problem #1: Poor selection
   A. Only one hot meal entree each day
   B. Only three sandwiches—PBJ, egg salad, and tuna
   C. Only one bread—white
   D. No low-fat milk products (milk, yogurt, LF cheeses, etc.)
   E. No options for those with restricted diets
II. Problem #2: High prices
   A. Soft drinks from $1.39 to $1.99 each
   B. Hamburgers now $3.19
   C. Average lunch now $7–$8
III. Problem #3: Inflexible staff
   A. Unwilling to change hours to meet McDuff's flexible work schedule
   B. Have not acted on suggestions
   C. Not willing to cater special events in building
IV. Solution #1: End lease and make food service a McDuff department
   A. Hire food service manager
   B. Use students enrolled in food service management program at
      Maryland Shore Community College
   C. Operate as nonprofit operation—just cover expenses
V. Solution #2: Hire outside restaurant to cater meals in to building
   A. Higher quality likely
   B. Initial interest by four nearby restaurants
      1. Sally Ann's
      2. Country Corner
      3. Mother's Palace
      4. Peaceful Platter
VI. Solution #3: Continue leasing space but change companies
   A. Initial interest by three vendors
      1. Excel—good regional reputation for quality
      2. APG—close by and local, with best price
      3. Good Food, Inc.—used by two other McDuff offices with good
         results
   B. Current contract over in two months
   C. Chance to develop contract not acceptable to current company

FIGURE 1–9
The outlining process: Later stage

Figure 1–8 shows the results of applying steps 1 and 2 to a writing project at McDuff, Inc., the company used throughout this book. Diane Simmons, office services manager at the Baltimore branch, plans to recommend a change in food service. She uses the brainstorming technique to record her major and minor points. First, she circles the six main ideas. As it happens, these ideas include three main problems and three possible solutions, so she labels them P#1 through P#3 (problems) and S#1 through S#3 (solutions). Second, she draws arrows between each main point and its related minor points. In this case, there is no material to be deleted. Although the result is messy, it prepares her for the next step of writing the formal outline.

Like Diane Simmons, you will face one main question as you plan your outline: What pattern of organization best serves the material? Chapter 3 presents an "ABC format" that applies to overall structure. Each document should start with an **A**bstract (summary), move to the **B**ody (discussion), and end with a **C**onclusion. However, outlines usually cover only the *body* of a document. Here are some common patterns to consider in outlining the middle part, or body, of a report or proposal. The examples all relate to a decision by McDuff, Inc., to revamp its photocopying system:

- **Chronological:** The writer wants to describe the step-by-step procedure for completing a major photocopying project, from receiving copy at the copy center to sending it out the next day.
- **Parts of an object:** The writer wants to provide a part-by-part description of the current photocopying machine, along with a list of parts that have been replaced in three years of service calls.
- **Simple to complex or vice versa:** The writer wants to describe problems associated with the current photocopying procedure, working from minor to major problems or vice versa.
- **Specific to general (inductive):** The writer wants to begin by listing about 20 individual complaints about the present photocopying machine, later arriving at three general groupings into which all the individual complaints fall.
- **General to specific (deductive):** The writer wants to lead off with some generalizations about photocopying problems at McDuff, followed by a description of incidents related to poor copies, missed deadlines, and high service bills.

■ *Outline Step 3: Draft a Final Outline*

Once related points are clustered, it is time to transform what you have done into a somewhat ordered outline. (See Figure 1–9, which continues with the report context used in Figure 1–8.) This step allows you to (1) refine the wording of your points and (2) organize them in preparation for writing the draft. Although you need not produce the traditional outline with Roman numerals, etc., some structure is definitely needed. Abide by these basic rules:

- Make sure every main point has enough subpoints so that it can be developed thoroughly in your draft.
- When you decide to subdivide a point, have at least two breakdowns (because *any* object that is divided will have at least two parts). This same rule applies to headings and subheadings in the final document. (In fact, a good outline will provide you with the wording for headings and subheadings. As such, the outline becomes the basis for a table of contents in formal documents.)
- For the sake of consistency, phrase your points in either topic or sentence form. Sentences give you a headstart on the draft, but they may lock you into wording that needs revision later. Most writers prefer the topic approach; topics take up less space on the page and are easier to revise as you proceed through the draft.

WRITING INITIAL DRAFTS

With your research and outline completed, you are ready to begin the draft. This stage in the writing process should go quickly *if* you have planned well. Yet many writers have trouble getting started. The problem is so widespread that it has its own name—"writer's block." If you suffer from it, you are in good company; some of the best and most productive writers often face the "block."

In business and industry, the worst result of writer's block is a tendency to delay the start of writing projects, especially proposals. These delays can lead to rushed final drafts and editing errors. Outlining and other planning steps are wasted if you fail to complete drafting on time. The suggestions that follow can help you to start writing and then to keep the words flowing.

■ Drafting Step 1: Schedule One- or Two-Hour Blocks of Drafting Time

Most writers can keep the creative juices flowing for an hour or two at a time *if* distractions are removed. Rather than writing for three or four hours with your door open and thus with constant interruptions, schedule an hour or two of un-interrupted writing time. Most other business can wait an hour, especially considering the importance of good writing to your success. Colleagues and staff members will adjust to your new strategy for drafting reports. They may even adopt it themselves.

■ Drafting Step 2: Do Not Stop to Edit

Later, you will have time to refine your writing; that time is not now. Instead, force yourself to get ideas from the outline to paper or computer screen as quickly as possible. Most writers have trouble getting back into their writing pace once they have switched gears from drafting to revising.

■ Drafting Step 3: Begin With the Easiest Section

In writing the body of the document, you need not move chronologically from beginning to end. Because the goal is to write the first draft quickly, you should start with the section that flows best for you. Later, you can piece together sections and adjust content.

■ Drafting Step 4: Write Summaries Last

As already noted, the outline used for drafting covers just body sections of the document. Only after you have drafted them should you write overview sections like summaries. You cannot summarize a report until you have actually completed it. Because most writers have trouble with the summary—a section that is geared mainly for decision-makers in the audience—they may get bogged down if they begin writing it prematurely.

REVISING DRAFTS

You may have heard the old saw, "There is no writing, only rewriting." In technical writing, as in other types of communication, careful revision breeds success. The term *revision* encompasses four tasks that transform early drafts into final copy:

1. Adjusting and reorganizing content
2. Editing for style
3. Editing for grammar
4. Editing for mechanics

Following are some broad-based suggestions for revising your technical prose. For more details, consult chapter 15, "Style in Technical Writing," or the Handbook at the end of the book. Also, chapter 4 examines the use of word processing during the revision process.

■ *Revision Step 1: Adjust and Reorganize Content*

In this step, go back through your draft to (1) expand sections that deserve more attention, (2) shorten sections that deserve less, and (3) change the location of sentences, paragraphs, or entire sections. The use of word processing has made this step considerably easier than it used to be.

■ *Revision Step 2: Edit First for Style*

The term *style* refers to changes that make writing more engaging, more interesting, more readable. Such changes are usually matters of choice, not correctness. For example, you might want to:

- Shorten paragraphs
- Rearrange a paragraph to place the main point first
- Change passive-voice sentences to active
- Shorten sentences
- Define technical terms
- Add headings, lists, or graphics

One stylistic error deserves special mention because of its frequency: long, convoluted sentences. As a rule, you should simplify a sentence if its meaning cannot easily be understood in one reading. Also, be wary of sentences that are so long you must take a breath before you complete them.

■ *Revision Step 3: Edit for Grammar*

You probably know your main grammatical weaknesses. Perhaps comma placement or subject-verb agreement gives you problems. Or maybe you have trouble distinguishing couplets like imply/infer, effect/affect, or complementary/complimentary. In this pass through the document for grammar, focus on the particular errors that have given you problems in the past.

■ *Revision Step 4: Edit for Mechanics*

Your last revision pass should be for mechanical errors such as misspelled words, misplaced pages, incorrect page numbers, missing illustrations, and errors in numbers (especially cost figures). Word-processing software can help prevent some of these errors, such as most misspellings, but computer technology has not eliminated the need for at least one final proofing check.

This four-stage revision process will produce final drafts that reflect well on you, the writer. Here are two final suggestions that apply to all stages of the process:

1. Depend on another set of eyes besides your own. One strategy is to form a partnership with another colleague, whether in a technical writing class or on the job. In this arrangement, you both agree that you will carefully review each other's writing. This "buddy system" works better than simply asking favors of friends and colleagues. Choose a colleague in whom you have some confidence and from whom you can expect consistent editing quality. However, never make changes suggested by another person unless you fully understand the reason for doing so. After all, it is *your* writing.

2. Remember the importance of completing each step separately. Revising in stages yields the best results.

WRITING IN GROUPS

Writing can seem like a lonely act at times. Your own experience in school may reinforce the image of the solitary writer—with sweat on brow—toiling away on research, outlines, and drafts. In fact, this description does *not* typify much writing in the working world outside college. Writing in teams is the rule rather than the exception in many professions and organizations.

Group writing (also called collaborative writing) can be defined in this way:

> **Group writing:** the effort by two or more people to produce one document, with each member *sharing* in the writing process. The term assumes all members actually help with the drafting process, as opposed to (1) the writing of a document by one person, after all group members have met to discuss the project, or (2) group editing of something written by one person. The team must have clear goals and effective leadership to achieve results. Group work can be done in person, over the phone, or through electronic mail (e-mail).

This section further defines group writing, highlighting benefits of the strategy and noting some pitfalls to avoid. Then you are given guidelines for using group writing in your classes and on the job.

Benefits and Drawbacks of Group Writing

Most organizations rely on people working together throughout the writing process to produce documents. The success of writing projects depends on information and skills contributed by varied employees. For example, a proposal writing team may include technical specialists, marketing experts, graphic designers, word processors, and technical editors. The company depends on all these individuals working together to produce the final product, a first-rate proposal.

In group writing, however, the whole is greater than the sum of the parts. In other words, benefits go beyond the collective specialties and experience of individual group members. Participants create *new* knowledge as they plan, draft, and edit their work together. They become better contributors and faster learners simply by being a part of the social process of a team. Discussion with fellow participants moves them toward new ways of thinking and inspires them to contribute their best. This collaborative effort yields ideas, writing strategies, and editorial decisions that result from the mixing of many perspectives.

Of course, group writing does have drawbacks. Most notably, the group must make decisions without falling into time traps that slow down the process. There must be procedures for getting everyone's ideas on the table *and* for reaching decisions on time. A leader with good interpersonal skills will help the group reach its potential, whereas an indecisive or autocratic leader will be an obstacle to progress. Good leadership rests at the core of every effective writing team.

In addition to good leadership, shared decision-making is at the heart of every successful writing team. Group writing is not one writer simply getting information from many people before he or she writes a draft. Nor is it one person writing a draft for the editorial red pen of individuals at higher levels. These two models may have their place in some types of company writing, but they do not constitute group writing. Instead, participants in a group must work together during the planning, drafting, and revising stages of writing. Although the degree of collaboration may vary, all forms of group writing differ considerably from the model of one person writing with only occasional help from others.

Guidelines for Group Writing

This section offers six pointers for group writing, to be used in this course and throughout your career. The suggestions concern the writing process as well as interpersonal communication.

■ Group Guideline 1: Get to Know Your Group

Most people are sensitive about strangers evaluating their writing. Before collaborating on a writing project, therefore, learn as much as you can about those with whom you will be working. Drop by their offices before your first meeting, or talk informally as a group before the writing process begins. In other words, first establish a personal relationship. This familiarity will help set the stage for the spirited dialogue, group criticism, and collaborative writing to follow.

■ *Group Guideline 2: Set Clear Goals and Ground Rules*

Every writing group needs a common understanding of its objectives and procedures for doing business. Either before or during the first meeting, these questions should be answered:

1. What is the group's main objective?
2. Who will serve as team leader?
3. What exactly will be the leader's role in the group?
4. How will the group's activities be recorded?
5. How will responsibilities be distributed?
6. How will conflicts be resolved?
7. What will the schedule be?
8. What procedures will be followed for planning, drafting, and revising?

The guidelines that follow offer suggestions for answering the preceding questions.

■ *Group Guideline 3: Use Brainstorming Techniques for Planning*

The term *brainstorming* means to pool ideas in a *nonjudgmental* fashion. In this early stage, participants should feel free to suggest ideas without criticism by colleagues in the group. This nonjudgmental approach does not come naturally to most people. Thus the leader may have to establish ground rules for brainstorming before the group proceeds.

Here is one sample approach to brainstorming:

Step 1. The group recorder takes down ideas as quickly as possible.
Step 2. Ideas are written on large pieces of paper affixed to walls around the meeting room so all participants can see how major ideas fit together.
Step 3. Members use ideas as springboards for suggesting others.
Step 4. The group takes some time to digest ideas generated during the first session before meeting again.

Results of a brainstorming session might look much like a nonlinear outline produced during a solo writing project (see Figure 1–8). The goal of both is to generate as many ideas as possible, which can be culled and organized later.

■ *Group Guideline 4: Use Storyboarding Techniques for Drafting*

Storyboarding helps propel participants from the brainstorming stage toward completion of a first draft. It also makes visuals an integral part of the document. Originating in the screenwriting trade in Hollywood, a storyboard is a sheet of paper that contains (1) one draft-quality illustration and (2) a series of sentences about one topic. (See Figure 1–10.) As applied to technical writing, the technique involves six main steps:

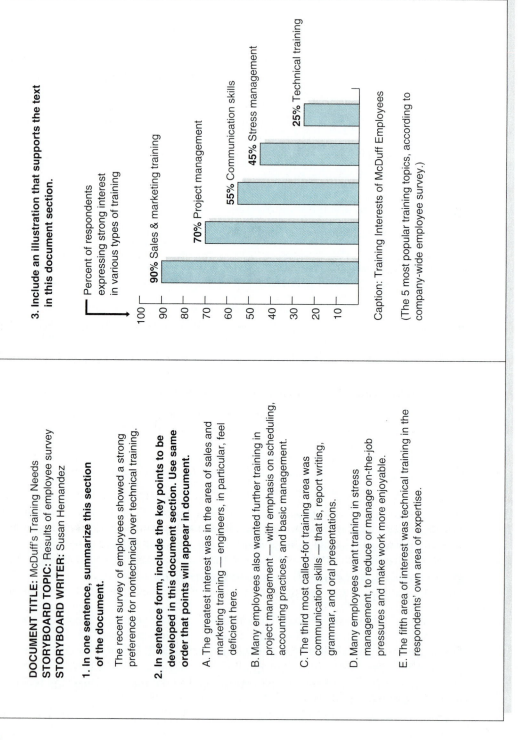

DOCUMENT TITLE: McDuff's Training Needs
STORYBOARD TOPIC: Results of employee survey
STORYBOARD WRITER: Susan Hernandez

1. In one sentence, summarize this section of the document.

The recent survey of employees showed a strong preference for nontechnical over technical training.

2. In sentence form, include the key points to be developed in this document section. Use same order that points will appear in document.

A. The greatest interest was in the area of sales and marketing training — engineers, in particular, feel deficient here.

B. Many employees also wanted further training in project management — with emphasis on scheduling, accounting practices, and basic management.

C. The third most called-for training area was communication skills — that is, report writing, grammar, and oral presentations.

D. Many employees want training in stress management, to reduce or manage on-the-job pressures and make work more enjoyable.

E. The fifth area of interest was technical training in the respondents' own area of expertise.

3. Include an illustration that supports the text in this document section.

Percent of respondents expressing strong interest in various types of training

90% Sales & marketing training
70% Project management
55% Communication skills
45% Stress management
25% Technical training

Caption: Training Interests of McDuff Employees

(The 5 most popular training topics, according to company-wide employee survey.)

FIGURE 1–10
Completed storyboard

Step 1. The group or its leader assembles a topic outline from ideas brought forth during the brainstorming session.

Step 2. All group members are given one or more topics to develop on storyboard forms.

Step 3. Each member works independently on the boards, creating an illustration and a series of subtopics for each main topic (see Figure 1–10).

Step 4. Members meet again to review all completed storyboards, modifying them where necessary and agreeing on key sentences.

Step 5. Individual members develop draft text and related graphics from their own storyboards.

Step 6. The group leader or the entire group assembles the draft from the various storyboards.

■ *Group Guideline 5: Agree on a Thorough Revision Process*

As with drafting, all members usually help with revision. Team editing can be difficult, however, as members strive to reach consensus on matters of style. Here are some suggestions for keeping the editing process on track:

■ Avoid making changes for the sake of individual preference.

■ Search for areas of agreement among group members, rather than those of disagreement.

■ Make only those changes that can be supported by accepted rules of style, grammar, and usage.

■ Ask the group's best all-round stylist to make a final edit.

This review will help produce a uniform document, no matter how many people work on the draft.

■ *Group Guideline 6: Use Computers to Communicate*

When team members are at different locations, computer technology can be used to complete some or all of the project. Team members must have personal computers and the software to connect their machines to a network, allowing members to send and receive information on-line. This section describes three specific computer applications that can improve communication among members of a group-writing project: e-mail, computer conference, and groupware. Chapter 4 ("Page Design") will discuss how the individual writer can use computers to plan, draft, and revise copy. Chapter 7 ("Letters and Memos") will discuss stylistic features of electronic mail.

■ **Electronic Mail (e-mail):** Individuals can send and receive messages from their office computers or from remote locations. Like written memos, e-mail messages usually include the date, sender, receiver, and subject. Unlike written memos, the style is often quite informal. Messages are sent at a time convenient for the sender and saved until a time that readers check their mail.

- **Computer Conference:** Members of a group can make their own comments and respond to comments of others on a specific topic or project. Computer conferences may be open to all interested users or open only to a particular group. For the purposes of group writing, the conference probably would be open only to members of the writing team. A leader may be chosen to monitor the contributions and keep the discussion focused. Contributions may be made over a long period, as opposed to a conventional face-to-face meeting wherein all team members are present at the same time. Accumulated comments in the conference can be organized or indexed by topic. The conference may be used to brainstorm and thus to generate ideas for a project, or it may be used for comments at a later stage of the writing project.
- **Groupware:** Team members using this software can work at the same time, or different times, on any part of a specific document. Groupware that permits contributions at the same time is called "synchronous"; that which permits contributions at different times is called "asynchronous." Because team members are at different locations, they may also be speaking on the phone at the same time they are writing or editing with synchronous groupware. Such sophisticated software gives writers a much greater capability than simply sending a document over a network for editing or comment. They can collaborate with team members on a document at the same time, almost as if they were in the same room. With several windows on the screen, they can view the document itself on one screen and make comments and changes on another screen.

Computers can be used to overcome many obstacles for writers and editors in different locations. Indeed, electronic communication can help to accomplish all the guidelines just noted. Specifically, (1) e-mail can be used by group members to get to know each other, (2) e-mail or a computer conference can be used to establish goals and ground rules, (3) synchronous, or real-time, groupware can help a team brainstorm about approaches to the project (and may, in fact, encourage more openness than a face-to-face brainstorming session), (4) computer conferences combined with groupware can approximate the storyboard process, and (5) either synchronous or asynchronous groupware can be used to approximate the editing process. Granted, such techniques lack the body language used in face-to-face meetings. Yet when personal meetings are not possible, computerized communication can provide a substitute that allows writers in different locations to work together to meet their deadline.

Team writing may play an important part in your career. If you use the preceding techniques, you and your team members will build on each other's strengths to produce top-quality writing.

CHAPTER SUMMARY

Technical writing refers to the many kinds of writing you will do in your career. In contrast to most academic writing, technical writing aims to get something done (not just to demonstrate knowledge), relays information from someone more knowledgeable about the topic (you) to someone less knowledgeable about it (the reader), and is read by people from mixed technical and decision-making levels.

For each writing project, you should complete a three-stage process of planning, drafting, and revising. Planning involves understanding your purpose, knowing the readers' needs, collecting information, and outlining major and minor points. In the drafting stage, you use the outline to write a first draft as quickly as possible—without stopping to make changes. Finally, the revision process requires that you adjust content and then edit for style, grammar, and mechanics.

Technical writing can be completed by you alone or by you as a member of a writing team. The latter approach is common in companies, for it exploits the strengths of the varied professionals within an organization. In group writing you work closely with your team during the planning, drafting, and revising processes.

ASSIGNMENTS

Your instructor will indicate whether assignments 1 through 5 should serve as the basis for class discussion, for a written exercise, or for both. Assignment 6 requires a written response.

1. **Features of Academic Writing.**

   Option A. Select an example of writing that you wrote for a college course other than this one. Then prepare a brief analysis in which you explain (1) the purpose of the writing sample, (2) the audience for which it was intended, and (3) the ways in which it differs from technical writing, as defined in this chapter.

   Option B. As an alternative to using your own example, complete the assignment by using the following example. Assume that the passage was written as homework or as an in-class essay in an environmental science class in college.

   > There are many different responses that are possible in the event toxic waste contamination is suspected or discovered at a site. First, you can simply monitor the site by periodically taking soil and/or water samples to check for contamination. This approach doesn't solve the problem and may not prove politically acceptable when contamination is obvious to the community, but it does help determine the extent of the problem. A second approach—useful when contamination is likely or proved—is to contain the toxic waste by sealing off the site in some fashion, such as by building barriers between it and the surrounding area or by "capping" it in some manner (as in the case of a toxic waste pit). Basically, this alternative depends on the ability to isolate the toxic substances effectively. A third strategy, useful when the contamination is liquified (as with toxic groundwater), is to pump the water from under the ground or from surface ponds and then transport it to treatment systems.
   >
   > A fourth method is appropriate when toxic substances need to be treated on-site, in which case they can be incinerated or they can be solidified at the site in some way. Then they can be placed in a landfill at the site. Fifth, waste can be hauled to another location where it can be incinerated or placed in some kind of secure landfill—when an off-site disposal approach is needed.

2. **Features of Technical Writing.**

    Option A.   Locate an example of technical writing (such as by borrowing it from a family member or an acquaintance who works in a technical profession). Then prepare a brief analysis in which you explain (1) the purpose for which the piece was written, (2) the apparent readers and their needs, (3) the way in which the example differs from typical academic writing, and (4) the relative success with which the piece satisfies this chapter's guidelines.

    Option B.   Using the following brief example of technical writing, prepare the analysis requested in Option A.

DATE: June 15, 1993
TO: Pat Jones, Office Coordinator
FROM: Sean Parker, Word-Processing Operator
SUBJECT: New Word-Processing Software

Introductory Summary
As you requested, I have examined the WordWonder word-processing software we are considering. On the basis of my observations, I recommend we secure one copy of Word-Wonder and test it in our office for two months. Then after comparing it to the other two packages we have tested, we can choose one of the three word-processing packages to use throughout the office.

Features of WordWonder
As we agreed, my quick survey of WordWonder involved reading the user's manual, completing the orientation disk, and meeting with a salesperson from the company. Here are the five features of the package that seemed most relevant to our needs:

1. **Formatting Flexibility:** WordWonder includes diverse "style sheets" to meet our needs in producing reports, proposals, letters, memos, articles, and even brochures. By engaging just one command on the keyboard, the user can change style sheets—whereby the program will automatically place text in a specified format.
2. **Mailers:** For large mailings, we can take advantage of WordWonder's "Mail Out" feature that automatically places names from mailing lists on form letters.
3. **Documentation:** To accommodate our staff's research needs, WordWonder has the capacity to renumber and rearrange footnotes as text is being edited.
4. **Page Review:** This package's "PagePeek" feature permits the user to view an entire written page on the screen. Without having to print the document, he or she can then see how every page of text will actually look on the page.
5. **Tables of Contents:** WordWonder can create and insert page numbers on tables of contents, created from the headings and subheadings in the text.

Conclusion
Though I gave WordWonder only a brief look, my survey suggests that it may be a strong contender for use in our office. If you wish to move to the next step of starting a two-month office test, just let me know. Then I will make arrangements with the manufacturer for us to receive a complimentary trial copy.

---

3. **Purpose and Audience.** The following examples deal with the same topic in four different ways. Using this chapter's guidelines on purpose and audience, determine the

main reason for which each excerpt was written and the technical level of the intended readers.

A. You can determine the magnitude of current flowing through a resistor by use of this process:
   - Connect the circuit (power supply, resistor, ammeter, voltmeter).
   - Set the resistor knob to a setting of "1."
   - Turn the voltage adjusting knob to the left until it stops rotating.
   - Switch the voltmeter to "on" and make sure it reads 0.00 volts.
   - Switch the power supply to "on."
   - Slowly increase the voltage on the voltmeter from 0 to 10 volts.
   - Take the reading from the ammeter to determine the amount of current flowing through the resistor.

B. After careful evaluation of several testers, I strongly recommend that Langston Electronics Institute purchase 100 Mantra Multitesters for use in our laboratories in Buffalo, Albany, and Syracuse.

C. Selected specifications for the Ames Multitester are as follows:
   - Ranges.................... 43
   - DC Voltage............... 0-125-250mV 1.25-2.5-10-5-125-500-1000V
   - AC Voltage............... 0-5-25-125-250-500-1000V
   - DC Current .... 0-25-50μA-2.5-5-25-50-250-500mA-10 amperes
   - Resistance .... 0-2K-20K-200K-20 Mega ohms
   - Decibels.................. −20 to +62 in db 8 ranges
   - Accuracy................. ±3% on DC measurements
                                ±4% on AC measurements
                                ±3% on scale length on resistance
   - Batteries .... one type AA penlight cell
   - Fuse .... 0.75A at 250V

   Note that the accuracy rate for the Ames is within our requirements of ±6%, and is considerably lower than the three other types of testers currently used by our staff.

D. Having used the Ames Multitester in my own home laboratory for the last few months, I found it extremely reliable during every experiment. In addition, it is quite simple to operate and includes clear instructions. As a demonstration of this operational ease, my 10-year-old son was able to follow the instructions that came with the device to set up a functioning circuit.

4. **Interview.** Interview a friend, relative, or recent college graduate who works as a technical professional or manager. Gather specific information on these topics:
   - The percentage of the workweek spent on writing
   - Types of documents that are written and their purpose
   - Specific types of readers of these documents

5. **Contrasting Styles.** Find two articles on the same topic in a professional field that interests you. One article should be taken from a newspaper or magazine of general interest, such as one you would find on a newsstand. The other should be from a magazine or journal written mainly for professionals in the field you have chosen. Now contrast the two articles according to purpose, intended audience, and level of technicality.

6. **Group Writing.** Your instructor will divide you into groups of three to five members for completing this assignment. Follow the guidelines for group writing in this chapter. (If your instructor decides that the storyboarding rule is impractical for this exercise, bypass that step.)

   Your objective is to write a brief evaluation of the teaching effectiveness of either the room in which your class is held or some other room or building of your instructor's choice. In following the tasks listed in this chapter, the group must establish criteria for evaluation, apply these criteria, and report on the results.

Your brief report should have three parts: (1) a one-paragraph summary of the room's effectiveness, (2) a list of the criteria used for evaluation, and (3) details of how the room met or did not meet the criteria you established.

Besides preparing the written report, be prepared to discuss the relative effectiveness with which the group followed this chapter's guidelines for group writing. What problems were encountered? How did you overcome them? How would you do things differently next time?

7. **Computerized Communication.** If your campus computer facilities permit, set up an e-mail system with members of a writing group to which you have been assigned by your instructor. Decide on a topic upon which you and your team members will comment. Each member should make four comments, or "postings." At least one posting should be an original comment, and at least one should be a response to another member's comment. Print the group's copy and submit it to your instructor. Depending on the instructions you have been given, this assignment may be independent or it may be related to a larger group-writing assignment.

2 | *Introduction to McDuff, Inc.*

Most employees at McDuff's corporate office in Baltimore depend on good writing for success in their jobs.

C hapter 1 defines technical writing and the writing process. This second chapter introduces you to McDuff, Inc., the fictional company referred to throughout this book. Then chapter 3 covers effective organization of information, and chapter 4 shows you techniques of page design. Together, this four-chapter package provides a foundation for your work in the rest of *Technical Writing: A Practical Approach.*

The first section of this chapter gives the rationale for using McDuff in this book. It also describes the main features of any company's "culture," with special emphasis on the search for quality among organizations in the 1990s. The rest of the chapter describes (1) the history of McDuff, (2) types of work done by the firm, (3) responsibilities of the corporate office, and (4) job descriptions of employees at the branch offices. The chapter concludes with a list of the most common writing tasks at McDuff, Inc. The color insert shows McDuff's worldwide locations and describes seven specific projects, with illustrations. These project sheets will be the basis for some assignments in later chapters.

MCDUFF, INC., AND THE WORKING WORLD

McDuff provides you with a window into the world of work. After graduation you may find yourself employed by an organization much like this company. All organizations—from profit-making companies like McDuff to nonprofit groups like colleges—have the same basic challenge: They must coordinate the efforts of diverse employees to accomplish common goals.

This section points out the relevance of McDuff material to the rest of the text. It also describes three generic features of any firm's culture and highlights one cultural feature—the drive for *quality*—that has attracted special attention. Today the quality revolution is having a large impact on every aspect of corporate life, including written and spoken communication.

Using McDuff, Inc., in This Book

McDuff, Inc., employs people in jobs much like those you will have. These employees encounter communication problems similar to ones faced by professionals in all walks of life. The use of McDuff, therefore, yields two main benefits for you as a student:

1. Real-World Context: McDuff provides an extended case study in modern technical communication. By placing you in actual working roles, the text prepares you for the writing and speaking tasks ahead in your career.

2. Continuity: The use of McDuff material will lend continuity to class assignments. Threading its way through all your course work will be the life of this one company. Intensive work with a realistic organization stresses connections among, and the cumulative nature of, all on-the-job assignments.

Be prepared to return to this chapter throughout the course. Here you will find information to help you with many later assignments. It will place you squarely in the typical working world of college graduates.

Elements of a Company's Culture

Unless you become self-employed, you will work for some sort of business enterprise or nonprofit organization after you leave college. For simplicity here, we'll use the term *company* or *firm* to refer to any organization where you may work. As noted in chapter 1, the writing you do in a company differs greatly from the writing you do in college. The stakes on the job are much higher than a grade on your transcript. Writing will directly influence your performance evaluations, your professional reputation, and your company's productivity and success in the marketplace. Given these high stakes, let's look at typical features of the organizations wherein you may spend your career.

Starting a job is both exciting and, sometimes, a bit intimidating. Although you look forward to practicing skills learned in college, you also wonder just how you will fare in new surroundings. Soon you discover that any organization you join has its own personality. This personality, or "culture," can be defined as follows:

> **Company culture:** current term for the main features of life at a particular company. A company's culture is influenced by the firm's history, type of business, management style, values, attitude toward customers, and attitude toward its own employees. Taken together, all features of a particular company's culture create a definable quality of life within the living organism of that company.

Before examining the culture at McDuff, let's look more closely at three features mentioned in the preceding definition: a firm's history, its type of business, and its management style.

■ *Feature 1: Company History*

A firm's origin often is central to its culture. On the one hand, the culture of a 100-year-old steel firm will depend on accumulated traditions to which most employees are accustomed. On the other hand, the culture of a recently established home electronics firm may depend more on the entrepreneurial spirit of its founders. Thus the facts, and even the mythology, of a company's origin may be central to its culture, especially if the person starting the firm remained at the helm for a long time. WalMart, for instance, possesses a culture very much connected to the dreams, aspirations, and open management style of its founder, the late Sam Walton. And IBM, while facing many changes in the computer marketplace during its history, still reflects the strong research-and-development orientation of its early leaders.

■ *Feature 2: Type of Business*

Culture is greatly influenced by a company's type of business. Many computer software firms, for example, are known for their flexible, nontraditional, and sometimes chaotic culture. Such firms encourage constant change and innovation. The industry's well-known competitiveness probably inspires this cultural trait. Some of the large computer hardware firms, however, have a culture focused more on tradition, formality, and custom. Yet both hardware and software companies share the same cultural trait of a high level of customer service.

■ *Feature 3: Management Style*

One major component of a company's culture is its style of leadership. Some companies run according to a rigid hierarchy, with decisions originating at the top. Other companies have fewer top-down pronouncements from upper management. Instead, they involve a wide range of employees in the decision-making process. As you might expect, most firms have a decision-making culture somewhere between these two extremes.

These three features give you some idea of what the culture can comprise in any company. A company's culture influences who is hired and promoted at the firm, how decisions are made, and even how reports and other company documents are drafted and reviewed. Let's examine one final feature—the emphasis on quality—which is becoming a part of the life of many companies throughout the world, including McDuff.

The Search for Quality

"Total quality management" and similar phrases have become buzzwords of the 1990s. Though overused, they are a good sign of industry's effort to give customers better goods and services. The many spokespersons in this movement agree that the search for quality must permeate every part of an organization. It especially applies to employee productivity and customer service. This section highlights several features of this quality revolution, with special emphasis on the relationship to a company's written and oral communications.

■ *Quality Feature 1: Putting the Customer First*

Most quality experts agree that any successful company must keep its finger on the pulse of customers, never taking them for granted. Putting the customer first starts with the design of products and services. It also is reflected in the efficiency and sincerity with which complaints are handled.

In the area of communication, this responsiveness to the customer can be seen in the way a writer strives to understand the reader's needs before a report is written. The reader may be a manager within the writer's company, or a customer outside the firm. From the quality perspective, both internal and external readers alike are "customers."

■ *Quality Feature 2: Stressing Teamwork Over Internal Competition*

Many quality experts believe we spend too much time pitting employees against each other, as in the way we distribute merit raises and give commissions on sales. These experts suggest that companies should focus attention on team goals, to promote the good of the entire company.

In communication, teamwork can translate into the need to increase the use of group writing (see chapter 1). Collaborative efforts help draw on specialties of writers, technical experts, editors, graphics specialists, and others to produce the best possible document. A culture that encourages group writing reduces the "lone ranger" mentality in communication. Employees work *together* for the good of the final written product.

■ *Quality Feature 3: Giving People the Freedom to Do Their Jobs*

Too often, management gets in the way of employee productivity by drawing a box around each person's responsibilities. Quality-centered firms, instead, give everyone more power to suggest changes that will improve the business. If employees believe their opinions are important to the overall plan of the company, they will be motivated to work harder and smarter.

In communication, this quality feature might mean that a good word processor should feel comfortable recommending improvements in writing style, rather than simply typing words without regard for the quality of the product. It might also mean that new hires, right out of school, would be encouraged to apply writing skills they learned in college to improve the company's report format, rather than simply following whatever format is handed to them by supervisors. Quality-based firms empower all employees to suggest improvements in the way the company does business.

■ *Quality Feature 4: Thinking Long Term, not Short Term*

Most quality experts agree that focusing on short-term profits works against the long-term well-being of organizations. Instead, companies should cultivate permanent, trusting relationships with suppliers and with customers. Although such relationships may not produce immediate profits or savings, they will pay off in the end by reducing cyclical ups and downs in profits. Another long-term strategy is to support company training of all types for *all* levels of employees. In particular, people can be cross-trained in other jobs to avoid burnout, to help them understand more about the firm's business, and to prepare them for career advancement.

In communication, investing in the long term can mean training employees to write and speak well. One such investment is to produce an up-to-date writing manual. This manual should specify guidelines for format and style and provide excellent models of all company documents. Most important, it should be followed by employees at all levels.

The concern for quality offers hope of changing the culture of companies throughout the world. It provides a framework within which you can view all your work as a professional—and as a communicator.

HISTORY OF MCDUFF, INC.

Just out of Georgia Tech in 1944, Rob McDuff spent several years as a civil engineer in the U.S. Army Corps of Engineers at the end of World War II. Then in 1947, he started a small engineering consulting firm in Baltimore, Maryland, where he had grown up. This firm's specialty was doing consulting work for construction firms and real estate developers. Specifically, McDuff, Inc., tested soils and then recommended foundation designs for structures that were being proposed.

In the early days, Rob McDuff did much of the fieldwork himself. He also analyzed the data, wrote reports, and did the marketing for new business. Work progressed well, and his new company earned a reputation for high-quality service. Now Rob McDuff nostalgically looks back upon those days as some of the most satisfying of his career. He had seized the opportunity to fulfill a dream that many people still have today: starting a business and then using skill, hard work, and imagination to make it grow.

From its founding in 1947 until about 1950, the company worked mostly for construction firms in the Baltimore area. Each job—whether a building, dam, or highway—involved tasks like those listed in Figure 2–1. The work was not glamorous. Yet it provided an important service. By the mid-1950s, the firm enjoyed a first-rate reputation. It had offices in Baltimore and Boston and about 80 employees.

McDuff, Inc., kept growing steadily, with a large spurt in the 1960s and another in the 1980s. The first was tied to increased oil exploration in all parts of the world. Oil firms needed experts to test soils, especially in offshore areas. The results of these projects were used to position oil rigs at locations where they could withstand rough seas. The second growth period was tied to environmental work

McDuff Sequence of Tasks
Typical Construction Job:
1947-1950

1. **Reviewing** whatever information was already on file about the construction site

2. **Visiting** the site to observe and record surface features such as rock formations or waste dumps

3. **Drilling** one or more borings (deep, cylindrical holes) into the earth, using special equipment to collect the dirt and rock samples from various depths

4. **Testing** the samples back in the office laboratory

5. **Analyzing** the laboratory data to come up with recommendations for foundation design

6. **Writing** a report that records project activities and specific recommendations for the client

7. **Observing** construction activities, like the pouring of concrete slabs, to be sure correct procedures are followed

FIGURE 2–1
McDuff sequence of tasks, typical construction job: 1947–1950

required by the federal government, state agencies, and private firms. McDuff became a major player in the waste-management business, consulting with clients about ways to store or clean up hazardous waste.

Today, as it nears its fiftieth birthday, McDuff, Inc., has about 1200 employees. There are nine offices in the United States and six overseas, as well as a corporate headquarters in Baltimore that is separate from the Baltimore branch office. McDuff in the 1990s performs a wide variety of work. What started as a consulting engineering firm has expanded into one that does both technical and nontechnical work for a variety of customers.

TYPES OF PROJECTS

Every company must improve its products and services to stay in business. McDuff is no exception. If it had stayed just with soils testing work, the company would be stagnant today. Periodic slowdowns in the construction and oil industries would have taken their toll. Fortunately, the company diversified. These are its seven main project areas today:

1. Soils work on land: Still the company's bread-and-butter work, these projects involve (a) taking soil samples, (b) testing samples in the laboratory, and (c) making design recommendations for foundations and other parts of office buildings, dams, factories, subdivisions, reservoirs, and mass transit systems.

When done well, this kind of work helps to prevent later problems, like cracks in building walls.

2. Soils work at sea: Now a smaller market than it was in the 1960s and 1970s, this geological and engineering work used to be done exclusively for oil and gas companies. It helped them to place offshore platforms at safe locations or to select drilling locations with the best chance of hitting oil. Now, however, McDuff also is hired by countries and states who want to preserve the ecologically sensitive offshore environment. By collecting and analyzing data from its ship, *Dolphin,* McDuff helps clients decide whether an offshore area should be preserved or developed.

3. Construction monitoring: Besides designing parts of structures, McDuff also helps with construction. Here are some services it offers during the construction process:

- Checking the quality of concrete being poured into structures like cooling towers for nuclear power plants
- Watching construction workers to make sure they follow proper procedures
- Testing the strength of concrete and other foundation materials, once they are put in place

4. Construction management: About 10 years ago, McDuff got into the business of actually managing projects other than its own jobs. Large construction companies hire McDuff to orchestrate all parts of a project so that it is completed on time. The work involves these activities:

- Establishing a schedule
- Observing the work of subcontractors
- Regularly informing contractors about job progress

5. Environmental management: In the early 1970s, Rob McDuff began to realize that garbage—all kinds of it—could mean big business for his firm. Suddenly the United States and other countries faced major problems caused by the volume of current wastes and by improper disposal of wastes since World War II. As McDuff's fastest growing market, environmental management work can involve one or more of these tasks:

- Testing surface soil and water for toxic wastes
- Drilling borings to see if surface pollution has filtered into the groundwater
- Designing a cleanup plan
- Supervising the cleanup
- Predicting the impact of proposed projects on the environment
- Analyzing the current environmental health of wetlands, beaches, national forests, lakes, and other areas

As Rob McDuff had hoped, managing wastes and determining the environmental impact of proposed projects proved to be excellent markets. Although there is growing competition, the company got into the business early enough to establish a good reputation for reliable, affordable work.

This color insert provides a glimpse into the world of McDuff, Inc. In addition to the map below, the insert contains information about seven specific projects completed by McDuff. Each page includes an illustration and project overview. The assignments throughout the book ask you to make use of information from the project sheets.

Worldwide Locations of McDuff, Inc., Offices

U.S. Locations

1. Corporate headquarters—Baltimore, Maryland
2. Baltimore, Maryland
3. Boston, Massachusetts
4. Atlanta, Georgia
5. Houston, Texas
6. Cleveland, Ohio
7. St. Paul, Minnesota
8. St. Louis, Missouri
9. Denver, Colorado
10. San Francisco, California

Non-U.S. Locations

11. Caracas, Venezuela
12. London, England
13. Munich, Germany
14. Nairobi, Kenya
15. Dammam, Saudi Arabia
16. Tokyo, Japan

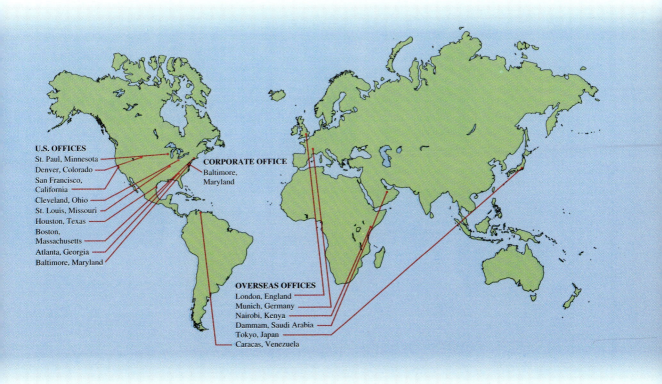

U.S. OFFICES
St. Paul, Minnesota
Denver, Colorado
San Francisco, California
Cleveland, Ohio
St. Louis, Missouri
Houston, Texas
Boston, Massachusetts
Atlanta, Georgia
Baltimore, Maryland

CORPORATE OFFICE
Baltimore, Maryland

OVERSEAS OFFICES
London, England
Munich, Germany
Nairobi, Kenya
Dammam, Saudi Arabia
Tokyo, Japan
Caracas, Venezuela

McDuff

Brief Project Description

McDuff worked from January through April, 1992, on field and laboratory work preceding construction of Sentry Dam. After submitting its geological and engineering report, the McDuff team worked with the water district on final dam design. It also gave some help during construction. The dam was completed in July of 1993. Because Sentry is a high-hazard dam—meaning that its failure would cause loss of life—safety was crucial.

Main Technical Tasks

- Drilled 20 soil borings at the site to sample soil and rock
- Drilled 15 test wells to find the depth to groundwater and to check on water seepage in the dam area
- Lab tested the soil, shale, and other samples from borings to evaluate the strength of material on which the dam would rest
- Designed an overflow spillway that would be anchored in strong bedrock, not weak shale
- Monitored water seepage in the foundation and dam during construction
- Certified the dam's safety after construction

Main Findings or Benefits

- Completed field and lab work on schedule and at budget
- Designed innovative concrete dam spillway that bypassed weak shale and connected with strong bedrock

Brief Project Description

During the spring of 1992, McDuff used its drillship *Dolphin* to examine the ocean floor over a 60-mile stretch off the California coast. We collected data on site and then tested and analyzed samples at our labs. After sending them our report on the study, we met with the client to help arrive at final conclusions and recommendations for further offshore use of the coastline.

Main Technical Tasks

- Kept *Dolphin* site for two months to map the seafloor, to drill borings, and to observe ocean habitats
- Used sonar to develop a profile of the surface of the ocean floor and its near-surface geology (return time of sound waves helped gauge the depth to the floor and to sediments below the floor)
- Drilled successfully for samples from *Dolphin's* drilling platform, often in difficult weather
- Analyzed samples from the borings to estimate geological age and stability of the ocean floor
- Viewed ocean life and geology firsthand at some locations, using a small submersible craft with a one-person crew

Main Findings or Benefits

- Concluded that most of the zone was too environmentally and geologically sensitive to be used for offshore drilling of oil and gas
- Found two locations where a pipeline might be safely placed, with minimum damage to ocean life and minimum risk of geological disturbance (such as earthquake or ocean avalanche)

PROJECT: *Monitored Construction of Grant Hospital*

CLIENT: *Floor County, Ohio*

McDuff

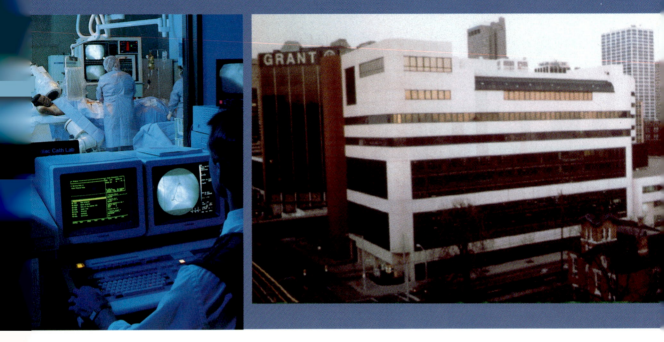

Brief Project Description

McDuff was hired to observe construction of Grant Hospital to make sure that all construction work was done according to agreed-on standards and legal specifications. McDuff had one or more employees on site continuously during the entire construction process.

Main Technical Tasks

- Checked quality of masonry, steel, and other materials used in the building's foundation and main structure
- Ascertained that heating, plumbing, and other systems were installed according to legal codes and according to contract
- Served as liaison between medical staff and construction personnel so that interior construction was done correctly
- Completed final "sign-off" for entire facility before it opened

Main Findings or Benefits

- Guaranteed client that all materials and procedures used in construction were up to contract standards
- Spent over 50 extra hours (without charge) consulting with doctors, nurses, and other technical staff about interior construction and placement of major pieces of equipment
- Billed client at 10% below proposed fee because of other work recently contracted with the county for the same calendar year

PROJECT: *Managed Construction of Nevada Gold Dome*

CLIENT: *City of Rondo, Nevada*

Brief Project Description

Rondo, Nevada, recently received permission to start a new football team, the Nevada Gamblers. The city is building a domed stadium, called the Gold Dome, for the new franchise. Because of concerns about scheduling the various construction firms and subcontractors involved in the project, the client hired McDuff to manage the entire project—from groundbreaking to occupancy.

Main Technical Tasks

- Established construction schedule that would allow for maximum overlapping of work by different subcontractors
- Held daily coordination meetings with a group of principals that represented every contractor on site that day
- Handled schedule delays by dealing immediately with contractors and subcontractors
- Used McDuff's patented "TimeTrack" scheduling software to stay aware of the entire flow of work at site

Main Findings or Benefits

- Had entire project completed two weeks before the deadline
- Saved client $100,000 by eliminating proposed overtime work that was not needed once scheduling was fine-tuned
- Arranged schedule so that team owners and staff members could complete a walkthrough of the facility at key points during the construction process

PROJECT: Examined Big Bluff Salt Marsh
CLIENT: State of Georgia

Brief Project Description

Georgia hired McDuff to explore the environmental quality of the Big Bluff Salt Marsh, located on Paradise Island off the Atlantic coast. The island is owned by the state. Developers have approached the state about buying island land for building condominiums and other tourist-related structures. McDuff's field work, lab work, and research resulted in a report about the level of development that the marsh can tolerate.

Main Technical Tasks

- Took inventory of wildlife and grasses throughout the marsh
- Consulted with wetlands experts around the U.S. about notable features of Big Bluff Salt Marsh
- Tested soil and water for current levels of pollution
- Researched two other Atlantic coast salt marshes where development has occurred to determine compatibility of development and marshes

Main Findings or Benefits

- Concluded that Big Bluff Salt Marsh serves as a nursery and feeding ground for fish caught commercially along the coast
- Learned from Environmental Protection Agency (EPA) that any major development of Big Bluff could be a violation of federal wetlands policy and thus could be challenged by the EPA
- Recommended that the state sell land next to marsh only if it would be used for low-impact activities—such as day trips by visitors—not for construction of homes and businesses

PROJECT:
Designed and
Installed Control
Panel for
Nuclear Plant

CLIENT:
Russian
Government

McDuff

Brief Project Description

Since the break-up of the old Soviet Union, some individual republics have sought Western assistance in updating their nuclear plants. McDuff's safety experts and mechanical engineers were hired to design a new control panel and to retrofit it into an existing plant. The panel was designed, manufactured, installed, and tested by McDuff—with the help of several subcontractors.

Main Technical Tasks

- Spent one week at site observing operators using old panel
- Hired ergonometric and nuclear power experts to help evaluate old panel design and to suggest features of new design
- Designed and manufactured panel
- Installed panel at Russian plant and observed one full week of testing, when panel was used at plant under simulated conditions
- Remained on site for three days after full power was resumed so we could continue training operators on use of new panel

Main Findings or Benefits

- Designed panel that international experts considered to be as safe as any currently in use
- Stayed on schedule, keeping the plant out of use only two weeks

PROJECT: *Designed and Taught Seminar in Technical Writing*
CLIENT: *Government of Germany*

McDuff

Brief Project Description

The government of Germany has greatly increased the number of technical experts in departments at the capital of Bonn. McDuff's Munich office was selected to design and teach an in-house technical writing seminar for 20 mid-level employees. They work in agriculture, health, and engineering. Though in different fields, the individuals write the same types of reports.

Main Technical Tasks

- Met with members of seminar and their managers to determine needs of the group
- Examined sample reports from all participants
- Studied the government's style guidelines
- Designed and taught a three-day seminar, using a manual of guidelines and in-house samples tailored for the group
- Evaluated actual on-the-job reports written by participants after the seminar

Main Findings or Benefits

- Received "very good" to "excellent" ratings from all 20 participants on the written critiques completed on the last day of the course
- Wrote a final report to client that documented improvement shown in participants' reports written after the seminar, as compared with those written before the seminar

6. Equipment development: Here the firm departs from its traditional emphasis on services and instead produces products. The ED group, as it is known, designs and builds specialized equipment, both for McDuff's own project needs and for its clients. As the company's newest and most innovative group, it takes on a variety of projects. For example, it is building mechanisms as diverse as a prototype for a new device to test water pollution levels, on the one hand, and a new instrument gauge to install in tractors, on the other. Although the ED group is based in the Baltimore corporate office, it is mobile enough to go to other offices—even in other countries—to complete projects.

7. Training: McDuff entered the training business about five years ago, when it realized that there was a good market for technical training in skills represented by the firm. Recently the Training Department also started offering nontechnical training in areas such as report writing, since the company employs several writers who are excellent trainers.

These seven project areas reflect McDuff's diversity. Though starting as a traditional engineering firm, McDuff has sought out new markets and become a scrappy competitor in many areas. Still an active company president at age 72, Rob McDuff likes to think that the entrepreneurial spirit thrives in this company he started out of his basement almost 50 years ago.

CORPORATE OFFICE

McDuff, Inc., has fifteen branch offices and a corporate headquarters. Though not a large company by international standards, it has become well known within its own fields. The company operates as a kind of loose confederation. Each office enjoys a good measure of independence. Yet some corporate structure is required for these purposes:

1. To coordinate projects that involve employees from several offices
2. To prevent duplication of the same work at different offices
3. To ensure fairness, consistency, and quality in the handling of human resources issues throughout the firm (salaries, benefits, work load, and so forth)

The corporate office gives special attention to problems related to international communications. Among its non-U.S. clients and employees, it must respond to differences in cultures and ways of doing business. This effort can mean the difference between success or failure in negotiating deals, completing projects, hiring employees, and so forth.

The corporate office in Baltimore is housed in a building across the street from the Baltimore branch office. In Rob McDuff's mind, this separation is important. He likes to keep the mostly "overhead" functions of the corporate office distinct from the mostly profit-generating functions of the branches. Also, he believes the physical separation is symbolic to offices outside of Baltimore, which already suspect that the large Baltimore branch office receives special treatment from corporate headquarters. Figure 2–2 shows an organization chart for this office. What follows is a brief description of the responsibilities of each service group shown on the chart:

■ *Service 1: Human Resources*

The Human Resources Department performs mostly personnel-related tasks. Its main work covers these four fields:

1. Employment: The office handles job advertisements, ensures that branch offices follow government guidelines that apply to hiring, gives legal advice on workers' compensation, and visits college campuses to recruit prospective graduates.

2. Benefits: Two staff members are responsible for handling company benefits. They send information to employees, check the accuracy of benefit deductions from paychecks, stay current about the newest benefits available for McDuff employees, and make recommendations to the corporate staff about benefit changes.

3. Safety: Given the company's interest in waste management, safety is crucial. The company hired a manager of safety in 1980 to complete these tasks:

- Educate employees about the importance of safe work practices
- Provide proper equipment and training
- Visit job sites to make safety checks
- Respond to questions by government agencies

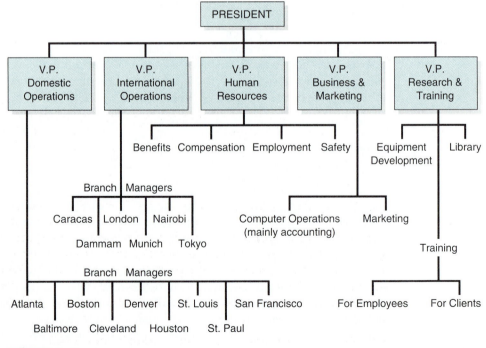

FIGURE 2–2
McDuff's corporate office

4. Compensation: The compensation expert has three main duties:

- Monitoring salaries at all offices to ensure some degree of uniformity within the job classifications
- Researching salary guidelines in all professions represented in the firm, to make sure the company's salaries are competitive
- Monitoring branch offices to make sure performance evaluation interviews are done each year for each employee, before salary decisions are made

Service 2: Project Coordination

With fifteen offices spread over a wide geographical area, the company sometimes falls into the trap of the left hand not knowing what the right hand is doing. Specifically, individual offices may not know what project resources exist in another office. If a large project requires experts from several offices, the corporate office can assemble the team. To make this system work, the office keeps an accurate record of current projects at each office.

Service 3: Marketing

By working closely with clients, the engineers and scientists at every branch help to secure repeat work from current clients. Yet their technical responsibilities keep them from spending much time on marketing for *new* clients. The corporate office, however, has a marketing and proposal-writing staff that works extensively on seeking new business. Also, it helps branch offices write proposals requested by current clients.

Service 4: Computer Operations

The corporate office houses the company's two mainframe computers. These units are used mainly for the firm's accounting data bases. Each branch office has terminals tied into the mainframes, giving them direct access to corporate data bases in Baltimore. Of course, each branch also uses its own stand-alone microcomputer systems for word processing and some other functions.

Service 5: Research

The term *research* at McDuff covers the services of the Equipment Development (ED) Group and the library. The ED lab is housed in the corporate office; it is the only group in the building that could be considered profit-generating. However, this young department is a start-up operation that does not make much money yet. It also retains the important "overhead" function of designing and building tools and other mechanisms that McDuff employees use on their projects. As for the McDuff library, two full-time librarians maintain a modest corporate collection that includes these items:

- Copies of all company reports and proposals since 1947
- Over 100 technical, business, and general periodicals
- Many reference works in technology and the sciences

McDuff employees around the world can receive information the same day they request it—either by phone or fax. In addition, each branch office usually keeps a small reference and periodical collection for its own use.

■ *Service 6: Training*

As noted earlier, McDuff now performs two types of training: in-house courses for its own employees, and external training for clients who need training in a number of technical and nontechnical subjects in which McDuff is proficient. The Training Department directs these efforts. Also, it helps McDuff employees find useful outside training or college courses.

BRANCH OFFICES

Each McDuff branch is unique in its particular combination of technical and nontechnical positions. Yet all fifteen branches include a common management structure, as shown in Figure 2–3. A branch manager, who reports to one of two corporate vice presidents, supervises a group of four or more department managers. These managers, in turn, supervise the technical and nontechnical employees at the branch.

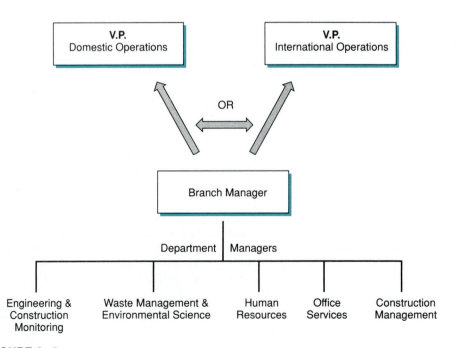

FIGURE 2–3
Branch office: Typical management structure

Branch positions below the manager can be grouped into four categories:

1. Technical professional
2. Nontechnical professional
3. Technical staff
4. Nontechnical staff

Figures 2–4 through 2–7 list some of the positions in these four groupings. Although all these employees are under the supervision of their respective branch managers, some interact closely with employees at the corporate level. For example, each human resources manager reports to his or her respective branch manager *and* works closely with the corporate vice president of human resources.

Most McDuff positions designated "professional" (Figures 2–5 and 2–6) require bachelor's degrees or higher in an appropriate field. Positions designated "staff" require at least a high-school education and sometimes more (such as a certificate program, vocational-technical training, or two-year college degree). Although these categories apply just to our fictional firm of McDuff, some may resemble jobs that exist in a real organization where you now work or will work.

Now that you have viewed McDuff's structure, you need to be made aware of a continuing management problem at the company—one common to many firms with widely spread offices and a corporate office. Branch employees often question whether the corporate people really understand and respond to their needs. Indeed,

Position	Typical Education (Minimum)	Main Duties
1. Department Manager	• B.S. in engineering or science • M.S. in engineering or science **or** M.B.A.	• Oversees entire technical department in engineering or science
2. Project Manager	• B.S. in engineering or science or engineering technology	• Oversees entire projects in engineering, waste management, construction, etc.
3. Research Engineer (Equipment Development)	• B.S. in engineering or design • M.S. or Ph.D. in engineering or design	• Designs new tools, mechanisms, or other equipment at ED lab at corporate office
4. Field Engineer	• B.S. in engineering or engineering technology	• Completes site work for projects and then completes remaining work at office
5. Field Scientist	• B.S. in biology, chemistry, environmental science, etc.	• Completes site work for hazardous waste projects and then completes remaining work back at office

FIGURE 2–4
Sample positions: Technical professionals

Position	Typical Education (Minimum)	Main Duties
1. Office Services Manager	• B.S./B.A. in business	• Oversees accounting, word processing, purchasing, physical plant, etc.
2. Human Resources Manager	• B.S./B.A. in liberal arts or in human resources	• Oversees benefits, safety, employment, compensation
3. Technical Writer	• B.S./B.A. in technical communication **or** in liberal arts	• Helps write and edit reports, proposals, and other branch documents
4. Marketing Specialist	• B.S./B.A. in business **or** in liberal arts	• Writes to and visits potential clients • Helps with proposals
5. Training Specialist	• B.S./B.A. in education **or** in liberal arts	• Works with corporate office to plan in-house training **and** external training for clients

FIGURE 2–5
Sample positions: Nontechnical professionals

Position	Typical Education (Minimum)	Main Duties
1. Field/Lab Technician	• Vo-tech or associate's degree (in technical field)	• Recovers samples from site • Completes lab tests
2. Computer Operator	• Vo-tech or associate's degree (in technical field)	• Inputs data
3. Field Hand	• High school diploma	• Operates and maintains equipment • Orders and picks up supplies
4. Research Technician	• Vo-tech or associate's degree (in technical field)	• Assists research engineers in work at the Equipment Development Lab
5. Warehouse Supervisor	• Vo-tech or associate's degree (in technical field)	• Keeps track of, and maintains, equipment in a branch office

FIGURE 2–6
Sample positions: Technical staff

Position	Typical Education (Minimum)	Main Duties
1. Word Processing Operator	• High school diploma **or** associate's degree	• Produces drafts of all company documents
2. Secretary	• High school diploma **or** associate's degree	• Handles paperwork for professional workers • Does some typing • Has some client contact
3. Receptionist	• High school diploma	• Oversees all of switchboard operation
4. Library Assistant	• High school diploma	• Helps librarian with cataloguing, ordering books, etc.
5. Training Assistant	• High school diploma	• Helps orchestrate training activities of all kinds

FIGURE 2–7
Sample positions: Nontechnical staff

it is easy to feel misunderstood and even neglected when you work hundreds or thousands of miles from the company's hub. Effective written and spoken communication can go a long way toward bridging the gaps between McDuff's local and central offices.

WRITING AT MCDUFF

Good writing is crucial to McDuff's existence. First, one of its main products is a written report. After the company completes a technical project, the project report stands as a permanent statement about, and reflection of, the quality of McDuff's work. Second, many company projects result from written proposals. Third, most routine activity within the firm is preceded or followed by memos, reports, in-house proposals, and manuals. As an employee at McDuff, you would be writing to readers in these groups:

- Superiors at your own McDuff branch
- Subordinates at your McDuff branch
- Employees at other branches or at the corporate office
- Clients
- Subcontractors and vendors

As pointed out in chapter 1, you often write to a mixed group of readers, all with different needs and backgrounds. Likewise, at McDuff, readers of the same document could come from more than one of the groups just listed. For example, assume McDuff's corporate training manager needs to send a memo to 20 employees officially confirming their attendance at an upcoming training seminar at the corporate office. Coming to Baltimore from all domestic offices, these employees need a seminar schedule as well as information about the course. Copies of the memo would have to be sent to (1) the participants' managers, who need to be reminded that they will be minus an employee for three days; (2) the vice president for research and training, who likes to be made aware of any company-wide training; and (3) a training assistant, who needs to make room arrangements for the seminar.

This memo is not unique. Most documents at McDuff and other companies are read by persons from different levels. Listed next are more examples of McDuff writing directed to diverse readers. Some of these projects resemble the examples and assignments in later chapters.

Examples of Internal Writing

1. Memo about changes in benefits—from a manager of human resources at a branch to all employees at that branch
2. Memo about changes in procedures for removing asbestos from buildings—from a project manager to field engineers and technicians
3. Orientation booklet on McDuff—from the manager of employment to all new employees at the firm
4. Internal proposal for funds to develop a new piece of equipment—from a technician to the Equipment Development lab manager
5. Draft of a project report—written by a project manager for review by a department manager (before being submitted to the client)
6. Long report on future markets for McDuff—from the vice president of business and marketing to all 1200 McDuff employees
7. Memo on new procedure for compensating domestic employees who work on overseas projects—from the corporate manager of compensation to all branch managers
8. Manual on new accounting procedures—from the corporate manager of computer operations to all branch managers
9. Article on an interesting environmental project at a national park—from a project manager to all employees who read the company's monthly newsletter
10. Trip report on a professional conference—from a biologist in the environmental science area to the corporate manager of training

Examples of External Writing

1. Sales letter—to potential client
2. McDuff brochure describing technical services—to potential client
3. Proposal—to potential client
4. Progress report—to client
5. Final project report—to client
6. Refresher letter—to previous client
7. Complaint letter—to supplier

8. Article on technical subject—for technical periodical
9. Training manual—for client
10. Affirmative-action report—for government

CHAPTER SUMMARY

This book uses the fictional firm of McDuff, Inc., to lend realism to your study of technical writing. The many McDuff examples and assignments give you a purpose, an audience, and an organizational context that simulate what you will face in your career.

Like other organizations where you might work, McDuff has developed its own personality, or "culture." A company's culture can be influenced by many features including its history, type of business, and management style. One particular feature that many organizations have in common today is an interest in improving the quality of their services or products.

This chapter looks specifically at the culture of McDuff, Inc. Though started as an engineering consulting firm with a narrow focus, McDuff is now an international company with 1200 employees, fifteen branches, and a corporate office. The firm works in seven main project areas: soils engineering on land, soils engineering at sea, construction monitoring, construction management, waste management (environmental science), equipment development, and training. The color insert gives summary information about specific McDuff projects in all seven areas.

McDuff's corporate office is in Baltimore, Maryland. It helps the branch offices in the areas of human resources, project coordination, marketing, computer operations, research, and training. Each of the fifteen offices is run by a branch manager. At each branch, employees are grouped into four categories: technical professionals, nontechnical professionals, technical staff, and nontechnical staff.

McDuff employees at all levels do a good deal of writing, both to superiors and subordinates within the organization and to clients and other outside readers. Documents often have multiple readers with different backgrounds, making writing even more challenging. McDuff employees who meet this challenge will have the best chance of doing valuable work for the company and succeeding in their careers.

ASSIGNMENTS

The following assignments present case studies at McDuff, Inc. They can be completed either as individual exercises or as group projects, depending on the directions of your instructor. Your instructor will also indicate if you are to prepare an oral or a written response.

Analyze the context of each case by considering what you learned in chapter 1 about the context of technical writing *and* what you learned in this chapter about McDuff. In particular, answer these five questions:

1. What is the purpose of the document to be written?
2. What result will you hope to achieve by writing it?

3. Who will be your readers and what will they want from your document?
4. What method of organization will be most useful?
5. What tone and choice of language will be most effective?

1. **Memo Changing Supplies Policy.** As the office services manager at the St. Louis office, you have a problem. In the last fiscal year, the office has used significantly more bond paper, computer paper, pens, mechanical pencils, eraser fluid, and file folders than in previous fiscal years. After going back through the year's projects, you can find no business-related reason why the office has bought $12,000 more of these items. Given that everyone has easy access to the supplies, you have concluded that some employees are taking them home. Putting the best face on it, you assume they may be "borrowing" supplies to complete company business they take home with them, then just keeping items at home. Putting the worst face on it, you wonder whether some employees are stealing from the company.

   After consulting with the branch manager and some other managers, you decide to restrict access to office supplies. Starting next month, these supplies must be signed out through secretaries in the various departments. First, you plan to meet with the secretaries to explain how to make the system work. Then on the following day, you will send a memo explaining the change to *all employees*.

   Would you change your approach in this memo if it were to be sent to a *specific* audience in the office? Why or why not? Answer this question with regard to the four employee groups shown in Figure 2–4 ("Technical Professionals"), Figure 2–5 ("Nontechnical Professionals"), Figure 2–6 ("Technical Staff"), and Figure 2–7 ("Nontechnical Staff").

2. **Letter Requesting Testimonials.** As a writer in the corporate marketing department, you spend a good deal of your time preparing materials to be used in sales letters, brochures, and company proposals. Yesterday you were assigned the task of asking about 20 customers if they will write "testimonial letters" about their satisfaction with McDuff's work. In all cases, these clients have used McDuff for many projects and have informally expressed satisfaction with the work. Now you are going to ask them to express their satisfaction in the form of a letter, which McDuff could use as a testimonial to secure other business.

   Your strategy is to write a "personalized" form letter to the 20 clients, and then follow it up with phone calls.

3. **Memo on Inventory Control.** For five years, you have supervised the supply warehouse at the Houston office. Your main job is to maintain equipment and see that it is returned after jobs are completed. When checking out equipment, each project manager is supposed to fill out part of a project equipment form that lists all equipment used on the job and the date of checkout. Upon returning the equipment, the project manager should complete the form by listing the date of return and any damage, no matter how small, that needs to be repaired before the equipment is used again. This equipment ranges from front-end loaders and pickup trucks to simple tools like hammers, wrenches, and power drills.

   Lately you have noticed that many forms you receive are incomplete. In particular, project managers are failing to record fully any equipment damage that occurred on the job. For example, if someone fails to report that a truck's alignment is out, the truck will not be in acceptable shape for the next project for which it could be used.

   Your oral comments to project managers have not done much good. Apparently, the project managers do not take the warehouse problem seriously, so you believe it is time to put your concerns in writing. The goal is to inform all technical professionals who manage projects that from now on the form must be correctly filled out. You have no "authority," as such, over the managers; however, you know that their boss would be very concerned about this problem if you chose to bring it to his or her attention.

At this point, you have decided to ask nicely one more time—this time in writing. You want your memo to emphasize issues of safety and profitability, as well as the need to follow a procedure that has helped you to maintain a first-rate warehouse.

4. **Memo Report on Flextime.** As branch manager of the Atlanta office, you have always tried to give employees as much flexibility as possible in their jobs—as long as the jobs got done. Recently you have had many requests to adopt flextime. In this arrangement, the office would end its standard 8:00 A.M. to 4:30 P.M. workday (with a half-hour lunch break). Instead, each employee would fit her or his eight-hour day within the following framework: 7:00 A.M. to 8:30 A.M. arrival, a half hour or full hour for lunch, and 3:30 P.M. to 5:30 P.M. departure.

Two conditions would prevail if flextime were adopted. First, each employee's supervisor would have to agree on the hours chosen, since the supervisor would need to make sure that departmental responsibilities were covered. Second, each employee would "lock in" a specific flextime schedule until another was negotiated with the supervisor. In other words, an employee's hours would not change from day to day.

Before you spend any more time considering this change, you want to get the views of employees. You decide to write a short memo report that (1) explains the changes being considered and the conditions (see previous paragraph); (2) solicits their views in writing, by a certain date; and (3) asks what particular work hours they would prefer, if given the choice. Also, you want your short report to indicate that later there may be department meetings and finally a general office meeting on the subject, depending on the degree of interest expressed by employees in their memos to you.

3 | *Organizing Information*

This McDuff committee discusses ways to organize information in its upcoming long report.

*T*om Kent asks the department secretary to hold his calls. Closing his door, he reaches for the report draft written by one of his staff members and sits down to read it. As a McDuff manager for 10 years, he has reviewed and signed off on every major report written by members of his department. Of all the problems that plague the drafts he reads, poor organization bothers him the most.

This problem is especially annoying at the beginning of a document and the beginning of individual sections. Sometimes he has no idea where the writer is going. His people don't seem to understand that they are supposed to be "telling a story," even in a technical report. Grammar and style errors are annoying to him, but organization problems are much more troublesome. They require extensive rewriting and time-consuming meetings with the report writer. Reaching for his red pen, Tom hopes for the best as he begins to read yet another report.

You, too, will face internal reviewers like Tom Kent when you write on the job. To help you avoid organization problems, this chapter offers strategies for organizing information as you plan, draft, and revise your writing. It builds on the discussion of the three stages of writing covered in chapter 1. Then the next chapter will complete your introduction to technical writing by showing you how to use effective page design to keep readers' attention.

IMPORTANCE OF ORGANIZATION

In a survey of engineering professionals, respondents named "organizing information" the most important topic for any undergraduate technical writing course

[Pinelli, T. E., M. Glassman, R. O. Barclay, and W. E. Oliu. 1989. *Technical communications in aeronautics: Results of an exploratory study—an analysis of profit managers' and nonprofit managers' responses.* (NASA TM-101626, p. 28.) Washington, DC: National Aeronautics and Space Administration]. This research is backed up by the experience of many communication consultants—including the author of this textbook, who for 15 years has helped companies improve their employees' writing. Overwhelmingly, these client firms have cited poor organization as the main writing problem among both new and veteran employees. That concern underlies all the suggestions in this chapter.

As you learned in chapter 1, your documents will be read by varied readers with diverse technical backgrounds. Chapter 2 displays this technical range within McDuff and refers to an even broader technical spectrum among McDuff's clients. Given this reader diversity, this chapter aims to answer one essential question: How can you best organize information to satisfy so many different people?

Figure 3–1 shows you three possible options for organizing information for a mixed technical audience, but only one is recommended in this book. Some writers, usually those with technical backgrounds themselves, choose Option A. They direct their writing to the *most* technical people. Other writers choose Option B. They respond to the dilemma of a mixed technical audience by finding the lowest common denominator—that is, they write to the level of the *least* technical person. Option A and B each satisfies one segment of readers at the expense of the others.

Option C is preferred in technical writing for mixed readers. It encourages you to organize documents so that **all** readers—both technical and nontechnical—get what they need. The rest of this chapter provides strategies for developing this option, describing general principles of organization and guidelines for organizing entire documents, individual document sections, and paragraphs.

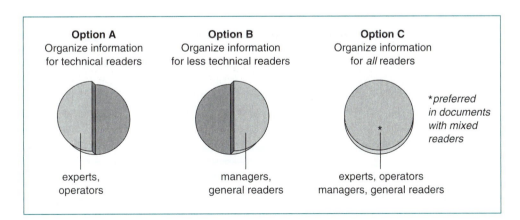

FIGURE 3–1
Options for organizing information

THREE PRINCIPLES OF ORGANIZATION

Good organization starts with careful analysis of your audience. Most readers are impatient and skip around as they read. Think about how you examine a weekly newsmagazine or an airline magazine. You are likely to take a quick look at articles of special interest to you; then you might read them more thoroughly, if there is time. That approach also resembles how *your* audience treats technical reports and other work-related documents. If important points are buried in long paragraphs or sections, busy readers may miss them. Three principles respond realistically to the needs of your readers:

■ *Principle 1: Write Different Parts for Different Readers*

The longer the document, the less likely it is that any of your readers will read it from beginning to end. As shown in Figure 3–2, they use a "speed-read" approach that includes these steps:

Step 1:   **Quick scan.** Readers scan easy-to-read sections like executive summaries, introductory summaries, introductions, tables of contents, conclusions, and recommendations. They pay special attention to the beginnings and endings of documents, especially those longer than a page or two, and to illustrations.

Step 2:   **Focused search.** Readers go directly to sections in the document body that will give them what they need at the moment. To find information quickly, they search for format devices like subheadings, listings, and white space in margins to guide their reading. (See chapter 4 for a discussion of page design.)

Step 3:   **Short follow-ups.** Readers return to the document, when time permits, to read or re-read important sections.

Your job is to write in a way that responds to this nonlinear, episodic reading process of your audience. Most important, you should direct each section to those in the audience most likely to read that particular section. Shift the level of tech-

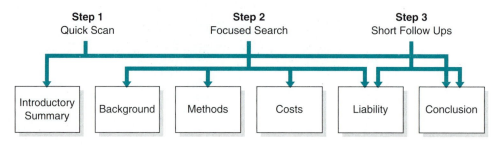

FIGURE 3–2
Speed-read approach of busy readers

nicality as you move from section to section *within* the document, to meet the needs of each section's specific readers. On the one hand, managers and general readers favor less technical language and depend most heavily on overviews at the beginning of documents. On the other hand, experts and operators expect more technical jargon and pay more attention than others to the body sections of documents.

Of course, you walk a thin line in designing different parts of the document for different readers. Although technical language and other stylistic features may change from section to section, your document must hang together as one piece of work. Common threads of organization, theme, and tone must keep it from appearing fragmented or pieced together.

This approach breaks the rules of nontechnical writing, which strives for dogged consistency throughout the same document. However, technical writing marches to the beat of a different drummer.

■ *Principle 2: Emphasize Beginnings and Endings*

Suspense fiction relies on the interest and patience of readers to ferret out important information. The writer usually drops hints throughout the narrative before finally revealing who did what to whom. Technical writing operates differently. Busy readers expect to find information in predictable locations without having to search for it. Their first-choice locations for important information are as follows:

- The beginning of the entire document
- The beginnings of report sections
- The beginnings of paragraphs

The reader interest curve in Figure 3−3 reflects this focus on beginnings. But the curve also shows that the readers' second choice for reading is the ends of documents, sections, and paragraphs. That is, most readers tend to remember best the first and last things they read. The ending is a slightly less desirable location

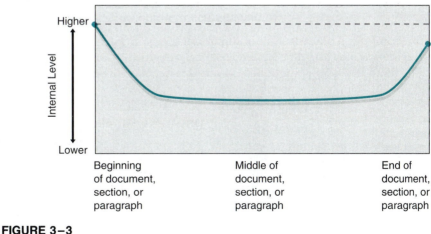

FIGURE 3−3
Reader interest curve

than the beginning because it is less accessible, especially in long sections or documents. Of course, some readers inevitably will read the last part of a document first, for they may have the habit of fanning pages when first seeing a document. Their thumb first locks on the last section of the report. Thus, although there is no guarantee that the first document section will be read first, you can be fairly sure that either the beginning or the ending will get first attention.

Emphasizing beginnings and endings responds to the reading habits and, indeed, the psychological needs of readers. At the beginning, they want to know where you're heading. They need a simple "road map" for the rest of the passage. In fact, if you don't provide something important at the beginnings of paragraphs, sections, and documents, readers will start guessing the main point themselves. It is in your best interest to direct the reader to what *you* consider most important in what they are about to read, rather than to encourage them to guess at the importance of the passage. At the ending, readers expect some sort of wrap-up or transition; your writing shouldn't simply drop off. The following paragraph begins and ends with such information:

> *The proposed word-processing software has two other features that will help our writers: a dictionary and a thesaurus.* When the built-in dictionary is engaged, it compares each word in a document with the same word in the system's dictionary. Differences are then highlighted so that they can be corrected by the operator of the system. The thesaurus also can help our writers by offering alternative word selections. When just the right word is escaping the writer, he or she can trigger the system to provide a list of related words or synonyms. *Both the dictionary and thesaurus are very quick and thus far superior to their counterparts in book format.*

The first sentence gives readers an immediate impression of the two topics to be covered in the paragraph. The paragraph body explores details of both topics. Then the last sentence flows smoothly from the paragraph body by reinforcing the main point about features of the dictionary and thesaurus.

Why does this top-down pattern, which seems so logical from the reader's perspective, frequently get ignored in technical writing? The answer arises from the difference between the way you complete your research or fieldwork and the way busy readers expect results of your work to be conveyed in a report. Figure 3–4 illustrates this difference. Having moved logically from data to conclusions and recommendations in technical work, many writers assume they should take this same approach in their report. They reason that the reader wants and needs all the supporting details before being confronted with conclusions and recommendations that result from these data.

Such reasoning is wrong. Readers want the results placed first, followed by details that support your main points. Of course, you must be careful not to give *detailed* conclusions and recommendations at the beginning; most readers want and expect only a brief summary. This overview will provide a framework within which readers can place the details presented later. Readers of technical documents want the "whodunit" answer at the beginning.

Recall the motto in chapter 1: *Write for your reader, not for yourself.* Now you can see that this rule governs the manner in which you organize information in everything you write.

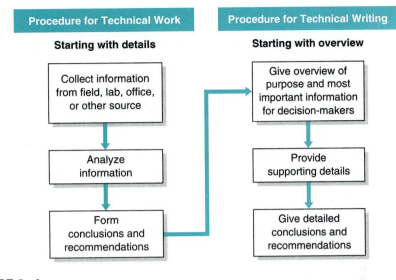

FIGURE 3-4
Technical *work* vs. technical *writing*

■ *Principle 3: Repeat Key Points*

You have learned that different people focus on different sections of a document. Sometimes no one carefully reads the *entire* report. For example, managers may have time to read only the summary, whereas technical experts may skip the leadoff sections and go directly to "meaty" technical sections with supporting information. These varied reading patterns require a *redundant* approach to organization—you must repeat important information in different sections for different readers.

For example, assume you are a McDuff employee in Atlanta and are writing a report to the 1996 Olympics staff on choosing sites for several practice fields. Having examined five alternatives, your report recommends one site for final consideration by the staff. Your 25-page report compares and contrasts all five alternatives according to criteria of land cost, nearness to other Olympics locations, and relative difficulty of grading the site and building the required facilities. Given this context, where will your recommendation appear in the report? Here are five likely spots:

1. Executive summary
2. Cost section in the body
3. Location section in the body
4. Grading/construction section of the body
5. Concluding section

Our assumption, you recall, is that few readers move straight through a report. Because they often skip to the section most interesting to them, you need to make main sections somewhat self-contained. In the Olympics report, that would mean placing the main recommendation at the beginning, at the end, and at one or more points within each main section. In this way, readers of all sections would encounter your main point.

What about the occasional readers who read all the way through your report, word for word? Will they be put off by the restatement of main points? No, they won't. Your strategic repetition of a major finding, conclusion, or recommendation gives helpful reinforcement to readers always searching for an answer to the "So what?" question as they read. Fiction and nonfiction may be alike in this respect—writers of both genres are "telling a story." The theme of this story must periodically reappear to keep readers on track.

Now we're ready to be more specific about how the three general principles of organization apply to documents, document sections, and paragraphs.

ABC FORMAT FOR DOCUMENTS

You have learned the three principles of organization: (1) write different parts of the document for different readers, (2) emphasize beginnings and endings, and (3) repeat key points. Now let's move from principles to practice. Here we will develop an all-purpose pattern of organization for writing entire documents. (The next major section covers document sections and paragraphs.)

Technical documents should assume a three-part structure that consists of a beginning, a middle, and an end. This book labels this structure the "ABC Format" (for **A**bstract, **B**ody, and **C**onclusion). Visually, think of this pattern as a three-part diamond structure, as shown in Figure 3–5:

- **Abstract:** A brief beginning component is represented by the narrow top of the diamond, which leads into the body.
- **Body:** The longer middle component is represented by the broad, expansive portion of the diamond figure.
- **Conclusion:** A brief ending component is represented by the narrow bottom of the diamond, which leads away from the body.

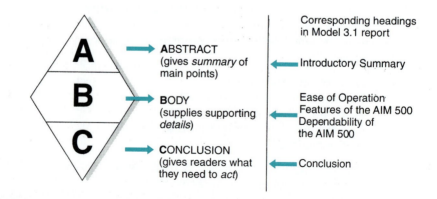

FIGURE 3–5
ABC format for all documents

Model 3–1 (pp. 74–75) includes a memo report that conforms to this structure. The following sections discuss the three ABC components in detail.

Document Abstract: The "Big Picture" for Decision-Makers

Every document should begin with an overview. As used in this text, *abstract* is defined as follows:

> **Abstract:** brief summary of a document's main points. Although its makeup varies with the type and length of the document, an abstract always includes (1) a clear purpose statement for the document and (2) the most important points for decision-makers. As a capsule version of the entire document, the abstract should answer readers' typical mental questions, such as the following: "How does this document concern me? What's the bottom line? So what?"

Abstract information is provided under different headings, depending on the document's length and degree of formality. Some common headings are "Summary," "Executive Summary," "Introductory Summary," and "Overview." The abstract may vary in length from a short paragraph to a page or so. Its purpose, however, is always the same: to provide decision-makers with highlights of the document.

For example, assume you are an engineer who has evaluated environmental hazards for the potential purchaser of a shopping mall site. The abstract information in your report should include a brief project summary, a statement of findings, and an indication of sections to follow. In effect, this summary should answer three questions:

1. What are the major risks at the site?
2. Are these risks great enough to warrant not buying the land?
3. What major sections does the rest of the report contain?

Here is how the summary might read:

> As you requested, we have examined the possibility of environmental contamination at the site being considered for the new Klinesburg Mall. Our field exploration revealed two locations with deposits of household trash, which can be easily cleaned up. Another spot has a more serious deposit problem of 10 barrels of industrial waste. However, our inspection of the containers and soil tests revealed no leaks.
>
> Given these limited observations and tests, we conclude that the site poses no major environmental risks and recommend development of the mall. The rest of this report details our field activities, test analyses, conclusions, and recommendations.

This general abstract, or overview, is mainly for decision-makers. Highlights must be brief, yet free of any possible misunderstanding. On some occasions, you may need to state that further clarification is included in the text, even though that point may seem obvious. For example, if your report concerns matters of safety, the overview may not be detailed enough to prevent or eliminate risks. In this case, state this point clearly so that the reader will not misunderstand or exaggerate the purpose of the abstract.

Later chapters in this book contain guidelines for writing the following specific types of abstracts:

- Introductory summaries for short reports (chapter 8)
- Executive summaries for formal reports (chapter 9)
- Introductory summaries for short proposals (chapter 10)
- Executive summaries for formal proposals (chapter 10)
- Abstracts of technical articles (chapter 13)

Document Body: Details for All Readers

The longest part of any document is the body. As used in this book, the body is defined as follows:

> **Body:** the middle section(s) of the document providing supporting information to readers, especially those with a technical background. Unlike the abstract and conclusion, the body component allows you to write expansively about items such as (1) the background of the project, (2) field, lab, office, or any other work upon which the document is based, and (3) details of any conclusions, recommendations, or proposals that might be highlighted at the beginning or end of the document. The body answers this main reader question: "What support is there for points put forth in the abstract at the beginning of the document?"

Managers may read much of the body, especially if they have a technical background and if the document is short. Yet the more likely readers are technical specialists who (1) verify technical information for the decision-makers or (2) use your document to do their jobs. In writing the body, use the following guidelines:

- **Separate fact from opinion.** Never leave the reader confused about where opinions begin and end. Body sections usually move from facts to opinions that are based on facts. To make the distinction clear, preface opinions with phrases such as "We believe that," "I feel that, "It is our opinion that," and the like. Such wording gives a clear signal to readers that you are presenting judgments, conclusions, and other nonfactual statements.
- **Adopt a format that reveals much structure.** Use frequent headings and subheadings to help busy readers locate important information immediately. (Chapter 4 covers these and other elements of page design.)
- **Use graphics whenever possible.** Use graphics to draw attention to important points. Today more than ever, readers expect visual reinforcement of your text, particularly in more persuasive documents like proposals. (Chapter 11 deals with graphical elements in technical documents.)

By following these guidelines, which apply to any document, you will make detailed body sections as readable as possible. They keep ideas from becoming buried in text and show readers what to do with the information they find.

Document Conclusion: Wrap-Up Leading to Next Step

Your conclusion deserves special attention. Good writers remember that the last thing read is often the first thing recalled. We define the conclusion component as follows:

> **Conclusion:** the final section(s) of the document bringing readers—especially decision-makers—back to one or more central points already mentioned in the body. Occasionally, it may include one or more points not previously mentioned. In any case, the conclusion provides closure to the document and often leads to the next step in the writer's relationship with the reader.

The conclusion component may have any one of several headings, depending on the type and length of the document. Possibilities include "Conclusion," "Closing," "Closing Remarks," and "Conclusions and Recommendations." The chapters in Part 2 of this text describe the options for short and long documents of many kinds. In general, however, every conclusion component must answer these sorts of questions:

- What major points have you made?
- What problem have you tried to solve?
- What should the reader do next?
- What will you do next?
- What single idea do you want to leave with the reader?

Because readers focus on beginnings and endings of documents, you want to exploit the opportunity to drive home your message—just as you did in the abstract. Format can greatly affect the impact you make on decision-makers. Although specific formats vary, most conclusions take one of these two forms:

- **Listings:** This format is especially useful when pulling together points mentioned throughout the document. Whereas the abstract often gives readers the big picture in narrative format, the conclusion may instead depend on listings of findings, conclusions, and/or recommendations. (Chapter 4 gives suggestions on using bulleted and numbered listings.)
- **Summary paragraph(s):** When a listing is not appropriate, you may want to write a concluding paragraph or two. Here you can leave readers with an important piece of information and make clear the next step to be taken.

Whichever alternative you choose, your goal is to return to the main concerns of the most important readers—decision-makers. Both the abstract and conclusion, in slightly different ways, should respond to the needs of this primary audience.

TIPS FOR ORGANIZING SECTIONS AND PARAGRAPHS

The ABC format pertains to the organization of entire documents. Yet the same "beginning-middle-end" strategy applies to the next smaller units of discourse— document sections and paragraphs. In fact, you can view the entire document as a series of interlocking units, with each responding to reader expectations as viewed on the reader interest curve in Figure 3–3.

Document Sections

As mentioned earlier, readers will move from the document abstract to the specific body sections they need to solve their problem or answer their immediate question. Just as they need abstracts and conclusions in the whole document, they need "mini-abstracts" and brief wrap-ups at the start and finish of each major section.

To see how a section abstract works, we must first understand the dilemma of readers. Refer to Model 3–2 on page 76, which contains one section from a long report. Some readers may read it from beginning to end, but others might not have the time or interest to do so in one sitting. Instead, they would look to a section beginning for an abstract, and then move around within that section at will. Thus the beginning must provide them with a map of what's ahead. Here are the two items that should be part of every section abstract:

1. **Interest grabber:** a sentence or more that captures the attention of the reader. Your grabber may be one sentence or an entire paragraph, depending on the overall length of the document.
2. **Lead-in:** a list, in sentence or bullet format, that indicates main topics to follow in the section. If the section contains subheadings, your lead-in may include the same wording as the subheadings and be in the same order.

The first part of the section gives readers everything they need to read on. First, you get their attention with a grabber. Then you give them an outline of the main points to follow so that they can move to the part of the section that interests them most. As in Model 3–2 on page 76, the section abstract immediately precedes the first subheading when subheads are used.

Sections also should end with some sort of closing thought, rather than just dropping off after the last supporting point has been stated. For example, you can (1) briefly restate the importance of the information in the section or (2) provide a transition to the section that follows. Model 3–2 (p. 76) takes the latter approach by suggesting the main topic for the next section. Whereas the section lead-in provides a map to help readers navigate through the section, the closing gives a sense of an ending so that readers are ready to move on.

Paragraphs

Paragraphs represent the basic building blocks of any document. Organizing them is not much different in technical writing than it is in nontechnical prose. Most paragraphs contain these elements:

1. **Topic sentence:** This sentence states the main idea to be developed in the paragraph. Usually it appears first. Do not delay or bury the main point, for busy readers may only read the beginnings of paragraphs. If you fail to put the main point there, they may miss it entirely.
2. **Development of main idea:** Sentences that follow the topic sentence develop the main idea with examples, narrative, explanation, or other details. Give the reader concrete supporting details, not generalizations.
3. **Transitional elements:** Structural transitions help the paragraph flow smoothly. Use transitions in the form of repeated nouns and pronouns, contrasting conjunctions, and introductory phrases.
4. **Closing sentence:** Most paragraphs, like sections and documents, need closure. Use the last sentence for a concluding point about the topic or a transitional point that links the paragraph with the one following it.

Model 3–3 on page 77 shows two paragraphs that follow this pattern of organization. The paragraphs are from a McDuff recommendation report. McDuff was hired to suggest ways for a hospital to modernize its physical plant. Each paragraph is a self-contained unit addressing a specific topic, while being linked to surrounding paragraphs (not shown) by theme and transitional elements.

This suggested format applies to many, but not to all, paragraphs included in technical documents. In one common exception, you may choose to delay statement of a topic sentence until you engage the reader's attention with the first few sentences. In other cases, the paragraph may be short and serve only as an attention grabber or a transitional device between several longer paragraphs. Yet for most paragraphs in technical writing, the beginning-middle-end model described here will serve you well. Remember these other points as well as you organize paragraphs in technical writing:

- *Length:* Keep the typical length of paragraphs at around 6 to 10 lines. Many readers won't read long blocks of text, no matter how well organized they may be. If you see that your topic requires more than 10 lines for its development, split the topic, and develop it in two or more paragraphs.
- *Listings:* Use short listings of three or four items to break up long paragraphs. Readers lose patience when they realize information could have been more clearly presented in listings. The next chapter offers detailed suggestions on using lists.
- *Use of Numbers:* Paragraphs are the worst format for presenting data of any kind, especially numbers that describe costs. Readers may ignore or miss data packed into paragraphs. Usually tables or figures would be a more clear and appropriate format. Also, be aware that some readers may think that cost data couched in paragraph form represent an attempt to hide important information.

This chapter mostly concerns the ordering of ideas within paragraphs, sections, and whole documents. Good organization helps make your writing successful. Organization alone, however, will not win the day. Readers also expect a visually appealing document. The next chapter describes technical devices for creating the best possible design of your pages.

CHAPTER SUMMARY

Good technical writing calls on special skills, especially in organization. Writers should follow three guidelines for organizing information: (1) write different parts of the document for different readers, (2) place important information at the beginnings and endings, and (3) repeat key points throughout the document.

This chapter recommends the "ABC format" for organizing technical documents. This format includes an **A**bstract (summary), a **B**ody (supporting details), and a **C**onclusion (wrap-up and transition to next step). The abstract section is particularly important because most readers give special attention to the start of a document.

Individual sections and paragraphs also require attention to organization. Sections need overviews and closing passages so that busy readers can find information quickly. Most paragraphs should contain a topic sentence, supporting details, transitional words and phrases, and a closing sentence that leads into the next paragraph.

ASSIGNMENTS

1. **Overall Organization.** Find an example of technical writing directed to more than one reader. Prepare a written or an oral report (your instructor's choice) that explains how well the excerpt follows this chapter's guidelines for organization.
2. **Evaluating an Abstract.** Read the following abstract and evaluate the degree to which it follows the guidelines in this chapter.

As one of the buyers for Randall Auto Parts, I constantly search for new products that I feel can increase our sales. I recently attended an electronics convention to see what new products were available. One product that caught my eye was the new Blaupunkt BMA5350B amplifier. I recommend that we adopt this amplifier into our line of car audio products.

This proposal supports my recommendation and includes the following sections:

1. Features of the Blaupunkt BMA5350B
2. Customer Benefits
3. Cost
4. Conclusions

3. **Section Organization.** As a graphics specialist at McDuff, you have written a recommendation report on ways to upgrade the graphics capabilities of the firm. One section of the report describes a new desktop publishing system, which you believe will make McDuff proposals and reports much more professional looking. Your report section describes technical features of the system, the free training that comes with purchase, and the cost.

Write a lead-in paragraph for this section of your report. If necessary, invent additional information for writing the paragraph.

4. **Paragraph Organization.** With the following list of related information, write a paragraph that follows the organizational guidelines in this chapter. Use all the information, change any of the wording when necessary, and add appropriate transitions. Assume that this paragraph is part of an internal McDuff document suggesting ways to improve work schedules.

- Four-day weeks may lower job stress—employees have long weekends with families and may avoid worst part of rush hour.
- A four-day, 10-hour-a-day workweek may not work for some service firms, where projects and clients need five days of attention.
- Standard five-day, eight-hour-a-day workweeks increase on-the-job stress, especially considering commuter time and family obligations.
- McDuff is considering a pilot program for one office, whereby the office would depart from the standard 40-hour workweek.
- There are also other strategies McDuff is considering to improve work schedules of employees.
- The 40-hour workweek came into being when many more families had one parent at home while the other worked.
- Some firms have gone completely to a four-day week (with 10-hour days).
- McDuff's pilot program would be for one year, after which it would be evaluated.

5. **Writing an Abstract.** The following short report lacks an abstract that states the purpose and provides the main conclusion or recommendation from the body of the report. Write a one- or two-paragraph abstract for this report.

DATE: June 13, 1993
TO: Ed Simpson
FROM: Jeff Radner
SUBJECT: Creation of an Operator Preventive Maintenance Program

The Problem
The lack of operator involvement in the equipment maintenance program has caused the reliability of equipment to decline. Here are a few examples:

- A tractor was operated without adequate oil in the crankcase, resulting in a $15,000 repair bill after the engine locked.
- Operators have received fines from police officers because safety lights were not operating. The bulbs were burned out and had not been replaced. Brake lights and turn-signal malfunctions have been cited as having caused rear-end collisions.
- A small grass fire erupted at a construction site. When the operator of the vehicle nearest to the fire attempted to extinguish the blaze, he discovered that the fire extinguisher had already been discharged.

When the operator fails to report deficiencies to the mechanics, dangerous consequences may result.

The Solution
The goal of any maintenance program is to maintain the company equipment so that the daily tasks can be performed safely and on schedule. Since the operator is using the equipment on a regular basis, he or she is in the position to spot potential problems before

they become serious. For a successful maintenance program, the following recommendations should be implemented:

- Hold a mandatory four-hour equipment maintenance training class conducted by mechanics in the motor pool. This training would consist of a hands-on approach to preventive
  maintenance checks and services at the operator level.
- Require operators to perform certain checks on a vehicle before checking it out of the
  motor pool. A vehicle checklist would be turned in to maintenance personnel.

The attached checklist would require 5 to 10 minutes to complete.

Conclusion

I believe the cost of maintaining the vehicle fleet at McDuff will be reduced when potential
problems are detected and corrected before they become serious. Operator training and the
vehicle pretrip inspection checklist will ensure that preventable accidents are avoided.
I will call you this week to answer any questions you may have about this proposal.

McDuff, Inc.
Fleet Maintenance Division
Vehicle Checklist
Pretrip Inspection

Inspected by: _____   Date: _____

Vehicle #: _____   Odometer: _____

Fluid Levels, Full/Low   Comments

_____   Engine Oil   _____

_____   Transmission Fluid   _____

_____   Brake Fluid   _____

_____   Power Steering   _____

_____   Radiator Level   _____

Before Cranking Vehicle

_____   Tire Condition   _____

_____   Battery Terminals   _____

_____   Fan Belts   _____

_____   Bumper and Hitch   _____

_____   Trailer Plug-in   _____

_____   Safety Chains   _____

After Cranking Vehicle

_____   Parking Brakes   _____

_____   Lights   _____

_____   All Gauges   _____

_____   Seat Belts   _____

_____   Mirrors/Windows/Wipers   _____

_____   Clutch   _____

_____   Fire Ext. Mounted and Charged   _____

_____   Two-Way Radio Working   _____

Additional Comments:_____

_____

Mc Duff, Inc.

MEMORANDUM

DATE: September 5, 1993
TO: Danielle Firestein
FROM: Barbara Ralston *BR*
SUBJECT: Recommendation for AIM 500 Fax

INTRODUCTORY SUMMARY

This memo presents my evaluation of the AIM 500 facsimile (fax) machine by Simko, Inc. The AIM 500 has served our department well for the past two years. If other departments need a fax machine, I highly recommend this model because it is easy to operate, has many useful features, and has been quite dependable.

EASE OF OPERATION

The AIM 500 is so easy to operate that a novice can learn to transmit a document to another location in about two minutes. Here's the basic procedure:

1. Press the button marked TEL on the face of the fax machine. You then hear a dial tone.
2. Press the telephone number of the person receiving the fax on the number pad on the face of the machine.
3. Lay the document facedown on the tray at the back of the machine.

At this point, just wait for the document to be transmitted—about 18 seconds per page to transmit. The fax machine will even signal the user with a beep and a message on its LCD display when the document has been transmitted. Other more advanced operations are equally simple to use and require little training. Provided with the machine are two different charts that illustrate the machine's main functions.

The size of the AIM 500 makes it easy to set up almost anywhere in an office. The dimensions are 13 inches in width, 15 inches in length, and 9.5 inches in height. The narrow width, in particular, allows the machine to fit on most desks, file cabinets, or shelves.

FEATURES OF THE AIM 500

The AIM 500 has many features that will be beneficial to our employees. In the two years of use in our department, the following features were found to be most helpful:

Automatic redial
Last number redial memory
LCD display

MODEL 3–1
ABC format in whole document

Preset dialing
Group dialing
Use as a phone

Automatic Redial. Often when sending a fax, the sender finds the receiving line busy. The redial feature will automatically redial the busy number at 30-second intervals until the busy line is reached, saving the sender considerable time.

Last Number Redial Memory. Occasionally there may be interference on the telephone line or some other technical problem with the transmissions. The last number memory feature allows the user to press one button to automatically trigger the machine to retry the number.

LCD Display. This display feature clearly shows pertinent information, such as error messages that tell a user exactly why a transmission was not completed.

Preset Dialing. The AIM 500 can store 16 preset numbers that can be engaged with one-touch dialing. This feature makes the unit as fast and efficient as a sophisticated telephone.

Group Dialing. Upon selecting two or more of the preset telephone numbers, the user can transmit a document to all of the preset numbers at once.

Use as a Phone. The AIM 500 can also be used as a telephone, providing the user with more flexibility and convenience.

DEPENDABILITY OF THE AIM 500

Over the entire two years our department has used this machine, there have been no complaints. We always receive clear copies from the machine, and we never hear complaints about the documents we send out. This record is all the more impressive in light of the fact that we average 32 outgoing and 15 incoming transmissions a day. Obviously, we depend heavily on this machine.

So far, the only required maintenance has been to change the paper and dust the cover.

CONCLUSION

The success our department has enjoyed with the AIM 500 compels me to recommend it highly. The ease of operation, many exceptional features, and record of dependability are all good reasons to purchase additional units. If you have further questions about the AIM 500, please contact me at extension 3646.

MODEL 3–1, *continued*

ADDITIONAL FEATURES OF MAGCAD

This report has presented two main advantages of the MagCad Drawing System: ease of correction and multiple use of drawings. However, there are two other features that make this sytem a wise purchase for McDuff's Boston office: the selective print feature and the cost.

Selective Printing

When printing a MagCad drawing, you can "turn off" specific objects that are in the drawing with a series of keystrokes. The excluded items will not appear in the printout of the drawing. That is, the printed drawing will reflect exactly what you have temporarily left on the screen, after the deletions. Yet the drawing that remains in the memory of the machine is complete and ready to be reconstructed for another printout.

The selective print feature is especially useful on jobs where different groups have different needs. For example, in a drawing of a construction project intended only for the builder, one drawing may contain only land contours and the building structures. If the same drawing is going to the paving company, we may need to include only land contours and parking lots. In each case, we will have used the selective print feature to tailor the drawing to the specific needs of each reader.

This feature improves our service to the client. In the past, we either had to complete several different drawings or we had to clutter one drawing with details sufficient for the needs of all clients.

Cost of MagCad

When we started this inquiry, we set a project cost limit of $12,000. The MagCad system stays well within this budget, even considering the five stations that we need to purchase.

The main cost savings occurs because we have to buy only one copy of the MagCad program. For additional work stations, we need pay only a $400 licensing fee per station. The complete costs quoted by the MagCad representative are listed below:

1. MagCad Version 5	$5,000
2. Licenses for five additional systems	1,500
3. Plotter	2,000
4. Installation	1,000
TOTAL	$9,500

With the $2,500 difference between the budgeted amount and the projected cost of the system, we could purchase additional work stations or other peripheral equipment. The next section suggests some add-ons we might want to purchase later, once we see how the MagCad can improve our responsiveness to client needs.

MODEL 3–2
ABC format in document section

Conversion to a partial solar heating and cooling system would upgrade the hospital building considerably. In fact, the use of modern solar equipment could decrease your utility bills by up to 50%, using the formula explained in Appendix B. As you may know, state-of-the-art solar systems are much more efficient than earlier models. In addition, equipment now being installed around the country is much more pleasing to the eye than was the equipment of ten years ago. The overall effect will be to enhance the appearance of the building, as well as to save on utility costs.

We also believe that changes in landscaping would be a useful improvement to the hospital's physical plant. Specifically, planting shade trees in front of the windows on the eastern side of the complex would block sun and wind. The result would be a decrease in utility costs and enhancement of the appearance of the building. Of course, shade trees will have to grow for about five years before they begin to affect utility bills. Once they have reached adequate height, however, they will be a permanent change with low maintenance. In addition, your employees, visitors, and patients alike will notice the way that trees cut down on glare from the building walls and add "green space" to the hospital grounds.

MODEL 3–3
ABC format in paragraphs

4 | *Page Design*

This McDuff manager reviews his report draft "on screen." At this stage he performs tasks such as adding white space, revising headings, positioning graphics, changing fonts, and transforming text to lists.

*T*he four chapters in Part 1 cover basics you need to know before moving to the applications in Parts 2 and 3. The first chapter describes the technical writing process, with emphasis on writing for your reader. The second introduces McDuff, Inc., the fictional company used throughout the book, and the third deals with organizing information. This chapter covers page design, another basic building block in technical writing. Here's an operating definition:

> **Page design:** a term that refers to formatting options used to create clear, readable, and visually interesting documents. Some of these options include judicious use of white space, headings, lists, and varied fonts.
> The term *page design* became an integral part of technical writing with the advent of word processing and desktop publishing (DTP). DTP refers to sophisticated hardware and software systems that individuals can use to write, edit, design, and print both text and graphics.

This chapter starts with brief sections on the history, benefits, and potential drawbacks of using computers to design pages. The rest of the chapter presents guidelines and examples for page design. As you read the material, remember that you can use this chapter during all three stages of the writing process: planning, drafting, and editing.

BRIEF HISTORY OF COMPUTERS AND WRITING

Personal computers provide the tools to write, edit, design, and print every part of a document. Today we take these tools for granted, but it was not long ago that legal pads and typewriters were all we had. Have you talked with anyone who wrote reports before the word processor? Let's take a brief look at the changes that have taken place, through the eyes of one person who lived and worked through it at the company used throughout this book, McDuff, Inc. As it happens, the origin and growth of McDuff, Inc., have paralleled the origin and growth of the use of computer technology in writing. (See Figure 4–1.)

Rob McDuff can well remember the precomputer era. Reports in the early days of the firm contained look-alike pages of text. Writers concentrated their efforts on producing accurate, readable technical writing, with little concern for devices like bulleted lists, white space, and graphics. The drafting staff supplied necessary engineering drawings, but these graphics were drawn by hand and usually separated from the text at the end of the report.

In those days, engineers and other professionals at McDuff wrote most first drafts in longhand. Then a secretary would type a first draft for editing. When final

Time Period	Technology Used at McDuff	Capabilities
Late 1940s	Basic typewriters	Straight typing—nothing fancy
Late 1950s	Advanced typewriters—with changeable font elements	Faster action, ability to switch fonts within document
Late 1960s	Advanced typewriters—with limited memory	Limited ability to save text for future printing on typewriter
Early 1970s	Advanced typewriters—with "daisy wheel" feature	Faster printing and wider range of font selection
Late 1970s	Basic word processors	Ability to change, save, and print text
Mid 1980s	Advanced word processors with laser printers	Sophisticated features and professional printing available at individual work station
Mid 1990s	Advanced word processors— networked within and among McDuff offices	Ability to design and print sophisticated documents *and* to communicate on-line among all McDuff offices

FIGURE 4–1
From typewriters to word processing at McDuff

changes were made by the writer, the secretary would type a final draft for one last review by the writers. Woe be to the engineers who made changes after this point, for the secretary would have to type a completely new draft. Remember—this was before word processing, and even before typewriters with correction keys.

What a difference a few decades have made. Figure 4–1 shows the transformation first brought about by advanced typewriters that kept text in a memory for later editing. Then word-processing systems allowed users to write and edit text on the screen. Finally, many firms graduated to systems that could produce features like these:

- Varied type fonts and type sizes
- Varied heading styles
- Multiple columns
- Innovative graphics incorporated into the text

Basic word processing—the use of computers to type and edit text—had been transformed into desktop publishing. That is, computers now could be used to produce text, design visuals, and prepare camera-ready copy without need of outside graphics shops.

This chapter shows the broad array of choices that access to the computer has brought to the business of page design. You'll learn how such writing techniques can add to the persuasive power of your text. The merging of message and design gives you a powerful tool for writing. The section that follows highlights the benefits of this tool, while also noting problems to avoid.

COMPUTERS IN THE WRITING PROCESS

Using a computer can speed up all three stages of writing: planning, drafting, and revising. Two main advantages are that (1) you can change text and graphics continuously and easily and (2) you have many typographical choices. Such choices give you the freedom to communicate your message in the most convincing way. Here are suggestions for using computers during all three writing stages.

Using Computers to Plan

As noted in chapter 1, the planning stage of writing requires that you determine your purpose, consider your readers' needs, collect information, and construct an outline. Computers are especially useful during the last two steps, collecting information and outlining.

Word processing can speed up the collection of information on your topic. The research procedure, detailed in chapter 13, may require that you assemble information from many sources. If you are taking notes next to your computer, you can enter them directly into your system. Each one can be placed on a separate ''page'' and then indexed and organized by topic, in preparation for writing the first draft. Of course, such computerized note-taking is practical only if you are reading sources next to a terminal. Some software programs are more helpful than others in making such note-taking functions easy to use.

You can also use computers to place ideas in outline form. Words and phrases in your outline can be added and moved quickly—without the constant scratching out and rewriting necessary with handwritten outlines. Clustering ideas and arranging points in logical order become as simple as touching the keyboard. Some software programs provide the hierarchal outline format on the screen, just awaiting your insertion of wording. For extensive outlines, you can set up a separate file for each main topic and, again, use the keyboard to shift back and forth among files as you build the underlying structure of your document.

In summary, the ability to write and edit on screen can help you collect information and construct outlines. Yet many writers prefer to complete research notes and outlines by hand. The old-fashioned approach still offers the advantage of writing nonlinear (i.e., messy) outlines. Outlined points and research notes can be stretched all over paper of any size or even pinned to the walls, giving you a "big picture" effect that small, ever-changing screens may not. Choose the best method for you, but at least be aware of what options computers offer.

Using Computers to Draft

Though helpful during the planning process, word processing shows its main strength as you write early drafts. Whether you touch-type or plunk away with two fingers, the word processor will keep you writing faster. For example, the common "word-wrap" feature automatically moves you from line to line without need of carriage returns. Most importantly, you add and delete text very quickly. You simply "scroll" forward or backward on the screen to the point where you want to make changes. Here are some other advantages of writing early drafts on the computer:

- Systems with multiple windows may keep your outline on the screen while you write the related draft section.
- When you cannot think of a word or phrase, you can type in an easily recognizable symbol and then return later to insert the word(s).
- When several alternative passages are possible, you can type in all of them, with brackets or other associated symbols. Then during the revision stage, you can make your final choice.

As noted in the chapter 1 discussion on collaborative writing, the computer helps you combine your work with that of others in team-writing efforts. Individual writers in the group can produce draft material at their separate terminals, which are part of one network. At that point, individual files can be "pulled up" on the screens of other group members, for revision and commentary. Later, separate files can be merged into one draft for revision. Thus both groups and individuals benefit from the computer at the drafting stage of writing.

Using Computers to Revise

Drafting requires that you get ideas onto paper (or screen) as quickly as possible, to avoid writer's block. When you shift to the revising stage, the aim is to select options and change text in whatever way best meets the readers' needs. As shown in Figure 4–2, computer systems offer these sorts of tools for revision:

- **Addition and deletion:** You can add or delete text by placing the cursor—an icon that shows where you are in the text—at the appropriate point. This technique eliminates the time-consuming process of scribbling changes into margins by hand, deleting words, and then retyping drafts.
- **Cut-and-paste:** This feature allows you to move sentences, paragraphs, graphics, or entire sections within one document, or among several. No longer must you literally cut and paste sections of typed or handwritten copy. Cut-and-paste allows you easily to revise text to produce the most logical arrangement of material.

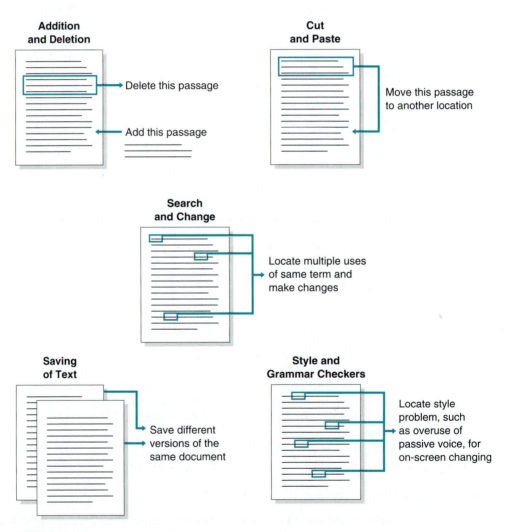

FIGURE 4–2
Sample changes during revising

- **Search and change:** Used correctly, this feature can save considerable time during revision. It helps you locate every instance of a particular word or phrase throughout the document. For example, assume that you want to check your draft to make certain that you used *data* consistently as a plural noun, not a singular noun. Simply enter the word *data* on the keyboard so that your system will scroll to each section of the text where the word is used. Then you can make necessary changes.
- **Saving of text:** Often you may be uncertain what version of a paragraph or section is most appropriate for the final draft. In this case you can save multiple versions in a file for later reference. A note or an asterisk on your screen will trigger your memory that another version of the text is available. Later you can select the best option and delete the others.
- **Style and grammar checkers:** Now there are editing programs that use the "search and change" function to locate awkward or incorrect passages in your writing. Once you see the highlighted passage, you can decide whether to change it. This software is sophisticated enough to identify errors or constructions like these: split infinitives ("to randomly select"), passive voice ("the recommendations were made"), clichés ("skyrocketing costs"), sexist pronouns ("an engineer should complete his reports on time"), and overuse of "to be" verbs (such as "is," "are," "was," and "were"). Another useful feature can analyze the "readability" of your style by, for example, determining the average number of words per sentence.

    Interactive software also helps you locate grammar, proofreading, and spelling errors. Examples include double words ("and and"), incorrect capitalization ("THey"), wrong spacing, incorrect punctuation, and subject-verb agreement errors. Another feature is a thesaurus. It can suggest alternative word choices and even give the etymology of words. Some programs allow you to add particular grammar or style rules that you want applied to your writing.

Computers allow changes as quickly as you can touch the keyboard. Besides speed, there is the advantage of convenience. Copy can be corrected until the last moment before printing. If errors are found, another draft can be printed. Ironically, it is the very ease of editing during word processing that can lead to misuse and inefficiency of computers during the writing process. The next section notes several problems and ways to avoid them.

Problems to Avoid

The following actual case typifies problems that can arise with computers during the writing process. At one engineering firm in the Southwest a few years ago, engineers usually produced their own report drafts in longhand or by cutting and pasting typed copy from previous reports. This hard copy was delivered to a central word-processing department, where operators produced first drafts for later editing. In theory, the operators were to see each report only two or three times—once during the initial inputting of information and another time or two when changes were incorporated into the final draft. In fact, however, writers constantly made changes that required multiple printing of the reports—often five or six times instead of two or three. As a result, productivity decreased, and word-processing expenses increased. The very ease of word processing had led to inefficiency in the writing process.

The company solved its productivity problem by forming a review committee of both users and operators, which resulted in the adoption of some new procedures. For example, the group agreed to limit the number of revision cycles, requiring writers to give more attention to each draft. In the same way, you can save time and take full advantage of computer technology in writing by following these guidelines:

- *Work in stages.* Note the suggestions in chapter 1 for completing stylistic, grammatical, and mechanical editing in stages—so that errors are caught at the appropriate time.
- *Limit number of printings.* Print as often as you need to print, but don't be wasteful. In other words, make sure you have edited enough to justify another printing. One hidden cost of writing today is the proliferation of word-processing drafts. Resources are squandered, and time is added to the writing process.
- *Use the buddy system.* Ask a colleague to help with the editing process so that another pair of eyes sees your work. Your closeness to the project may keep you from spotting errors, no matter how much technology is at your command.
- *Know the limitations of style programs.* Remember that software programs to correct spelling and other mechanical errors will not eliminate all errors. You still need to watch for problems with words like *effect/affect, complement/compliment,* and *principal/principle,* as well as misspellings of personal names and other proper nouns. You might want to use the "search and change" feature of your system to locate trouble spots.

If used with common sense, computers can be a powerful tool as you plan, draft, and revise documents. You will spend more time preparing the content of writing and less time on mechanics.

However, there is one more hazard brought about by this new technology. Computers allow far more typographical variations than most readers want or need. For example, systems give you the option of using one or all of the following techniques for typographical emphasis: underlining, bold, full caps, italics, and increased type size. Too much variety in one document can confuse rather than clarify a passage. The next two sections provide guidelines for effective page design.

GUIDELINES FOR PAGE DESIGN

As one expert says, often you will write for readers who are "in a hurry, frustrated, and bored, and who would prefer to get the information needed from text *in any other way but reading*" (my emphasis) [Schriver, K. A. Fourth Quarter, 1989. Document design from 1980 to 1989: Challenges that remain. *Technical Communication,* p. 319]. Most readers dislike solid text. Your challenge is to respond to this prejudice by making pages interesting to the eye.

Good organization, as pointed out in the last chapter, can fight readers' indifference by giving information when and where they want it. But to keep readers interested, you must use effective page design—that is, the physical appearance of each page in your document. Each page needs the right combination of visual elements to match the needs of readers and the purpose of the document.

This section briefly describes these elements of page design: (1) white space, (2) headings, (3) lists, and (4) in-text emphasis. Computers have made these features and others easy to introduce into documents. The next section covers aspects of document design dealing with type selection and font size. Together these two parts of the chapter give you the guidelines and examples you need to make text visually appealing to your readers.

White Space

The term *white space* simply means the open places on the page with no text or graphics—literally, the *white* space. Experts have learned that readers are attracted to text because of the white space that surrounds it, as with a newspaper ad that includes a few lines of copy in the middle of a white page. Readers connect white space with important information.

In technical writing you should use white space in a way that (1) attracts attention, (2) guides the eye to important information on the page, (3) relieves the boredom of reading text, and (4) helps readers organize information. Here are some opportunities for using white space effectively:

1. Margins: Most readers appreciate generous use of white space around the edges of text. Marginal space tends to "frame" your document, so the text doesn't appear to push the boundaries of the page. Good practice is to use 1" to 1 1/2" margins, with more space on the bottom margin. When the document is bound on the left, as in the Figure 4–3, also place more space in the left margin than at the bottom—to account for space lost in the binding process.

2. Columns: Long lines can be an obstacle to keeping the readers' attention. Eyes get weary of overly long lines, so some writers add double columns to their design options. This "book look," as shown in Figure 4–4, uses white space between columns to break up text and thus reduce line length.

3. Line space: When choosing single, double, or 1 1/2 line spacing, consider the document's length and degree of formality. Letters, memos, short reports, and other documents read in one sitting are usually single spaced, with one line space between paragraphs. Longer documents, especially if they are formal, are usually 1 1/2-space or double-spaced, without extra spacing between paragraphs. (See Figure 4–5.)

4. Right-justified versus ragged edge: To justify or not to justify lines is often the question. In right-justified copy, all lines are the same length—as on this textbook page. In ragged-edge copy, lines are variable length. Some readers prefer ragged-edge copy because it adds variation to the page, making reading less predictable for the eye. Yet many readers like the professional appearance of right-justified lines, especially in formal documents. Both views have merit.

As a rule, (1) use ragged edge on densely packed, single-spaced documents and (2) use either ragged or justified margins on 1 1/2-space and double-spaced

FIGURE 4–3
Use of white space:
margins

FIGURE 4–4
Use of white space:
columns

documents, depending on reader preference or company style (see Figure 4–5). However, only use justified margins if your word-processing software maintains uniform spacing between words and between letters within words. It is unnerving to read justified text with inconsistent spacing at these points.

5. Paragraph length: New paragraphs give readers a chance to regroup, as one topic ends and another begins. These shifts also have a visual impact. The amount of white space produced by paragraph lengths can shape reader expectations. For example, two long paragraphs suggest a heavier reading burden than do three or four paragraphs of differing lengths. Most readers skip long paragraphs, so vary paragraph lengths and avoid putting more than 8 or 10 lines in any one paragraph (see Figure 4–6).

6. Paragraph indenting: The argument goes on about indenting or not indenting the first lines of paragraphs. As with ragged-edge copy, most readers prefer indented paragraphs because the extra white space creates visual variety. Reading text is hard work for the eye. You should take advantage of any opportunity to snag the reader (see Figure 4–5).

7. In-text graphics: Any illustration within the text needs special attention. Chapter 11 provides a complete discussion of graphics, but this section discusses their placement for visual appeal. Here are some pointers:

- Make sure there is ample white space between any in-text graphic and the text. If the figure is too large to permit adequate margins, reduce its size.
- When you have the choice, place in-text graphics near the top of the page. That position gives them the most attention.

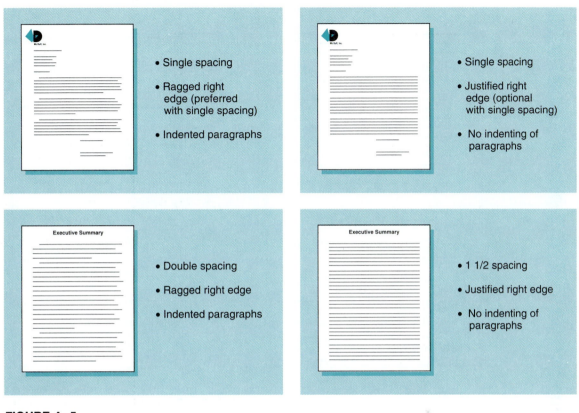

FIGURE 4–5
Use of white space: lines

- When a graphic doesn't fit well on a page with text, place it on its own page where there will be adequate space. Normally it appears on the page following the first reference to it.
- Pay special attention to page balance when graphics will be included on multi-column pages, two-page spreads, or both.
- Draw rough sketches of the layout for the entire document so that you can use white space consistently and persuasively from start to finish.

8. Heading space and lines: White space helps the reader connect related information immediately. Always have slightly more space above a heading than below it. That extra space visually connects the heading with the material into which it leads. In a double-spaced document, for example, you would add a third line of space between the heading and the text that came before it.

In addition, some writers add a horizontal line across the page above headings, to emphasize the visual break. The next section will cover other aspects of headings.

In summary, well-used white space can add to the persuasive power of your text. As with any design element, however, it can be overused and abused. Make

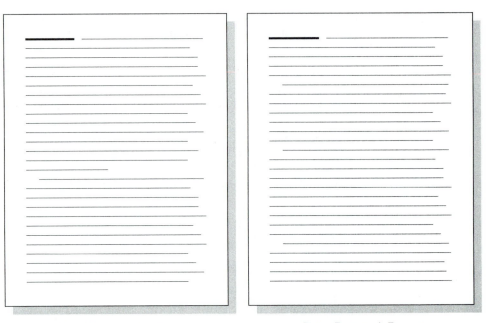

Poor Paragraph Format   Better Paragraph Format

FIGURE 4–6
Use of white space: paragraphs

sure there is a *reason* for every decision you make with regard to white space on your pages.

Headings

Headings are brief labels used to introduce each new section or subsection of text. They serve as (1) a signpost for the reader who wants to know the content, (2) a ''grabber'' to entice readers to read documents, and (3) a visual ''oasis'' of white space where the reader gets relief from text.

As a general rule, every page of any document over one page should have at least one heading. Readers need these markers to find their way through your writing. Models throughout this text show how headings can be used in short and long documents. Of course, heading formats differ greatly from company to company and even from writer to writer. With all the typographical possibilities of word processing, there is incredible variety in typeface, type size, and the use of bold, underlining, and capitals. Here are some general guidelines that apply:

1. Use your outline to create headings and subheadings. A well-organized outline lists major and minor topics. With little or no change in wording, they can be converted to headings and subheadings within the document. As with outlines, you need to follow basic principles of organization.

- First, if you have one subheading, you must have at least one more at that same level—anything that is divided has at least two parts.

- Second, the number of subheadings should be one indication of the relative length or importance of the section. Be consistent with your approach to headings throughout the document.

2. Use substantive wording. Headings give readers an overview of what content will follow. They entice readers into your document; they can determine whether readers—especially those who are hurried and impatient—will read or skip over the text. Strive to use concrete rather than abstract nouns, even if the heading must be a bit longer. Note the improvements in the following revised headings:

Original:	"Background"
Revised:	"How the Simmons Road Project Got Started" or "Background on Simmons Road Project"
Original:	"Discussion"
Revised:	"Procedure for Measuring Toxicity" or "How to Measure Toxicity"
Original:	"Costs"
Revised:	"Production Costs of the FastCopy 800" or "Producing the FastCopy 800: How Much?"

3. Maintain parallel form in wording. Headings of equal value and degree should have the same grammatical form, as shown in the following:

A) *Headings With Parallel Form*
    Scope of Services
    Schedule for Fieldwork
    Conditions of Contract

B) *Headings That Lack Parallel Form*
    Scope of Services
    How Will Fieldwork Be Scheduled?
    Establish Contract Conditions

You don't have to be a grammar expert to see that the three headings in "B" are in different forms. The first is a noun phrase, the second is a question, and the third is an action phrase beginning with a verb. Because such inconsistencies distract the reader, you should make headings in each section uniform in wording.

4. Establish clear hierarchy in headings. Whatever typographical techniques you choose for headings, your readers must be able to distinguish one heading level from another. Visual features should be increasingly more striking as you move up the ranking of levels. Figure 4–7 shows several heading options that reflect such distinctions. Here are specific guidelines for using such typographical distinctions:

- *Use larger type size for higher-level headings.* You want readers to grasp quickly the relative importance of heading levels as they read your document. Type size fixes this relative importance in their minds so that they can easily find their way through your material both the first time and upon rereading it. The examples in Figure 4–7 reflect two and sometimes three different type sizes in some of the examples. The incremental upgrading of type size helps readers determine the relative importance of the information.

1. Four levels, formal report

LEVEL-1 HEADING

xxxxxxxxxxxxxxxxxxxxxxxxxxxxxxxxxxxxxxxxxxxxxxxxxxxxxxxxxxxxxxxxxxxxxxxx xxxxxxxxxxxxxxxxxxxxxxxxxxxxxxxx

LEVEL-2 HEADING

xxxxxxxxxxxxxxxxxxxxxxxxxxxxxxxxxxxxxxxxxxxxxxxxxxxxxxxxxxxxxxxxxxxxxxxx xxxxxxxxxxxxxxxxxxxxxxxxxxxxxxxxxxxxxxxxxxxxxxxxxxxxxxxxxxxxxxxxxxxxxxxx xxxxxxxxxxxxx

Level-3 Heading

xxxxxxxxxxxxxxxxxxxxxxxxxxxxxxxxxxxxxxxxxxxxxxxxxxxxxxxxxxxxxxxxxxxxxxxx xxxxxxxxxxxxxxxxxxxxxxxxxxxxxxxxxxxxxxxxxxxxxxxxxxxxxxxxxxxxxxxxxxxxxxxx xxxxxxxxxxxxxxxxxxxxxxxxxxxxxxxxxxxx

Level-4 Heading. xxxxxxxxxxxxxxxxxxxxxxxxxxxxxxxxxxxxxxxxxxxxxxxxx xxxxxxxxxxxxxxxxxxxxxxxxxxxxxxxxxxxxxxxxxxxxxxxxxxxxxxxxxxxxxxxxxxxxxxxx

2. Three levels, formal report: Option A

Level-1 Heading

xxxxxxxxxxxxxxxxxxxxxxxxxxxxxxxxxxxxxxxxxxxxxxxxxxxxxxxxxxxxxxxxxxxxxxxx xxxxxxxxxxxxxxxxxxxxxxxxxxxxxxxxxxxxxxxxxxxxxxxx

Level-2 Heading

xxxxxxxxxxxxxxxxxxxxxxxxxxxxxxxxxxxxxxxxxxxxxxxxxxxxxxxxxxxxxxxxxxxxxxxx xxxxxxxxxxxxxxxxxxxxxxxxxxxxxxxxxxxxxxxxxxxxxxxxxxx

Level-3 Heading. xxxxxxxxxxxxxxxxxxxxxxxxxxxxxxxxxxxxxxxxxxxxxxxx xxxxxxxxxxxxxxxxxxxxxxxxxxxxxxxxxxxxxxxxxxxxxxxxxxxxxxxxxxxxxxxxxxxxxxxx

3. Three levels, formal report: Option B

LEVEL-1 HEADING

xxxxxxxxxxxxxxxxxxxxxxxxxxxxxxxxxxxxxxxxxxxxxxxxxxxxxxxxxxxxxxxxxxxxxxxx xxxxxxxxxxxxxxxxxxxxxxxxxxxxxxxxxxxxxxxxxxxxxxxxxxxxxxxxxxx

LEVEL-2 HEADING

xxxxxxxxxxxxxxxxxxxxxxxxxxxxxxxxxxxxxxxxxxxxxxxxxxxxxxxxxxxxxxxxxxxxxxxx xxxxxxxxxxxxxxxxxxxxxxxxxxxxxxxxxxxxxxxxxxxxxxxx

Level-3 Heading

xxxxxxxxxxxxxxxxxxxxxxxxxxxxxxxxxxxxxxxxxxxxxxxxxxxxxxxxxxxxxxxxxxxxxxxx xxxxxxxxxxxxxxxxxxxxxxxxxxxxxxxxxxxxxxxxx

FIGURE 4–7
Some heading options

4. Three levels, informal (letter or memo) report

Level-1 Heading

xxxxxxxxxxxxxxxxxxxxxxxxxxxxxxxxxxxxxxxxxxxxxxxxxxxxxxxxxxxxxxxxxxxx
xxxxxxxxxxxxxxxxxxxxxxxxxxxxxxxxxxxxxxxxxx

Level-2 Heading

xxxxxxxxxxxxxxxxxxxxxxxxxxxxxxxxxxxxxxxxxxxxxxxxxxxxxxxxxxxxxxxxxxxx
xxxxxxxxxxxxxxxxxxxxxxxxxxxxxxxxxxxxxxxxxxxxxxx

Level-3 Heading. xxxxxxxxxxxxxxxxxxxxxxxxxxxxxxxxxxxxxxxxxxxxxx
xxxxxxxxxxxxxxxxxxxxxxxxxxxxxxxxxxxxxxxxxxxx

---

5. Two levels, informal (letter or memo) report: Option A

LEVEL-1 HEADING

xxxxxxxxxxxxxxxxxxxxxxxxxxxxxxxxxxxxxxxxxxxxxxxxxxxxxxxxxxxxxxxxxxxx
xxxxxxxxxxxxxxxxxxxxxxxxxxxxxxxxxxxxxxxxxxxxxxxxx

Level-2 Heading

xxxxxxxxxxxxxxxxxxxxxxxxxxxxxxxxxxxxxxxxxxxxxxxxxxxxxxxxxxxxxxxxxxxx
xxxxxxxxxxxxxxxxxxxxxxxxxxxxxxxxxxxxxxxxxxxx

---

6. Two levels, informal (letter or memo) report: Option B

Level-1 Heading

xxxxxxxxxxxxxxxxxxxxxxxxxxxxxxxxxxxxxxxxxxxxxxxxxxxxxxxxxxxxxxxxxxxx
xxxxxxxxxxxxxxxxxxxxxxxxxxxxxxxxxxxxxxxxxxxx

Level-2 Heading. xxxxxxxxxxxxxxxxxxxxxxxxxxxxxxxxxxxxxxxxxxxxxxxx
xxxxxxxxxxxxxxxxxxxxxxxxxxxxxxxxxxxxxxx

---

7. One level, informal (letter or memo) report: Option A

LEVEL-1 HEADING

xxxxxxxxxxxxxxxxxxxxxxxxxxxxxxxxxxxxxxxxxxxxxxxxxxxxxxxxxxxxxxxxxxxx
xxxxxxxxxxxxxxxxxxxxxxxxxxxxxxxxxxxxxxxxxxxxxxx

---

8. One level, informal (letter or memo) report: Option B

Level-1 Heading

xxxxxxxxxxxxxxxxxxxxxxxxxxxxxxxxxxxxxxxxxxxxxxxxxxxxxxxxxxxxxxxxxxxx
xxxxxxxxxxxxxxxxxxxxxxxxxxxxxxxxxxxxxxxxxxxx

FIGURE 4–7, *continued*

- *Use heading position to show ranking.* In formal documents, your high-level headings can be centered. The next two or three levels of headings are at or off the left margin, as shown in Figure 4–7. Be sure these lower-level headings also use other typographical techniques, such as size, to help the reader distinguish levels.
- *Use typographical techniques to accomplish your purpose.* Besides type size and position, previously mentioned, you can vary heading type with features such as these:

    Uppercase and lowercase

    Bold type

    Underlining

    Changes in type font

With this embarrassment of riches, writers must be careful not to overdo it and create "busy" pages of print. Use only those features that will look good on the page and provide an easy-to-grasp hierarchy of levels for the reader.

- *Consider using decimal headings for long documents.* Decimal headings include a hierarchy of numbers for every heading and subheading listed in the table of contents. Many an argument has been waged over their use. People who like them say that they help readers find their way through documents and refer to subsections in later discussions. People who dislike them say that they are cumbersome and give the appearance of "bureaucratic" writing.

    Unless decimal headings are expected by your reader, use them only with formal documents that are fairly long. Include at least three heading levels. Following is the normal progression of numbering in decimal headings for a three-level document:

```
1.0   xxxxxxxxxxxxxx
        1.1   xxxxxxxxxxx
                1.1.1   xxxxxxxxxx
                1.1.2   xxxxxxxxxx
        1.2   xxxxxxxxxxx
                1.2.1   xxxxxxxxxx
                1.2.2   xxxxxxxxxx
2.0   xxxxxxxxxxxxxx
        2.1   xxxxxxxxxxx
                2.1.1   xxxxxxxxxx
                2.1.2   xxxxxxxxxx
        2.2   xxxxxxxxxxx
3.0   xxxxxxxxxxxxxx
```

Listings

Technical writing benefits from the use of lists. Readers welcome your efforts to cluster items into lists for easy reading. In fact, almost any group of three or more related points can be made into a bulleted or numbered listing. Here are some points to consider as you apply this important feature of page design:

1. **Typical uses:** Lists emphasize important points and provide a welcome change in format. Because they attract more attention than text surrounding them, they are usually reserved for these uses:

   Examples

   Reasons for a decision

   Conclusions

   Recommendations

   Steps in a process

   Cautions or warnings about a product

   Limitations or restrictions on conclusions

2. **Number of items:** The best lists are those that subscribe to the rule of short-term memory. That is, people can retain no more than five to nine items in their short-term memory. A listing over nine items may confuse rather than clarify an issue. Consider placing 10 or more items in two or three groupings, or grouped lists, as you would in an outline. This format gives the reader a way to grasp information being presented.

3. **Use of bullets and numbers:** The most common visual clues for listings are numbers and bullets (enlarged dots or squares like those used in the following listing). Here are a few pointers for choosing one or the other:

   - *Bullets:* Best in lists of five or fewer items, unless there is a special reason for using numbers.
   - *Numbers:* Best in lists of over five items *or* when needed to indicate an ordering of steps, procedures, or ranked alternatives. Remember that your readers normally will *infer* sequence or ranking in a numbered list.

4. **Format on page:** Every listing should be easy to read and pleasing to the eye. These specific guidelines cover practices preferred by most readers:

   - *Indent the listing.* Although there is no standard list format, readers prefer lists that are indented farther than the standard left margin. Five spaces is adequate.
   - *Hang your numbers and bullets.* Visual appeal is enhanced by placing numbers or bullets to the left of the margin used for the list, as done with the items here.
   - *Use line spaces for easier reading.* When one or more listed items contain over a line of text, an extra line space between listed items can enhance readability.
   - *Keep items as short as possible.* Depending on purpose and substance, lists can consist of words, phrases, or sentences—like the list you are reading. Whichever format you choose, pare down the wording as much as possible to retain the impact of the list format.

5. **Parallelism and lead-ins:** Make the listing easy to read by keeping all points grammatically parallel and by including a smooth transition from the lead-in to the listing itself. (The term *lead-in* refers to the sentence or fragment preceding the listing.) "Parallel" means that each point in the list is in the same grammatical form, whether a complete sentence, verb phrase, or noun phrase. If you change form in the midst of a listing, you take the chance of upsetting the flow of information.

EXAMPLE:

"To complete this project, we plan to do the following:

- Survey the site
- Take samples from the three boring locations
- Test selected samples in our lab
- Report on the results of the study"

The listed items are in verb form (note introductory words "survey," "take," "test," "report"). An alternative would be to put them in noun form, with a slightly different lead-in.

EXAMPLE:

"To complete the project, we will perform the following activities:

- Surveying the site
- Taking samples from the three boring locations
- Testing selected samples in our lab
- Reporting on the results of the study"

6. **Punctuation and capitalization:** Although there are acceptable variations on the punctuation of lists, preferred usage includes a colon before a listing, no punctuation after any of the items, and capitalization of the first letter of the first word of each item. Refer to the alphabetized Handbook under "Punctuation: Lists" for alternative ways to punctuate lists.

7. **Overuse:** With listings it *is* possible to have too much of a good thing. Too many lists on one page can create a distracting, fragmented effect. One rule of thumb is to use no more than one list per page. Two lists on the same page force the reader to decide which one deserves attention first.

In-Text Emphasis

Sometimes you want to emphasize an important word or phrase within a sentence. Computers give you these options: underlining, boldface, italics, and caps. The least effective are FULL CAPS and *italics,* for both are difficult to read within a paragraph and distracting to the eye. The most effective highlighting techniques are <u>underlining</u> or **boldface;** they add emphasis without distracting the reader. This writer's preference is <u>underlining</u>, which draws attention to the word without the overstatedness of bold copy (which can also be hard to read).

Whatever typographical techniques you select, use them sparingly. They can create a "busy" page that leaves the reader confused about what to read. Excessive in-text emphasis also detracts from the impact of headings and subheadings, which should be receiving significant attention.

SELECTION OF FONTS

Besides page format, you have something else in your word-processing bag of tricks: changes in the size and type of font you use in the text itself.

Size of Type

Traditionally, type size has been measured in "points" (72 points to an inch). When you go to your font-selection menu on your computer screen, the sizes may be listed as such: 9, 10, 12, 14, 18, and 24. Other features of your software will allow you to expand type even farther, for specialty uses.

Despite these many options, most technical writing is printed off the desktop in 10- or 12-point type. When you are choosing type size, however, be aware that the actual size of letters varies among font types. Some 12-point type appears larger than other 12-point type. Differences stem from the fact that your selection of a font affects (1) the thickness of the letters, (2) the size of lowercase letters, and (3) the length and style of the parts of letters that extend above and below the line. The following examples show the differences in three common fonts. Note that the typeface used in setting the text of this book is 10-point Meridien.

New Century Schoolbook
9 point
10 point
12 point
14 point
18 point
24 point

Times Roman
9 point
10 point
12 point
14 point
18 point
24 point

Helvetica
9 point
10 point
12 point
14 point
18 point
24 point

Before selecting your type size, run samples on your printer so that you are certain of how your copy will appear in final form.

FIGURE 4–8
Font types   Serif Type   Nn▷ extra lines
                                          (serifs)

                  Sans Serif Type   Nn

Font Types

Font types vary tremendously. Most word-processing systems give you more choices than you will ever use. Generally, these types are classified into two groups:

- Serif fonts: Characters have "tails" at the ends of the letterlines.
- Sans serif fonts: Characters do not have tails. (See Figure 4–8.)

If you are able to choose your font, the obvious advice is to use the one that you know is preferred by your readers. A phone call or a look at documents generated by your reader may help you. If you have no reader-specific guidelines, here are three general rules to follow:

1. *Use serif fonts for regular text in your documents.* The tails on letters make letters and entire words more visually interesting to the reader's eye. In this sense, they serve the same purpose as ragged-edge copy—helping your reader move smoothly through the document.

2. *Consider using another typeface—sans serif—for headings.* Headings benefit from a clean look that emphasizes the white space around letters. Sans serif type helps attract attention to these elements of organization within your text.

3. *Avoid too many font variations in the same document.* There is a fine line between interesting font variations and busy and distracting text. But there *is* a line. Your rule of thumb might be to use no more than two fonts per document, one for text and another for headings and subheadings.

CHAPTER SUMMARY

This chapter shows you how to apply principles of page design to your assignments in this class and your on-the-job writing. The term *page design* refers to the array of formatting options you can use to improve the visual effect of your document.

The last several decades have seen an incredible change in the way documents are produced. Today individual writers working at their personal computers write, edit, design, and print sophisticated documents. Used judiciously, computers are an effective tool during the planning, drafting, and revising stages.

Effective page design requires that you use specific elements such as white space, headings, listings, and in-text emphasis. White space draws attention to adjacent items. Headings quickly lead the reader to important points and subpoints. Listings emphasize related groups of points. And conservative use of in-text emphasis, like underlining and bold print, can draw attention to items within sentences and paragraphs.

Another strategy for page design is to change the size and type of fonts in your documents. Like other strategies, this one must be used with care so that your document does not become too "busy." Page design remains a technique for highlighting content, not a substitute for careful organization and editing.

ASSIGNMENTS

1. **Group Evaluation of Page Design.** Working in small groups, analyze the effectiveness of the page design of the document in Model 4–1 on pages 101–102. Your instructor will indicate whether you should prepare a written or an oral report of your findings. Give specific support for your praise or criticism. (Assignment 4 uses this same memorandum for a writing exercise.)

2. **Individual Evaluation of Page Design.** Visit your library and locate an example of technical writing, such as a government document or a company's annual report. Use the guidelines in this chapter to analyze the document's page design. Your instructor will indicate whether your report should be oral or written.

3. **Individual Practice in Page Design.** As a manager at McDuff, you have just finished a major report to a client. It gives recommendations for transporting a variety of hazardous materials by sea, land, and air. The body of your report contains a section that defines the term *stowage plan* and describes its use. Given your mixed technical and nontechnical audience, this basic information is much needed. What follows is the *text* of that section. Revise the passage by applying any of this chapter's principles of page design that seem appropriate—such as adding headings, graphics, lists, and white space. If you wish, you also can make changes in organization and style. *Optional:* If your class has access to e-mail, transmit your version to another student to receive his or her response.

In the chemical shipping industry, a stowage plan is a kind of blueprint for a vessel. It lists all stowage tanks and provides information about tank volume, tank coating, stowed product, weight of product, loading port, and discharging port. A stowage plan is made out for each vessel on each voyage and records all chemicals loaded. The following information concerns cargo considerations (chemical properties and tank features) and some specific uses of the stowage plan in industry.

The three main cargo considerations in planning stowage are temperature, compatibility, and safety. Chemicals have physical properties that distinguish them from one another. To maintain the natural state of chemicals and to prevent alteration of their physical properties, a controlled environment becomes necessary. Some chemicals, for example, require firm temperature controls to maintain their physical characteristics and degree of viscosity (thickness) and to prevent contamination of the chemicals by any moisture in the tanks. In addition, some chemicals, like acids, react violently with each other and should not be stowed in adjoining, or even neighboring, tanks. In shipping, this relationship is known as chemical compatibility.

The controlled environment and compatibility of chemicals have resulted in safety regulations for the handling and transporting of these chemicals. These regulations originate with the federal government, which bases them on research done by the private manufacturers. Location and size of tanks also determine the placement of cargo. A ship's tanks are arranged with all smaller tanks around the periphery of the tank grouping and all larger tanks in the center. These tanks, made of heavy steel and coated with zinc or epoxy, are highly resistant to most chemicals, thereby reducing the chance of cargo contamination. Each tank has a maximum cargo capacity, and the amounts of each chemical are matched with the tanks. Often chemicals to be discharged at the same port are staggered in the stowage plan layout so that after they are discharged the ship maintains its equilibrium.

The stowage plan is finalized after considering the cargo and tank characteristics. In its final form, the plan is used as a reference document with all information relevant to the loading/discharging voyage recorded. If an accident occurs involving a ship, or when questions arise involving discharging operations, this document serves as a visual reference and brings about quick decisions.

---

4. **Group Practice in Page Design.** Working in small groups, prepare a redesigned version of the memorandum in Model 4–1 (pp. 101–102). If your class is being held in a computer lab, present your group's version on screen. If you are not using a lab, present your version on an overhead transparency.

Mc Duff, Inc.

MEMORANDUM

DATE: August 19, 1993
TO: Randall Demorest, Dean *RD*
FROM: Kenneth Payne, Professor and Head
SUBJECT: BSTW Advisory Board

What? Lunch meetings between Advisory Board members and me
Why?   To get more Board support for the BSTW degree program
Who?   Each individual member at a separate luncheon
When? Fall 1993
How?   Allocation of $360 to pay for the lunches

Rationale

When we seek support for the college, we have to (1) make people feel that they will get something in return and (2) make them feel comfortable about us and our organization. As businesses have demonstrated, one way we can accomplish these goals is by taking potential donors to lunch.

As you and I have discussed, the B.S. in Technical Writing degree program (BSTW) needs to strengthen ties to its Advisory Board. We must ask Board participants to provide tangible support for the program *and* give them meaningful involvement in the work we are doing.

Method

The immediate need is to involve members of the Advisory Board in the coming year's program. I want to do this in two ways:

1. Plan carefully for a fall Board meeting
2. Discuss with each of them individually what we want to accomplish this year

Cost

To do the second item mentioned, I request an allocation of $360 so that I can take each member to lunch for an extended one-on-one discussion. I plan to discuss the needs of our program and each member's capabilities to support it.

continues

MODEL 4–1
Page design in memorandum

Specifics

Each member of the Board will be asked individually to consider the following ways to contribute:

1. Continuing support for the internship program
2. Participation in the research project we began a year ago
3. Cooperative work experiences for BSTW faculty, possibly during the summer of 1994
4. Financial support for the following items:
   - The college's membership as a sponsoring organization in the Society for Technical Communication
   - Contribution's—financial or otherwise—to library holdings in technical writing
   - Usability testing laboratory
   - A workshop series bringing to the campus some outstanding technical communicators (for example, Edward Tufte, expert in graphics; JoAnn Hackos, President of STC and an expert in quality management; and William Horton, expert in on-line documentation)

Benefits

What are Board members going to get from this?

Long range: A better BSTW program, which will produce better technical writers for them to hire

Immediately: Meaningful involvement in the program

Specifically: Training opportunities for their personnel through the workshops mentioned

My tentative plan for those workshops is to provide a one-day seminar for our students and a second seminar for employees of Advisory Board members. (We will allow them a number of participants based on how much they contribute to the workshops.)

Response Needed

Please let me know as soon as possible if money is available for the lunches. I hope to begin scheduling meetings within a week.

MODEL 4–1, *continued*

PART II

5 | *Patterns of Organization*

Definition, description, and other patterns of organization form the body of technical documents. These patterns must be clearly presented to a diverse audience, including hands-on technicians and supervisors.

P art 1 gave you some basic background on technical writing and on McDuff, Inc., the example company used throughout this book. Building on that foundation, Part 2 focuses on specific types of technical writing. This chapter and the next one cover seven common patterns for organizing information. Then chapters 7–10 discuss the following formats for entire documents: letters and memos, short reports, long reports, and proposals and feasibility studies.

If you have already taken a basic composition course, you will see similarities between patterns studied in that course and those described in this chapter. Indeed, technical writing uses the same building blocks as all other good writing. What follows are separate sections on five of these main patterns: argument, definition, description, classification/division, and comparison/contrast. They are roughly arranged by the frequency with which they are used in technical writing, starting with the most common pattern. Each section contains these four parts:

1. Introduction about the pattern
2. Short case studies from McDuff
3. Writing guidelines
4. Extended example at the end of the chapter, with marginal comments

Then the assignments challenge you to use the chapter's five patterns in the context of some short letter or memo reports.

ARGUMENT

Good argument forms the basis for *all* technical writing. Some people have the mistaken impression that only recommendation reports and proposals argue their case to the reader, and that all other writing should be objective rather than argumentative. The fact is, every time you commit words to paper you are "arguing" your point. This text uses the following broad-based definition for argument:

> **Argument:** the strategies you use in presenting evidence to support your point *and* to support your professional credibility, while still keeping the reader's goodwill.

Thus even the most uncontroversial document, like a trip report, involves argument in the sense defined here. That is, a trip report would present information to support the fact that you accomplished certain objectives on a business trip. It also would show the reader, usually your boss, that you worked hard to accomplish your objectives. In fact, you hope that every document you write becomes a written argument for your own conscientiousness as a professional.

The strongest form of argument—called "persuasion"—tries to convince your reader to adopt a certain point of view or pursue a certain line of action (see chapter 10, "Proposals and Feasibility Studies"). In other words, persuasion seeks obvious *changes* in opinions or actions, whereas argument only presents evidence or logic to support a point of view. To help you use argument correctly, the next section gives some short case studies from McDuff, guidelines for applying argument, and an annotated example.

Short Cases From McDuff: Argument

Under the definition just given, many in-house and external documents could be considered argument. Here are three McDuff examples:

■ Argument Case 1: Fire Control in the Everglades

As a fire-control expert in McDuff's Atlanta office, you just completed a project for the state of Florida. You spent two weeks in the Everglades examining the likelihood for major fires this season. Last season, major fires and westerly winds caused soot and smoke to drift to Florida's tourist area. Your investigation suggests that fires are likely and, in fact, necessary this season. Regular, contained swamp fires help keep down underbrush and thus reduce the chance for much bigger fires. Limited fires will not damage wildlife or major trees, whereas major fires will. Now you must write a report to Florida authorities arguing that controlled fires in the Everglades are desirable, despite the general feeling that all fires should be prevented.

■ Argument Case 2: Back Pain in the Boston Office

The engineers, programmers, scientists, and other office workers in McDuff's Boston branch spend a lot of time in their chairs. Although all the office furniture is new and expensive, you and your Boston branch colleagues have experienced regular back pain since the new chairs arrived. Unfortunately, the furniture was ordered through the corporate office, so your complaint cannot be handled in a routine, informal way in your own office. As the branch manager, you have mentioned the problem to your boss, the vice president for domestic operations. Predictably, he asked you to "put it in writing." Although this memo report must explain the problem, you do not want to convey a complaining attitude. Instead, you must thoroughly and objectively document the problems associated with the arrival of the new chairs.

■ Argument Case 3: Quality Control in the Labs

As McDuff's quality-control manager, you operate out of the corporate office in Baltimore. Though friendly with all the branch managers, you have the sensitive job of making sure that the quality of McDuff's products and services remains high. During a recent road trip to every office, you noticed problems at the labs. In particular, testing procedures were not being followed exactly, equipment was sometimes not properly cleaned, and samples from the field were not always labeled clearly. Now you must write a memo report to all lab supervisors noting these lapses.

All three cases involve argument as defined in this chapter. That is, you must collect evidence to support your point, while still keeping the goodwill of your readers.

Guidelines for Writing Argument

Argumentative writing has a long tradition, from the rhetoric of ancient times to the political debate of today. This section describes five guidelines about argument that apply to your on-the-job writing.

■ Argument Guideline 1: Use Evidence Correctly

Whether writing a proposal to a customer or a memo to your boss, you often move from specific evidence toward a general conclusion supported by the evidence. Called "inductive reasoning," this approach to argument requires that you follow some accepted guidelines. Here are three, along with brief examples:

- **Use points the reader can grasp.** For example, as the fire-control expert in Case 1, you would need to educate readers about how limited fires can benefit swamp ecosystems. Your report must use evidence that could be understood by a mixed audience of park technicians and bureaucrats.
- **Use points that are a representative sample.** For example, as the quality-control expert in Case 3, you would want to have collected evidence from *all* labs mentioned in the memorandum, before sending the memo calling for corrective action. As the writer in Case 2, you would want to have collected information from enough colleagues to support your point about the office furniture.
- **Have enough points to justify your conclusion.** Again from Case 3, you would want to have made *many* observations, not two or three, to justify the memo's conclusions about poor lab procedures.

■ *Argument Guideline 2: Choose the Most Convincing Order for Points*

Chapter 3 points out that readers pay most attention to beginnings and endings, not the middle. Thus you should include your stronger points at the beginning and ending of an argumentative sequence, with weaker points in the middle. It follows that the strongest point usually appears in one of these two locations:

- **Last**—in arguments that are fairly short
- **First**—in arguments that are fairly long

Readers finish short passages in one sitting. You can expect them to stay attentive as you lead up to your strongest point. For long passages, however, the audience may take several sittings to read through your argument. For this reason, place the most important evidence at the beginning of longer arguments, where readers will see it before they get distracted.

For Case 2, assume you plan to write a fairly short memo about the 45 new office chairs, as a follow-up to your phone call to your manager at the corporate office. Your main points are that (1) employees have had medical problems leading to frequent sick leave, (2) productivity has suffered among those who are at work, (3) a local expert on health in the workplace has confirmed design flaws in the chairs, and (4) you have contacted a representative of the chair company, who has agreed to trade your chairs for a better model, for a net cost to McDuff of only $2000.

You decide that your fourth point is the strongest because it suggests a feasible solution. The next strongest is the first point because it deals with employee health. Given the brevity of your memo, therefore, you decide to place the medical point first, and the suggested solution last. In between will occur the point about reduced productivity and the expert's views. This arrangement holds the best chance for gaining your reader's support.

■ *Argument Guideline 3: Be Logical*

Some arguments rest on a logical structure called a syllogism. When it occurs formally (which is rare), a syllogism may look something like this:

Major Premise:	All chairs with straight backs are unacceptable for prolonged office work.
Minor Premise:	The chairs at the Boston office have straight backs.
Conclusion:	The chairs at the Boston office are unacceptable for prolonged office work.

In fact, syllogisms usually appear in an implied, less-structured fashion than the sample shown here. For example, recall that your McDuff chair report would include a testimonial from an outside expert in office furniture design. That person could have confirmed the major premise, leading to your statement of the minor premise and the conclusion. Indeed, this example shows that a syllogism is effective—that is, logical—*only* if the premises (1) can be verified as true and (2) will be accepted by the reader.

■ *Argument Guideline 4: Use Only Appropriate Authorities*

In the technical world, experts often supply evidence that will win the day in reports, proposals, and other documents. Using Case 2 again, your argument would rely on opinions by the expert in workplace health. This testimonial, however, would work only if you could verify that (1) there is a legitimate body of research on the health effects of chair design and (2) the expert you have chosen has credentials that readers will accept.

You will encounter a similar rule in chapter 13, which covers proper documentation in research writing. Such papers also require that you cite sources that will be accepted as valid in the field about which you are writing.

■ *Argument Guideline 5: Avoid Argumentative Fallacies*

Guidelines 1–4 mention a few ways you can err in arguing your point. Listed next are some other fallacies that can damage your argument. Accompanying each fallacy is an example from Case 2:

- **Ad hominem** (Latin: "To the man"): arguing against a person rather than discussing the issue. (Example: suggesting that the chairs should be replaced because the purchasing agent at the Baltimore office, who ordered the chairs, is surly and incompetent.)
- **Circular reasoning:** failing to give any reason for why something is or is not true, other than stating that it is. (Example: proposing that the straight-backed chairs should be replaced simply because they are straight-backed chairs.)
- **Either/or fallacy:** stating that only two alternatives exist—yours and another one that is much worse—when in fact there are other options. (Example: claiming that if the office does not purchase new chairs next week, 15 employees will quit.)
- **False analogy:** suggesting that one thing should be true because of its similarity to something else, when in fact the two items are not enough alike to justify the analogy. (Example: suggesting that the chairs in the Boston office should be replaced because the Atlanta office has much more expensive chairs.)
- **Hasty generalization:** forming a generalization without adequate supporting evidence. (Example: proposing that all Jones office furniture will be unacceptable, simply because the current chairs carry the Jones trademark.)
- **Non sequitur** (Latin: "It does not follow"): making a statement that does not follow logically from what came before it. (Example: claiming that if the office chairs are not replaced soon, productivity will decrease so much that the future of the office will be in jeopardy—even though there is no evidence that the chairs would have this sort of dramatic effect on business.)

■ **Post hoc ergo propter hoc** (Latin: "After this, therefore because of this"): claiming a cause-effect relationship between two events simply because one occurred before the other. (Example: associating the departure of an excellent secretary with the purchase of the poorly designed chairs a week earlier—even though no evidence supports the connection.)

Example of Argument

This section presents the complete memo report resulting from Case 2 discussed previously (see Model 5–1, pp. 135–136). As the branch manager in McDuff's Boston office, you are writing to your supervisor in the Baltimore corporate office. From your brief phone conversation, you expect that he supports the recommendation in your memo. Yet you know that he will show the memo to a few other corporate colleagues, including the procurement officer who ordered the chairs. Thus you cannot assume support. You must present good arguments, in a tactful manner.

DEFINITION

During your career, you will use technical terms known only to those in your profession. As a civil engineer, for example, you would know that a "triaxial compression test" helps determine the strength of soil samples. As a computer professional, you would know the meaning of RAM (random access memory), ROM (read only memory), and LAN (local area network). When writing to readers unfamiliar with these fields, however, you would need to define technical terms.

Good definitions can support findings, conclusions, and recommendations throughout your document. They also keep readers interested. Conversely, the most organized, well-written report will fall on deaf ears if it includes terms that readers do not grasp. "Define your terms!" is the frustrated exclamation of many a reader. For your reader's sake, then, you need to be asking questions like these about definitions:

■ How often do you use them?
■ Where should they be placed?
■ What format should they take?
■ How much information is enough and how much is too much?

To answer these questions, the following sections give guidelines for definitions and supply an annotated example. First, here are some typical contexts for definitions within McDuff, Inc.

Short Cases From McDuff: Definitions

These cases may help you visualize the use of definitions in your own on-the-job writing.

■ *Definition Case 1: New Power Plant*

As a McDuff technician helping to build a power plant, you often use the term *turbine* in speech and writing. Obviously your co-workers and clients understand the term. Now, however, you are using it in a plant brochure to be sent to consumers.

For this general audience, you decide to define "turbine" as follows: "a machine that receives energy from a moving fluid and then uses this energy to make a shaft rotate." You accompany the definition with an illustration so that the unfamiliar audience can visualize how the turbine works.

■ *Definition Case 2: Health Bulletin to Employees*

As a health and benefits specialist in McDuff's corporate office, you have watched the company's health insurance costs increase dramatically over the last five years. Much of this cost has not been passed on to the employee, resulting in a drain on profits. You have convinced management to support a major campaign to encourage healthy habits by employees. The campaign might include these free benefits: in-house quit-smoking seminars, stress-reduction sessions, and annual physicals.

Your first project, however, is a health memorandum to all employees. In it you must (1) explain the nature of the cholesterol problem and (2) invite employees to free cholesterol screening tests sponsored by the company. Your memo must provide clear definitions for terms like "good" cholesterol and "bad" cholesterol.

■ *Definition Case 3: Forestry Report*

As a forestry and agriculture expert with McDuff's Denver office, you have coordinated a major study for the state of Idaho. Your job has been to recommend ways that a major forested region can still be used for timber with little or no damage to the region's ecological balance.

Although the report will go first to technical experts in Idaho's Department of Natural Resources, you have been told that it must be understood by other potential readers like politicians, bureaucrats, and citizen groups. Thus you have decided to include a glossary that defines terms such as these:

- **Silvaculture** (cultivation of forests to supply a renewable crop of timber)
- **Shelterwood cutting** (removal of only mature trees in several cuttings over a few decades)
- **Seed-tree cutting** (removal in one cutting of all but a few trees, which are left to regenerate the forest)
- **Clearcutting** (removal of all trees in an area in one cutting)*

These three cases all require well-placed definitions for their readers. When in doubt, insert definitions! Readers can always skip over ones they do not need.

---

*Adapted from ENVIRONMENTAL SCIENCE: AN INTRODUCTION, 2/E by G. Tyler Miller, Jr. © 1988 by Wadsworth, Inc. Reprinted by permission of the publisher.

Guidelines for Writing Definitions

Once you know definitions are needed, you must decide on their format and location. Again, consider your readers. How much information do they need? Where is this information best placed within the document? To answer these and other questions, here are five working guidelines for writing good definitions.

■ Definition Guideline 1: Keep It Simple

Occasionally the sole purpose of a report is to define a term. Most times, however, a definition just clarifies a term in a document with a larger purpose. Your definitions should be as simple and unobtrusive as possible. Always present the simplest possible definition, with only that level of detail needed by the reader.

For example, in writing to a client on your land survey of her farm, you might briefly define a transit as "the instrument used by land surveyors to measure horizontal and vertical angles." The report's main purpose is to present property lines and total acreage, not to give a lesson in surveying, so this sentence definition is adequate. Choose from these three main formats (listed from least to most complex) in deciding the form and length of definitions:

- **Informal definition:** a word or brief phrase, often in parentheses, that gives only a synonym or other minimal information about the term
- **Formal definition:** a full sentence that distinguishes the term from other similar terms and that includes these three parts: the term itself, a class to which the term belongs, and distinguishing features of the term
- **Expanded definition:** a lengthy explanation that begins with a formal definition and is developed into several paragraphs or more

Guidelines 2–4 show you when to use these three options and where to put them in your document.

■ Definition Guideline 2: Use Informal Definitions for Simple Terms Most Readers Understand

Informal definitions appear right after the terms being defined, often as one-word synonyms in parentheses. They give just enough information to keep the reader moving quickly. As such, they are best used with simple terms that can be adequately defined without much detail.

Here is a situation in which an informal definition would apply. McDuff has been hired to examine a possible shopping-mall site. The buyers, a group of physicians, want a list of previous owners and an opinion about the suitability of the site. As legal assistant at McDuff, you must assemble a list of owners in your part of the group-written report. You want your report to agree with court records, so you decide to include real-estate jargon such as "grantor" and "grantee." For your nontechnical readers, you include parenthetical definitions like these:

> All **grantors** (persons from whom the property was obtained) and **grantees** (persons who purchased the property) are listed on the following chart, by year of ownership.

This same McDuff report has a section describing creosote pollution found at the site. The chemist writing the contamination section also uses an informal definition for the readers' benefit:

> At the southwest corner of the mall site, we found 16 barrels of **creosote** (a coal tar derivative) buried under about three feet of sand.

The readers do not need a fancy chemical explanation of creosote. They only need enough information to keep them from getting lost in the terminology. Informal definitions perform this task nicely.

■ *Definition Guideline 3: Use Formal Definitions for More Complex Terms*

A formal definition appears in the form of a sentence that lists (1) the **term** to be defined, (2) the **class** to which it belongs, and (3) the **features** that distinguish the term from others in the same class. Use it when your reader needs more background than an informal definition provides. Formal definitions define in two stages:

- First, they place the term into a *class* (group) of similar items.
- Second, they list *features* (characteristics) of the term that separate it from all others in that same class.

In the list of sample definitions that follows, note that some terms are tangible (like "pumper") and others are intangible (like "arrest"). Yet all can be defined by first choosing a class and then selecting features that distinguish the term from others in the same class.

Term	Class	Features
An arrest is	restraint of persons	that deprives them of freedom of movement and binds them to the will and control of the arresting officer.
A financial statement is	a historical report about a business	and is prepared by an accountant to provide information useful in making economic decisions, particularly for owners and creditors.
A triaxial compression test is	a soils lab test	that determines the amount of force needed to cause a shear failure in a soil sample.
A pumper	is a fire-fighting apparatus	used to provide adequate pressure to propel streams of water toward a fire.

This list demonstrates three important points about formal definitions. First, the definition itself must not contain terms that are confusing to your readers. The definition of "triaxial compression test," for example, assumes readers will understand the term "shear failure" that is used to describe features. If this assumption were incorrect, then the term would need to be defined. Second, formal definitions may be so long that they create a major distraction in the text. (See Guideline 5 for alternative locations.) Third, the class must be narrow enough so that you will not have to list too many distinguishing features.

■ Definition Guideline 4: Use Expanded Definitions for Supporting Information

Sometimes a parenthetical phrase or formal sentence definition is not enough. If readers need more information, use an expanded definition with this three-part structure:

- **An overview at the beginning**—which includes a formal sentence definition and a description of the ways you will expand the definition
- **Supporting information in the middle**—perhaps using headings and lists as helpful format devices for the reader
- **Brief closing remarks at the end**—reminding the reader of the definition's relevance to the whole document

Here are seven ways to expand a definition, along with brief examples:

1. **Background and/or history of term**—expand the definition of "triaxial compression test" by giving a dictionary definition of "triaxial" and a brief history of the origin of the test
2. **Applications**—expand the definition of "financial statement" to include a description of the use of such a statement by a company about to purchase controlling interest in another
3. **List of parts**—expand the definition of "pumper" by listing the parts of the device, such as the compressor, the hose compartment, and the water tank
4. **Graphics**—expand the description of the triaxial compression test with an illustration showing the laboratory test apparatus
5. **Comparison/contrast**—expand the definition of a term like "management by objectives" (a technique for motivating and assessing the performance of employees) by pointing out similarities and differences between it and other management techniques
6. **Basic principle**—expand the definition of "ohm" (a unit of electrical resistance equal to that of a conductor in which a current of one ampere is produced by a potential of one volt across its terminals) by explaining the principle of Ohm's Law (that for any circuit the electric current is directly proportional to the voltage and inversely proportional to the resistance)
7. **Illustration**—expand the definition of CAD/CAM (Computer-Aided Design/Computer-Aided Manufacturing—computerized techniques to automate the design and manufacture of products) by giving examples of how CAD/CAM is changing methods of manufacturing many items, from blue jeans to airplanes

Obviously, long definitions might seem unwieldy within the text of a report, or even within a footnote. For this reason, they often appear in appendices, as noted in the next guideline. Readers who want additional information can seek them out, whereas other readers will not be distracted by digressions in the text.

■ Definition Guideline 5: Choose the Right Location for Your Definition

Short definitions are likely to be in the main text; long ones are often relegated to footnotes or appendices. However, length is not the main consideration. Think first about the *importance* of the definition to your reader. If you know that decision-makers reading your report will need the definition, then place it in the text—even if it is fairly lengthy. If the definition only provides supplementary information, then it can go elsewhere. You have these five choices for locating a definition:

1. **In the same sentence as the term,** as with an informal, parenthetical definition
2. **In a separate sentence,** as with a formal sentence definition occurring right after a term is mentioned
3. **In a footnote,** as with a formal or expanded definition listed at the bottom of the page on which the term is first mentioned
4. **In a glossary at the beginning or end of the document,** along with all other terms needing definition in that document
5. **In an appendix at the end of the document,** as with an expanded definition that would otherwise clutter the text of the document

Example of Expanded Definition

Expanded definitions are especially useful in reports from technical experts to nontechnical readers. McDuff's report writers, for example, often must explain environmental, structural, or geological problems to concerned citizens or nontechnical decision-makers. Figure 5–1 gives a definition that might appear in a report to a county government about the placement of a landfill.

DESCRIPTION

Technical descriptions, like expanded definitions, require that you pay special attention to *details*. In fact, you can consider a description to be a special type of definition that focuses on parts, functions, or other features. It emphasizes *physical* details. Descriptions can appear in any part of a document, from the introduction to the appendix. As a rule, however, detailed descriptions tend to be placed in the technical sections.

To help you write accurate descriptions, this section offers case studies, guidelines for writing descriptions, and two annotated examples of equipment descriptions.

Starts with for-
mal sentence
definition—
including term,
class, and fea-
tures.

Indicates way
definition will be
developed (de-
scription of
acute and
chronic types).

Gives examples
of first type.

Gives examples
of second type.

> A **toxic substance** is a chemical that is harmful to people or other living
> organisms. The effects from exposure to a toxic substance may be acute or chronic.
> Acute effects are those that appear shortly after exposure, usually to a large
> concentration or dose over a short time. Examples are skin burns or rashes, eye
> irritation, chest pains, kidney damage, headache, convulsions, and death.
>
> Effects that are delayed and usually long-lasting are called "chronic" effects. They
> may not appear for months or years after exposure and usually last for years.
> Examples are cancer, lung and heart disease, birth defects, genetic defects, and
> nerve and behavioral disorders. Chronic effects often occur as a result of prolonged
> exposure to fairly low concentrations or doses of a toxin. However, they may occur
> as the delayed effects of short-term exposure to high doses.

FIGURE 5-1
Definitions in a report

Excerpted from LIVING IN THE ENVIRONMENT: AN INTRODUCTION TO ENVIRONMENTAL SCI-
ENCE, 6/E by G. Tyler Miller, Jr. © 1990 by Wadsworth, Inc. Reprinted by permission of the publisher.

Short Cases From McDuff: Descriptions

Descriptions often appear as supporting information in the document body or in
appendices. Here are three situations from McDuff in which a detailed description
would add to the effectiveness of the complete document.

■ Description Case 1: Possible Harbors

McDuff's San Francisco office has been hired to recommend possible locations for
a new swimming and surfing park in northern California. Written to a county
commission (five laypersons who will make the decision), your report gives three
possible locations and your reasons for selecting them. Then it refers to appendices
that give brief physical descriptions of the sites.

Specifically, your appendices to the report describe (1) surface features,
(2) current structures, (3) types of soils gathered from the surface, (4) water quality, and (5) aesthetic features, such as quality of the ocean views. Language is kept
nontechnical, considering the lay background of the commissioners.

■ Description Case 2: Sonar Testing Equipment

A potential McDuff client, Rebecca Stern, calls you in your capacity as a geologist
at McDuff's Baltimore office. She wants information about the kind of sonar equipment McDuff uses to map geological features on the seafloor.

This client has a strong technical background, so you write a letter with a
detailed technical description of the McDuff system. The body of the letter describes

the locations and functions of (1) the seismic source (a device, towed behind a boat, that sends the sound waves) and (2) the receiver (a unit, also towed behind the boat, that receives the signals).

◼ *Description Case 3: Asbestos Site*

McDuff's Cleveland office was hired to examine asbestos contamination in a large high school built in 1949. As a member of the investigating team, you found asbestos throughout the basement in old pipe coverings. Your final report to the school board provides conclusions about the level of contamination and recommendations for removal. An appendix gives a detailed technical description of the entire basement, including a map with a layout of the plumbing system.

Guidelines for Writing Descriptions

Now that you know how descriptions fit into entire documents, here are some simple guidelines for writing accurate, detailed descriptions. Follow them carefully as you prepare assignments in this class and on the job.

◼ *Description Guideline 1: Remember Your Readers' Needs*

The level of detail in a technical description depends on the purpose a description serves. Give readers precisely what they need—but no more. In the harbor description in Case 1, the commissioners do not want a detailed description of soil samples taken from borings. That level of detail will be reserved for a few sites selected later for further study. Instead, they want only surface descriptions. Always know just how much detail will get the job done.

◼ *Description Guideline 2: Be Accurate and Objective*

More than anything else, readers expect accuracy in descriptions. Pay close attention to details. (As noted previously, the *degree* of detail in a description depends on the *purpose* of the document.) In the building-floor description in Case 3, for example, you would want to describe every possible location of asbestos in the school basement. Because the description will become the basis for a cost proposal to remove the material, accuracy is crucial.

Along with accuracy should come objectivity. This term is more difficult to pin down, however. Some writers assume that an objective description leaves out all opinion. This is not the case. Instead, an objective description may very well include opinions that have these features:

- They are based on your professional background.
- They can be justified by the time you have had to complete the description.
- They can be supported by details from the site or object being described.

For example, your description of the basement pipes mentioned in Case 3 might include a statement like this: "Because there is asbestos wrapping on the exposed pipes above the boiler, my experience suggests that asbestos wrapping probably also exists around the pipes above the ceiling—in areas that we were not able to view." This opinion does not reduce the objectivity of your description; it is simply a logical conclusion based on your experience.

◼ *Description Guideline 3: Choose an Overall Organization Plan*

Like other patterns discussed in this chapter, technical descriptions usually make up only parts of documents. Nevertheless, they must have an organization plan that permits them to be read as self-contained, stand-alone sections. Indeed, a description may be excerpted later for separate use.

Following are three common ways to describe physical objects and events. In all three cases, a description should move from general to specific. That is, you begin with a view of the entire object or event. Then in the rest of the description, you focus on specifics. Headings may be used, depending on the format of the larger document.

1. **Description of the parts:** For many physical objects, like the basement floor and coastal scene in the previous cases, you will simply organize the description by moving from part to part.
2. **Description of the functions:** Often the most appropriate overall plan relies on how things work, not on how they look. In the sonar example, the reader was more interested in the way that the sender and receiver worked together to provide a map of the seafloor. This function-oriented description would include only a brief description of the parts.
3. **Description of the sequence:** If your description involves events, as in a police officer's description of an accident investigation, you can organize ideas around the major actions that occurred, in their correct sequence. As with any list, it is best to place a series of many activities into just a few groups. Four groups of 5 events each is much easier for readers to comprehend than a single list of 20 events.

◼ *Description Guideline 4: Use "Helpers" Like Graphics and Analogies*

The words of a technical description need to come alive. Because your readers may be unfamiliar with the item, you must search for ways to connect with their experience and with their senses. Two effective tools are graphics and analogies.

Graphics respond to the desire of most readers to see pictures along with words. As readers move through your part-by-part or functional breakdown of a mechanism, they can refer to your graphic aid for assistance. The illustration helps you too, of course, in that you need not be as detailed in describing locations and dimensions of parts when you know the reader has easy access to a visual. Note how the diagrams in Models 5–2 and 5–3 on pages 137–140 give meaning to the technical details in the verbal descriptions.

Analogies, like illustrations, give readers a convenient handle for understanding your description. Put simply, an analogy allows you to describe something unknown or uncommon in terms of something that is known or more common. A brief analogy can sometimes save you hundreds of words of technical description. This paragraph description contains three analogies:

> McDuff, Inc., is equipped to help clean up oil spills with its patented product, Sea-Clean. This highly absorbent chemical is spread over the entire spill by means of a helicopter, which makes passes over the spill much like a lawn mower would cover the complete surface area of a lawn. When the chemical contacts the oil, it acts like sawdust coming in contact with oil on a garage floor. That is, the oil is immediately absorbed into the chemical and physically transformed into a product that is easily collected. Then our nearby ship can collect the product, using a machine that operates much like a vacuum cleaner. This machine sucks the SeaClean (now full of oil) off the surface of the water and into a sealed container in the ship's hold.

First analogy

Second analogy

Third analogy

■ Description Guideline 5: Give Your Description the "Visualizing Test"

After completing a description, test its effectiveness by reading it to someone unfamiliar with the material—someone with about the same level of knowledge as your intended reader. If this person can draw a rough sketch of the object or events while listening to your description, then you have done a good job. If not, ask your listener for suggestions to improve the description. If you are too close to the subject yourself, sometimes an outside point of view will help refine your technical description.

Examples of Description

This section contains two descriptions from McDuff documents, one written at a U.S. office and the other produced overseas. The first example includes a description of physical parts and a brief operating procedure. (For a thorough discussion of process descriptions and instructions, see chapter 6.) The second includes only a description of physical parts. Both show the importance of graphics in technical descriptions.

■ Description 1: Blueprint Machine

Donna Millsly, human resources coordinator at McDuff's St. Paul office, has been asked to assemble an orientation guide for new secretaries, office assistants, and other members of the office staff. The manual will contain descriptions and locations of the most common pieces of equipment in the office, sometimes with brief instructions for their use.

One section of her manual includes a series of short equipment descriptions organized by general purpose of the equipment. Model 5−2 on pages 137−138 includes a description of the office blueprint machine, which gets used by a wide variety of employees. Because the blueprint machine has an operating procedure that is not self-evident, the description includes a brief procedure that shows the manner in which the machine should be used.

■ *Description 2: Desk Stapler*

Fahdi Ahmad, the director of procurement at McDuff's office in Saudi Arabia, is changing his buying procedures. He has decided to purchase some basic office and audiovisual supplies from companies in nearby developing nations, rather than from firms in industrialized countries. For one thing, he thinks this move will save a lot of money. For another, he believes it will help the company get more projects from these nations, for McDuff will become known as a firm that pumps back some of its profits into the local economies.

As a first step in this process, Fahdi is sending possible suppliers a set of detailed descriptions of basic products it needs, such as pencils, photocopying paper, staplers, tape dispensers, and transparencies for overhead projectors. Fahdi decides to describe the typical desk stapler in moderate detail so that potential suppliers will know that they must come up with a product that has the basic features of staplers manufactured in developed countries. Model 5−3 on pages 139−140 contains the description included in his report. Later, if need be, he could submit more detailed specifications listing exact measurements, metal grades, and so forth.

CLASSIFICATION/DIVISION

In technical writing, you often perform these related tasks: (1) grouping lists of items into categories, a process called "classification," or (2) separating an individual item into its parts, a process called "division." In practice, the patterns usually work together. So that you can see classification and division in action, this section starts with some case studies from McDuff. Then it presents basic guidelines and an example for classification, followed by basic guidelines and an example for division.

Short Cases From McDuff: Classification/Division

The first case shows how you might use both classification and division in the same context. The other cases show only one pattern at work.

■ *Classification/Division Case 1: Civil-Engineering Projects*

A major client needs some detailed information before deciding between McDuff and a competitor for a big project. Specifically, the firm wants a detailed description of the civil-engineering projects completed by the Caracas office over the last 10 years.

- First, you *divide* the office's civil-engineering capabilities into these five groupings, based on type of project: (1) city planning, (2) construction, (3) transportation, (4) sanitation, and (5) hydraulics. (In each case, you describe the type of work done. For example, hydraulics projects concern the use of water.)
- Second, you collect descriptions of the 214 projects completed.

- Third, you *classify* the 214 projects into the five groupings mentioned. Thus the combined process of division and classification gives you a way to organize details for the client.

Classification/Division Case 2: Birds Near a Sub Base

The federal government wants to expand a submarine base off the U.S. southeastern coast. Plans went along fine until last year, when a major environmental group raised questions about the effect the expansion would have on bird nesting areas.

The concerned organization, BirdWatch, hired McDuff to determine the environmental impact of increased base traffic on the bird populations of a nearby national seashore. As project director, you determine that 61 species of birds on the island could be affected by the expansion. For the purposes of your report, you classify the 61 species by physical features into the following groupings: (1) loons; (2) grebes; (3) gulls and terns; (4) cranes, rails, and coots; and (5) ducks, geese, and swans. These classifications, often used in ornithological work, will help readers understand the significance of your environmental impact statement.

Classification/Division Case 3: Employee Grievances

Rob McDuff wants to change the procedure by which employees can register grievances. He prefers that a committee be appointed to hear all grievances. Before he makes his final decision, however, he wants you to submit a summary of the types of grievances filed since the company was founded in 1947. Your painstaking research uncovers 116 separate grievances for which there is paperwork. For easier reading, your report to the company president classifies the 116 incidents into four main categories, by type: (1) firing or discharge from the firm; (2) the use of seniority in layoffs, promotions, or transfers; (3) yearly performance evaluations; and (4) overtime pay and the issue of required overtime at some offices.

Guidelines for Classification

Classification helps you (and of course your reader) make sense out of diverse but related items. The process of outlining a writing assignment requires the use of classification. Outlining (as described in chapter 1) forces you to apply both classification and division to organize diverse information into manageable "chunks." The guidelines presented here provide a three-step procedure for classifying any group of related items.

Classification Guideline 1: Find a Common Basis

Classification requires that you establish your groupings on one main basis. This basis can relate to size, function, purpose, or any other factor that serves to produce logical groupings. For example, here are the bases for groupings established in the three cases cited:

Case 1 basis: the *type of work* being done in the civil-engineering field at McDuff (such as sanitation or city planning)

Case 2 basis: the *physical features* of the marsh or ocean birds found at the site (note that the classifications in this case are not original but rather are ones commonly used in ornithology)

Case 3 basis: the *purpose for the grievance* (that is, the grievances can be grouped under four main reasons why the employees made formal complaints to the company)

■ Classification Guideline 2: Limit the Number of Groups

In chapter 4, you learned that readers prefer groupings of under nine items—the fewer the better. This principle of organization, as well as the appropriateness of the basis, should guide your use of classification. Strive to select a basis that will result in a limited number of groupings.

In Case 1, another basis for classifying the 214 civil-engineering projects would be by country in which the project was located, rather than type of work. Assume that after applying this project-site basis, you end up with 18 different classifications (that is, countries) for the 214 projects. Given this unwieldy number of groupings, however, you should either (1) avoid using this basis or (2) reduce the number of classifications by grouping countries together, perhaps by continent. In other words, the number of classifications affects the degree to which this pattern of organization succeeds with the reader.

■ Classification Guideline 3: Carefully Classify Each Item

The final step is to place each item in its appropriate classification. If you have chosen classifications carefully, this step is no problem. For example, assume that as McDuff's finance expert, you want to identify sources for funding new projects. Therefore, you need to collect the names of all commercial banks started in Pennsylvania last year. A reference librarian gives you the names of 42 banks. To impose some order, you decide to classify them by charter. The resulting groups are as follows:

1. National banks (those chartered in all states)
2. Non-Federal Reserve state banks (exclusively Pennsylvania banks that are not members of the Federal Reserve System)
3. Federal Reserve state banks (exclusively Pennsylvania banks that are members of the Federal Reserve System)

With three such classifications that do not overlap, you can put each of the 42 banks in one of the three classifications.

Example of Classification

Now apply the three-part classification strategy to a detailed problem at McDuff. You are on a committee to write an outline summary of McDuff's safety practices. Since much of the company's work involves hazardous materials, you want to place special emphasis on techniques used to protect workers from toxic chemicals, polluted air, asbestos fibers, and so forth. This outline will be used as an attachment to sales letters and proposals sent to potential clients, such as the federal government and various state governments. Good organization is crucial so that the reader can locate information quickly. At its first meeting, your committee comes up with this list of topics to be included in the outline:

1. Headgear provided by company
2. Complete physicals done yearly
3. Partial physicals done at six-month intervals
4. Blood monitoring throughout work at hazardous sites
5. State-of-the-art face masks and oxygen equipment
6. Twenty-hour safety program during employment orientation
7. Certified safety official on site during hazardous projects
8. Pre-employment drug testing
9. Random drug testing of all workers in dangerous jobs
10. State-of-the-art hand and body protection equipment
11. Biweekly training sessions on safety-related topics
12. Pre-employment complete physical
13. Monthly in-house newsletter with safety column
14. Incentive awards for employees with safety suggestions
15. Company membership in national safety organizations

Next the committee (1) selects an appropriate basis (the *time* at which these safety precautions should take place), (2) establishes three main groupings, (3) classifies each of the 15 items into one of the groups, and (4) assembles the following outline for use in McDuff sales literature. (Numbers in parentheses refer to the topics just listed.)

I. Procedures before employment
   A. Twenty-hour safety program during orientation (6)
   B. Pre-employment drug testing (8)
   C. Pre-employment complete physical (12)
II. Procedures during a project
   A. Headgear provided by company (1)
   B. Blood monitoring throughout work at hazardous sites (4)
   C. State-of-the-art face masks and oxygen equipment (5)
   D. Certified safety official on site during hazardous projects (7)
   E. State-of-the-art hand and body protection equipment (10)
III. Procedures that take place periodically during one's employment at McDuff
   A. Complete physicals done yearly (2)
   B. Partial physicals done at six-month intervals (3)
   C. Random drug testing of all workers in dangerous jobs (9)
   D. Biweekly training sessions on safety-related topics (11)
   E. Monthly in-house newsletter with safety column (13)
   F. Incentive awards for employees with safety suggestions (14)
   G. Company membership in national safety organizations (15)

Guidelines for Division

Division begins with an entire item that must be *broken down* or *partitioned* into its parts, whereas classification begins with a series of items that must be *grouped* into related categories. Division is especially useful when you need to explain a complicated piece of equipment to an unfamiliar audience. Follow these three guidelines for applying this pattern:

■ *Division Guideline 1: Choose the Right Basis for Dividing*

Like classification, division means you must find a logical reason for establishing groups or parts. Assume, for example, that you are planning to teach a McDuff training seminar in project management. In dividing this four-day training seminar into appropriate segments, it seems clear to you that each day should cover one of these crucial parts of managing projects: meeting budgets, scheduling staff, completing written reports, and seeking follow-up work from the client. In this case, the principle of division seems easy to employ.

Yet other cases present you with choices. If, for example, you were planning a training seminar on report writing, you could divide it in these three ways, among others: (1) by *purpose of the report* (for example, progress, trip, recommendation), (2) by *parts of the writing process* (for example, brainstorming, outlining, drafting, revising), or (3) by *report format* (for example, letter report, informal report, formal report). Here you would have to choose the basis most appropriate for your purpose and audience.

■ *Division Guideline 2: Subdivide Parts When Necessary*

As with classification, the division pattern of organization can suffer from the "laundry list" syndrome. Specifically, any particular level of groupings should have from three to seven partitions—the number that most readers find they can absorb. When you go over that number, consider reorganizing information or subdividing it.

Assume that you want to partition a short manual on writing formal proposals. Your first effort to divide the topic results in nine segments: cover page, letter of transmittal, table of contents, executive summary, introduction, discussion, conclusions, recommendations, appendices. You would prefer fewer groupings, so you then establish three main divisions: front matter, discussion, and back matter, with breakdowns of each. In other words, your effort to partition was guided by every reader's preference for a limited number of groupings.

■ *Division Guideline 3: Describe Each Part With Care*

This last step may seem obvious. Make sure to give equal treatment to each part of the item or process you have partitioned. Readers expect this sort of parallelism, just as they prefer the limited number of parts mentioned in Guideline 2.

Example of Division

You are a McDuff manager helping to prepare a client's report. The main task is to describe major types of offshore oil rigs. You decide to partition the subject on the basis of *environmental application,* which gives you five main types. In the following list, notice that each of the five is described in a parallel fashion: the rig's purpose, type of structure, and design with respect to wave strength.

Five Main Types of Offshore Oil Rigs

1. **Platform rig:** Generally for drilling at water depths of less than 1000 ft/ Permanent structure supported by steel and concrete legs driven into ocean floor/ Designed to withstand waves of about 50 ft
2. **Submersible rig:** Generally for drilling at water depths of less than 100 ft/ Temporary structure supported by large tanks that, when filled, go to the ocean floor and thus form the foundation for the columns extending to the deck structure above/ Designed to withstand waves of about 30 ft
3. **Semisubmersible rig:** Generally for drilling in extreme water depths of 1000 ft or more/ Temporary structure with large water-filled pontoons that keep it suspended just below the water surface, with anchors or cables extending to the ocean bottom/ Designed to withstand extreme wave heights of 90 ft or more
4. **Drillship:** Generally for drilling at water depths between 100 and 1000 ft/ Drilling operations take place through opening in the middle of the ship/ Designed to operate in relatively low wave heights of less than 30 ft
5. **Jack-up rig:** Generally for drilling at water depths between 25 and 500 ft/ Temporary structure with platformlike legs that can be jacked up and down to rest on the ocean surface, much like the jacking system used to elevate a car when changing a tire/ Designed to withstand hurricane-force winds that produce waves of over 50 ft

COMPARISON/CONTRAST

Many writing projects obligate you to show similarities or differences between ideas or objects. (For our purposes, the word *comparison* emphasizes similarities, whereas the word *contrast* emphasizes differences.) In the real world of career writing, this technique especially applies to situations wherein readers are making buying decisions. To help you write effective comparisons, this section presents several cases from McDuff, puts forth some simple guidelines, and gives an annotated example.

Short Cases From McDuff: Comparison/Contrast

One of the most common patterns at McDuff, comparison/contrast is used in many in-house and external reports. Examples of each are noted here.

■ *Comparison/Contrast Case 1: Jamaica Hotels*

As an architect for McDuff, you have had the good fortune of traveling to Jamaica three times on business in the last year. Each time you stayed at a different hotel on Montego Bay. Now your boss, Janet Scarsdale, is going there for an extended business trip. Before making her travel arrangements, she asks you for a memorandum that compares and contrasts main features of the three hotels. Presumably, she will use the information to decide where she will stay.

■ *Comparison/Contrast Case 2: Security Systems*

One of your jobs as McDuff's corporate purchasing agent is to advise managers. The manager of the London branch has asked that you send him a summary of similarities and differences among five high-quality infrared detectors. He wants to install several in a new computer lab, to detect movement of any after-hours intruders and then to send a signal to local police.

■ *Comparison/Contrast Case 3: Poisons*

Your job in McDuff's toxicology laboratory brings you in contact with dozens of poisons. One client has asked you to compare and contrast the relative dangers of five poisonous chemicals found in the well water at the site where the client firm wants to build a warehouse. Your report will be used in making the decision to build or not to build.

Guidelines for Comparison/Contrast

The three cases just cited show you real contexts in which you will use the comparison/contrast pattern of organization. Now here are guidelines for writing effective comparisons and contrasts.

■ *Comparison/Contrast Guideline 1: Remember Your Purpose*

When using comparison/contrast in on-the-job writing, your purpose usually falls into one of these two categories:

1. **Objective:** essentially an unbiased presentation of features wherein you have no real "axe to grind"
2. **Persuasive:** an approach wherein you compare features in such a way as to recommend a preference

You must constantly remember your main purpose. You will either provide raw data that someone else will use to make decisions *or* you will urge someone toward your preference. In either case, the comparison must show fairness in dealing with all alternatives. Only in this way can you establish credibility in the eyes of the reader.

■ *Comparison/Contrast Guideline 2: Establish Clear Criteria and Use Them Consistently*

In any technical comparison, you must set clear standards of comparison and then apply them uniformly. Otherwise, your reader will not understand the evaluation or accept your recommendation (if there is one).

For example, assume you are a McDuff field supervisor who must recommend the purchase of a bulldozer for construction sites. You have been asked to recommend just *one* of these models: Cannon-D, Foley-G, or Koso-L. After background reading and field tests, you decide upon three main criteria for your comparison: (1) pushing capacity, (2) purchase details, and (3) dependability. These three factors, in your view, are most relevant to McDuff's needs. Having made this decision about criteria, you then must discuss all three criteria with regard to *each* of the three bulldozers. Only in this way can readers get the data needed for an informed decision.

■ *Comparison/Contrast Guideline 3: Choose the "Whole-by-Whole" Approach for Short Comparison/Contrasts*

This strategy requires that you discuss one item in full, then another item in full, and so on. Using the bulldozer example, you might first discuss all features of the Cannon, then all features of the Foley, and finally all features of the Koso. This strategy works best if individual descriptions are quite short so readers can remember points made about the Cannon bulldozer as they proceed to read sections on the Foley and then the Koso machines.

Keep these two points in mind if you select the whole-by-whole approach:

- Discuss subpoints in the same order—that is, if you start the Cannon description with information about dependability, begin the Foley and Koso discussions in the same way.
- If you are making a recommendation, move from least important to most important, or vice versa, depending on what approach will be most effective with your reader. Busy readers usually prefer that you start with the recommended item, followed by the others in descending order of importance.

■ *Comparison/Contrast Guideline 4: Choose the "Part-by-Part" Approach for Long Comparison/Contrasts*

Longer comparison/contrasts require readers to remember much information. Thus readers usually prefer that you organize the comparison around major criteria, *not* around the whole items. Using the bulldozer example, your text would follow this outline:

   I. Pushing Capacity
      A. Cannon
      B. Foley
      C. Koso

II. Purchase Details
    A. Cannon
    B. Foley
    C. Koso
III. Dependability
    A. Cannon
    B. Foley
    C. Koso

Note that the bulldozers are discussed in the same order within each major section. As with the whole-by-whole approach, the order of the items can go either from most important to least important or vice versa—depending on what strategy you believe would be most effective with your audience.

■ *Comparison/Contrast Guideline 5: Use Illustrations*

Comparison/contrasts of all kinds benefit from accompanying graphics. In particular, tables are an effective way to present comparative data. For example, your report on the bulldozers might present some pushing-capacity data you found in company brochures.

Example of Comparison/Contrast

Using the bulldozer example, Model 5–4 on pages 141–142 presents a sample of the part-by-part pattern. Remember that such a comparison would be only part of a final report—in this case, one that recommends the Cannon-D. The complete report would include an introductory summary at the beginning and a list of conclusions and recommendations at the end (see chapter 8). This example could be written in either a whole-by-whole or part-by-part manner. The latter is used here to emphasize the importance of the three criteria for comparison.

CHAPTER SUMMARY

This chapter examines five common patterns of organization that make up reports, proposals, and correspondence. By studying these patterns, you will be better prepared to write the documents covered later in this book.

    Argument is used when the writer needs to support points with evidence. (Persuasion, the strongest form of argument, occurs when the writer seeks to change the reader's opinions or actions.) To write effective arguments, follow these guidelines:

1. Use evidence correctly.
2. Choose the most convincing order for points.
3. Be logical.
4. Use only appropriate authorities.
5. Avoid argumentative fallacies.

Definitions occur in technical writing in one of three forms: informal (in parentheses), formal (in sentence form with term, class, and features), and expanded (in a paragraph or more). These main guidelines apply:

1. Keep it simple.
2. Use informal definitions for simple terms most readers understand.
3. Use formal definitions for more complex terms.
4. Use expanded definitions for supporting information.
5. Choose the right location for your definition.

Description, like definition, depends on detail and accuracy for its effect. Careful descriptions usually include a lengthy itemizing of the parts of a mechanism or the functions of a term. Follow these basic guidelines for producing effective descriptions:

1. Remember your readers' needs.
2. Be accurate and objective.
3. Choose an overall organization plan.
4. Use "helpers" like graphics and analogies.
5. Give your description the "visualizing test."

Classification/division patterns help you organize groups of related items (classification) and break down an item into its parts (division). The guidelines for classification are as follows:

1. Find a common basis (for grouping diverse items).
2. Limit the number of groups.
3. Carefully classify each item.

Similarly, the guidelines for division are as follows:

1. Choose the right basis for dividing.
2. Subdivide parts when necessary.
3. Describe each part with care.

Comparison/contrast gives you an organized way to highlight similarities and differences in related items—whether you are simply presenting data or are attempting to argue a point. These main writing rules apply:

1. Remember your purpose.
2. Establish clear criteria and use them consistently.
3. Choose the "whole-by-whole" approach for short comparison/contrasts.
4. Choose the "part-by-part" approach for long comparison/contrasts.
5. Use illustrations.

ASSIGNMENTS

Part 1: Short Assignments

The following short assignments can be completed either orally or in writing. The first three can be either group or individual assignments; the last two are exclusively group assignments. Your instructor will give you specific directions for each one.

1. **Argument.** Analyze the argumentative effectiveness, or lack thereof, in the following passage [NOTE: The "leave bill" mentioned in point "d" was signed into law after the CEO wrote this response.]:

> After a good deal of thought, I have decided not to accept the committee's recommendation to allow up to six weeks of unpaid annual leave to employees for family emergencies, such as sickness of family members. My reasons are as follows:
>
> a. The proposal obviously would cause the company to lose many customers, for we definitely would not have the staff to cover our daily operations in the office and in the field.
>
> b. Just as the policy of flextime hours has caused some companies to fail to respond adequately to phone calls early in the morning, the leave plan would keep the office uncovered during days when an excessive number of employees were taking leave.
>
> c. Because almost all employees would occasionally miss some days of work, their salary would decrease and they would be less able to pay their bills. In effect, then, the leave time policy would hurt their families.
>
> d. Family leave has been supported by some members of Congress, and we all know how little that organization knows about how to operate business. In fact, I heard the other day that our local congressman, who supported the leave bill, bounced 17 checks during the House banking scandal several years ago. That will tell you something about his knowledge of business activities.
>
> e. One of our competitors, Jonquil Engineering, adopted a family leave policy just two years ago, and I have just learned that the firm's stock dropped 10% recently.
>
> f. Family leave has been supported by the Americans for Family, an organization whose president is Arlin Thomas. And we all know about his antics in the media. In the last few years, he has supported any cause that has come his way.

2. **Definition.** Using the guidelines in this chapter, discuss the relative effectiveness of the following short definitions. Speculate on the likely audience the definitions are addressing. [Excerpted from Turner, R. S., R. B. Cook, H. Van Miegroet, D. W. Johnson, J. W. Elwood, O. P. Bricker, S. E. Lindberg, and G. M. Hornberger. September 1990. Watershed and lake processes affecting surface water acid-base chemistry (NAPAP Report 10). In *Acidic Deposition: State of Science and Technology.* National Acid Precipitation Assessment Program, 722 Jackson Place, NW, Washington, D.C.]

> a. Watershed—the geographic area from which surface water drains into a particular lake or point along a stream.
>
> b. Acid mine drainage—runoff having high concentrations of metals and sulfate and high levels of acidity resulting from the oxidation of sulfide minerals that have been exposed to air and water by mining activities.
>
> c. Steady-state model—a model in which the variables under investigation are assumed to reach equilibrium and are independent of time.
>
> d. Acidic lake or stream—a lake or stream in which the acid neutralizing capacity is less than or equal to zero.
>
> e. Biomass—the total quantity of organic matter in units of weight or mass.
>
> f. Detritus—dead and decaying organic matter originating from plants and animals.
>
> g. Hydrology—the science that deals with the waters of the earth—their occurrence, circulation, and distribution; their chemical and physical properties; and their relationship to living things.
>
> h. Plankton—plant or animal species that spend part or all of their lives carried passively by water currents.
>
> i. Mineral weathering—dissolution of rocks and minerals by chemical and physical processes.

3. **Description.** Write a description of a piece of equipment or furniture located in your classroom or brought to class by your instructor—for example, a classroom chair, an overhead projector, a screen, a three-hole punch, a mechanical pencil, or a computer floppy disk. Write the description for a reader totally unfamiliar with the item.

4. **Classification/Division.** Divide into groups of four or five students, as selected by your instructor. Then, as a group, take an inventory of the career plans of each member of your group. Finally, using career plans as your basis, classify the group participants into a limited number of groupings. Be sure to adjust the size and focus of categories so that you have at least two students in each grouping. Present the group results to the entire class.

5. **Comparison/Contrast.** Divide into groups of three to five students, as determined by your instructor. Select two members who will give the other group members information about their (a) academic career, (b) extracurricular activities, and (c) work experience. Using the information gathered during this exercise, develop an outline of either a whole-by-whole or part-by-part comparison/contrast.

Part 2: Longer Assignments

These assignments test your ability to use patterns of organization covered in this chapter. To lend realism, they are placed in the context of short reports within McDuff, Inc. Follow these guidelines for each assignment:

- Write each exercise in the form of a *memo report* (if it is directed within McDuff, Inc.) or a *letter report* (if it is directed to an outside reader).
- Follow organization and design guidelines given in chapters 3 and 4, especially with regard to the ABC format (Abstract/Body/Conclusion) and the use of headings. Chapter 8 gives thorough format guidelines for short reports, but such detail is not necessary to complete the assignments here.
- For each assignment, fill out a copy of the Planning Form at the end of the book. Invent any audience analysis information not included in the following so that you have a "real" person to whom you are writing.

6. **Argument for New Wastewater Specialist.** You are a wastewater specialist and manager of a four-person crew with McDuff's Houston office. Recently you have become concerned about the amount of overtime worked by you and your group. Here are your main concerns, from most to least important. First, you are worried that excessive overtime might lead to errors by exhausted workers. The four employees on your crew often work at dangerous sites with poisonous chemicals and polluted water. Mistakes could lead to worker exposure to chemicals or errors in recording data or collecting samples. Second, three of your four field-workers have complained about their 50–55-hour weeks and the excessive time spent out of town on projects. Although they like the overtime pay, they would prefer to average just an extra five hours a week. You are concerned that they may quit to work at a competing firm. Third, you are convinced that there is enough work to support a fifth field employee, and still have three or four hours of overtime per employee per week. (Of course, you realize that there is always the risk that a future work slowdown would mean laying off an additional worker.)

Given these concerns, you want to write a short report to the branch manager, Elmore Lindley, describing the problems. You also want to suggest that McDuff hire another field-worker, or at least give the matter some study. Present this information to Lindley, using the argumentative strategies described in this chapter.

7. **Technical Definitions in Your Field.** Select a technical area in which you have taken course work or in which you have technical experience. Now assume that you are employed as an outside consulting expert, acting as a resource in your particular area

to a McDuff manager not familiar with your specialty. (For example, a food-science expert might provide information related to the dietary needs of oil workers working on an offshore rig for three months; a business or management expert might report on a new management technique; an electronics expert might explain the operation of some new piece of equipment that McDuff is considering buying; a computer programmer might explain some new piece of hardware that could provide supporting services to McDuff; and a legal expert might define ''sexism in the workplace'' for the benefit of McDuff's human resources professionals.

For the purpose of this report, develop a context in which you would have to define terms for an uninformed reader. Incorporate *one expanded definition* and *at least one sentence definition* into your report.

8. **Description of Equipment in Your Field.** Select a common piece of laboratory, office, or field equipment with which you are familiar. Now assume that you must write a short report to your McDuff supervisor, who wants this report to contain a thorough physical description of the equipment. Later he or she plans to incorporate your description into a training manual for those who need to know how to use, and perform minor repairs on, the equipment. For the body of your description, choose either a part-by-part physical description and/or a thorough description of functions.

9. **Description of Position in Your Field.** Interview a friend or colleague about the specific job that person holds. Make certain it is a job that you yourself have *not* had. On the basis of data collected in the interview, write a thorough description of the person's position—including major responsibilities, reporting relationships, educational preparation, experience required, etc.

Now place this description in the context of a letter report to the manager of human resources at McDuff. Assume she has hired you, a technical consultant to McDuff, to submit a letter report that contains the description. She is preparing to advertise such an opening at McDuff but needs your report to write the job description and the advertisement. Because she has little firsthand knowledge of the position about which you are writing, you should avoid technical jargon.

10. **Classification/Division in Technical Courses.** For this assignment, you will need a catalog from your college or university or that of another school. First, select a number of courses that have not yet been classified except perhaps by academic department. Using course descriptions in the catalog or any other information you can find, *classify* the courses using an appropriate basis (for example, the course level, topic, purpose in the department, prerequisites, etc.). Also *partition* one of the courses by using information gathered from a course syllabus or an interview with someone familiar with the course.

The context for this exercise is a memo report written by you, a technical training specialist at McDuff's St. Louis office. Assume that the courses are taught at St. Louis Tech, a nearby college. A group of McDuff's St. Louis employees has expressed interest in further education in the area you have investigated. Your classification of the courses and your in-depth partition of one particular course will help these employees decide whether to consider enrolling.

11. **Comparison/Contrast for the Purchasing Agent.** For this memo report, select a category in the list following or another category approved by your instructor. Then choose three specific types or brand-name products that fit within the category and for which you can find data. Write either a whole-by-whole or a part-by-part comparison/contrast. Place this comparison/contrast pattern within the context of a short report from you, as a McDuff employee, to the company purchasing agent. You may or may not include a preference for one item or the other. However, understand that the purpose of your report is to give the reader the information needed to order one of the items in your comparison/contrast.

Categories

1. Briefcases
2. Calculators
3. Compact disks
4. Credit cards
5. Electric pencil sharpeners
6. Electric sanders or saws
7. Lawn mowers
8. Mortgages
9. Personal computers
10. Photocopiers
11. Pickup trucks
12. Professional journals in a technical field
13. Refrigerators
14. Software programs
15. Retirement plans
16. Surveying methods
17. Telephone systems or companies
18. Types of savings accounts
19. Vacuum cleaners
20. VCRs

Mc Duff, Inc.

MEMORANDUM

DATE: February 25, 1993
TO: Kerry F. Camp, Vice President of Domestic Operations
FROM: Your Name, Boston Branch Manager
SUBJECT: Problems with Office Chairs

I enjoyed talking to you yesterday and look forward to seeing you at the Environmental Science Convention in New York. In the meantime, I am writing to ask your help with the chair problem I mentioned in our conversation. This memo will give you some background on our difficulties with the chairs. Also, I have presented a solution for you to consider.

SPECIFIC PROBLEMS WITH OFFICE CHAIRS

About six months ago, we received new office furniture ordered by the corporate office. In many ways, it has been a considerable improvement over the 20-year-old furniture it replaced. The desks, cabinets, and credenzas have been especially well received. Besides being quite practical, they give a much more professional appearance to our office.

As I mentioned in our conversation, the chairs (Model 223) that accompany the new desks have not worked out as well as the other furniture. Here are the two main problems.

Increased Employee Illness

Starting about a month after the chairs arrived, I received complaints from ten employees about back pains they claimed were related to the new chairs. Taking a wait-and-see approach, I asked managers to have chairs adjusted and to give the furniture more break-in time. In the second month, however, I heard from a dozen additional employees about the chairs and noticed a 25 percent increase in sick days. Upon further inquiry, I learned that most managers attributed the increased sick time to back problems related to the office chairs.

Decreased Productivity

The increased sick leave presented a serious enough concern. Adding to the problem, however, was roughly a 10 percent decrease in productivity last month among employees who were not sick. I measured this decrease by three criteria:

- Number of pages produced by the word-processing staff
- Number of billable hours logged by the professional staff
- Number of client calls made by the marketing staff

In my seven years managing this office, I have never witnessed so large a drop with no clear cause.

Side annotations:

Approaches problem tactfully.

Mentions points of agreement— before moving toward argumentative issues.

Presents evidence clearly— including statistic.

Cites reliable authority.

States supporting statistic.

Gives basis for statistical information.

MODEL 5–1
Memo report

continues

CONFIRMATION OF PROBLEM

Concerned about the health problems and lower productivity, I sought a recognized consultant in the area of workplace health and office design. Dr. Stacy Y. Stephens, professor of ergonomics at East Boston College, was referred to me when I called the local office of the Occupational Safety and Health Administration.

Dr. Stephens visited our office on January 17. She spent two days (1) interviewing and observing users of the new chairs, (2) examining the chairs in detail, and (3) meeting with me to offer her conclusions. At her suggestion, I called the supplier, Jones Office Furniture, to ask about the availability of other Jones chairs with different features. According to Dr. Stephens, the Jones Model 623 is much better suited to normal office use. Specifically, the Model 623 has these advantages:

- Excellent lower-lumbar support, compared to the inadequate support in the straight-backed Model 223
- Twelve inches of vertical adjustment, as opposed to the six inches available in Model 223
- Ball casters for easier rolling, as opposed to Model 223 casters that move with difficulty on our office carpeting
- Adjustable back spring that provides limited movement backward, as opposed to the rigid Model 223

Considering Dr. Stephens' suggestions, I decided to seek more information from Jones Office Furniture Company.

PROPOSED SOLUTION

Last week, I met twice with Mr. Dan McCartney, the Boston sales representative for Jones Office Furniture. In response to our problem, he has offered this proposal. Jones will trade our 45 Model 223 chairs for 45 new Model 623s, for only an extra $2000 from us. Under this arrangement, he would be (1) crediting us for the full purchase of our four-month-old Model 223s and (2) discounting the new chairs about $3000. I believe Jones is making this offer because it truly wants to retain our goodwill — and our future business.

CONCLUSION

As noted, this problem affects our employees' well-being and our office productivity. Would you please present my proposed solution to Ben Garner, vice president of business and finance, so that he can approve the $2000 funding? We could have the new chairs within a week after approval.

Thanks for your help, Kerry. With relatively small investment, we should be able to solve our medical and productivity problems.

MODEL 5–1, *continued*

Cites authority and gives her credentials.

Mentions this expert's opinion *last,* because it is the strongest evidence in this short report.

Gives *specific* points to support expert's views.

Sets up clear contrast between two chair types, to support chair exchange.

Describes alternative clearly — giving necessary details about cost, etc.

Ends with request for action.

Closes with tone that encourages agreement on *mutual* problem.

TECHNICAL DESCRIPTION: BLUEPRINT MACHINE

A blueprint machine is a piece of office equipment used to make photographic repro-
ductions of architectural plans, technical drawings, and other types of figures. The blueprint
shows as white lines on blue paper. This description covers the physical parts of the machine
and a brief operation procedure.

Physical Description

The blueprint machine is contained in a metal cabinet measuring 36" wide, 10" tall, and
18" deep. Typically, the machine is placed on a cabinet containing blueprint supplies. The
most evident features of the machine are the controls and the paper path.

Blueprint Machine

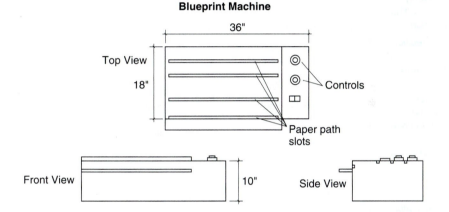

Controls: The controls for the machine are located on the top right-hand side of the case.

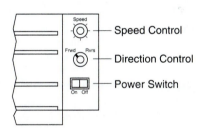

There are two knobs and one power switch. The topmost knob controls the speed of the paper
feed. The lower knob changes the direction of the paper. The power switch turns the machine
on and off.

continues

MODEL 5–2
Detailed description

Integrates pro-
cedure with il-
lustration.

Paper Path: The top of the machine has four horizontal slots. These slots are the openings to the paper path.

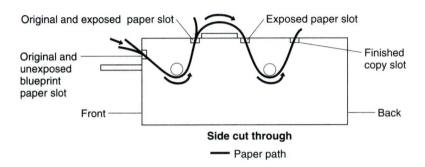

Side cut through
— Paper path

The first slot, beginning at the front of the machine, is the feed slot for the original and the unexposed blue print paper. The second slot is the discharge for the original and the exposed print paper. The third slot is the feed for developing the exposed blueprint paper. The fourth slot is the discharge for the finished blueprint copy.

Operation of the Blueprint Machine

Operating the blueprint machine involves setting the controls and making copies.
Setting the Controls:
1. Turn on the machine.
2. Set the paper path to forward.
3. Set the speed for the type of paper used.

Includes brief
instructions that
help expand
upon description
of mechanism.

Making Copies:
1. Place the original, face side up, on top of the blueprint paper, yellow side up.
2. Feed the two pieces of paper into the original feed slot.
3. Separate the original from the copy as they emerge from the discharge slot.
4. Feed the exposed blueprint paper into the developer slot.
5. Collect the copy as it emerges from the discharge slot.
6. Turn power off when desired copies have been made.

Conclusion

Blueprint machines are found in all McDuff offices and are used frequently. If a wide variety of employees understands how these machines function, our office will continue to run smoothly.

MODEL 5–2, *continued*

TECHNICAL DESCRIPTION DESK STAPLER

The stapler is a standard piece of office equipment used to drive a U-shaped piece of metal (staple) through papers, thus binding them together. Although staplers are made in different shapes and sizes, the description here is of the standard desk-use stapler. It is usually 6 to 8 in. long and made out of lightweight strips of metal. The major parts are the head, ejector, and base. The following description covers these main components.

Stapler Head

The head is the part of the stapler that, when compressed, drives the staple from the ejector through the papers. The stapler head has two main parts—the outer casing and the U-shaped driver.

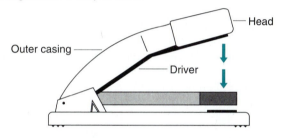

The outer casing is made of lightweight sheet metal. On a typical stapler it is about 7 in. long, 1 in. wide, and 3/4 in. high. A U-shaped piece with the open portion facing downward, it provides a sheath that wraps around the driver.

The driver is a metal piece that fits inside the outer casing. On a typical stapler it is about 6 1/2 in. long, 1/2 in. wide, and 1/4 in. high. Like the outer casing, it is U-shaped with the open portion facing downward. The driver is connected to the outer casing at the base by a metal pin. This piece, made of heavier metal than the outer casing, serves to drive the staple out of the ejector.

Ejector

The ejector houses the staples that bind the pieces of paper together. Like the outer casing, it is made of U-shaped metal but with the open part of the U facing upward. This piece is about the same size as the driver. Inside the U of the ejector is a tray onto which the row of staples fits. These staples are forced to the front of the tray by a tension-mounted spring.

continues

MODEL 5–3
Detailed description

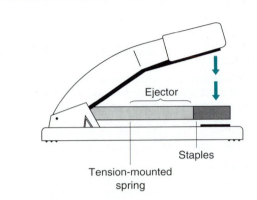

Ejector

Staples

Tension-mounted
spring

Base

The base is the heaviest and largest part of the stapler. It is about 8 in. long, 1 1/2 in. wide, and 1 in. thick. The base is tapered somewhat so that the edges closest to the bottom are longer than the edges near the top of the base, thus giving the entire stapler as much stability as possible. On the bottom of the base there are two rubber pieces, one at each end, that are riveted to the base and that cushion

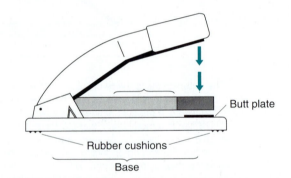

Butt plate

Rubber cushions

Base

the stapler from the surface on which it rests. On the top side of the base there is a butt plate that is used to flatten the staples when they are driven from the ejector through the paper. The head and ejector are connected to the base with a pin located at the rear of the stapler.

Conclusion

The standard stapler is a light-duty machine used mainly for paper binding. Because of its structural simplicity, it usually lasts a long time and can endure a significant amount of abuse (heavy pounding by some users, dropping onto floors, use with excessive amount of paper, etc.).

MODEL 5–3, *continued*

PART-BY-PART PATTERN COMPARISON OF BULLDOZERS

This is a comparison of the Cannon-D, Foley-G, and Koso-L bulldozers. The major criteria considered are pushing capacity, purchase details, and dependability.

Pushing Capacity

Both the Cannon-D and Foley-G have an excellent pushing capacity that is more than adequate for our U.S. construction projects. The Cannon has a pushing capacity of 1300 tons per hour (TPH) over 500 ft, whereas the Foley has a pushing capacity of 1100 TPH over 500 ft. Both figures are mentioned in brochures from their respective companies. Furthermore, they are confirmed by recent tests reported in the trade journal, Bulldozer Unlimited.

The Koso-L, however, lags behind its two competitors, being able to push only 1000 TPH over 500 ft. Furthermore, this figure was found only in the company's sales brochures, with no verification available in the trade journals I researched.

Purchase Details

The main elements of the purchase—price, warranty, and cost of extended warranty—vary considerably among the three machines. The basic purchase prices are as follows:

1. $250,000 for the Cannon-D
2. $300,000 for the Foley-G
3. $310,000 for the Koso-L

Concerning warranties, Cannon provides a complete parts-and-service warranty for 90 days, with an additional parts warranty on the drivetrain for another 12 months. Foley offers the same initial parts-and-service warranty, but it has a longer parts warranty on the drivetrain—18 months. Koso has a complete parts-and-service warranty for six months, with no additional parts warranty on the drivetrain.

Beyond these warranties that come with the machines, all three companies offer the exact option of an additional two-year warranty that covers parts in the drivetrain. This additional warranty must be acquired at time of purchase and costs $4000 for the Cannon, $4000 for the Foley, and $7000 for the Koso.

Dependability

Bulldozer downtime can cost the firm a good deal of money, either in project delays or in the added expense of renting another bulldozer, so dependability is an important criterion in comparing the three models. Although the trade journals I consulted contained no model-by-model comparison of dependability of the three bulldozers, I did find occasional references to features that affect reliability. In

continues

MODEL 5–4
Part-by-part comparison/contrast: three bulldozers

Uses subhead-
ings because of
large amount of
information to
relate.

Introduces Fo-
ley's main fea-
ture (technical
innovation).

But reinforces
Cannon prefer-
ence by noting
that sealed
tracks are not
fully tested.

Keeps option
open for buying
Koso *later*—but
makes it clear
that more data
are needed.

addition, I sought out anecdotal evidence from nearby construction firms that have used the machines. Here is what I discovered.

Canon-D. This machine is an established commercial bulldozer with a good reputation for reliability in the field. The "D" model is the latest version of a machine that began with the "A" model in 1954, so the company has had years to refine its technology. Interestingly, several years ago Cannon decided not to add sealed tracks to its list of features, even though this sealed approach is one of the newest track innovations in the industry. Even so, the Cannon-D generally will operate for 10,000 hours without major maintenance in the tracks. It remains to be seen if the decision about sealed tracks was the correct one.

Foley-G. While the Foley also has a good reputation for reliability in the field, the firm began making bulldozers only in 1970—16 years after Cannon started. Foley has the general reputation of being quicker than Cannon to introduce new technology into its bulldozer line. For example, 2 years ago it started using sealed tracks that lubricate themselves automatically. The track assembly should not need major maintenance for 15,000 hours, a 5000-hour improvement over Cannon. Of course, the newness of this advance suggests that there hasn't yet been enough time to judge the long-term effectiveness of sealed tracks.

Koso-L. This machine is definitely the "new guy on the block." Unlike the American-made Cannon and Foley, the Koso is produced in Korea by a company that got into business rather recently—1986. The Koso-L uses fairly traditional technology, not having incorporated sealed tracks into its design, for example. Despite the recent entry into the market, Koso has already built a strong reputation for reliability on projects in other countries. However, there are not yet enough data about performance on construction projects in the United States.

MODEL 5–4, *continued*

6 | *Process Descriptions and Instructions*

This vendor depends on clearly written instructions to repair the photocopying machine at McDuff's Atlanta office.

L ike other McDuff offices, McDuff's Denver office installed an electronic-mail system in the last few years. It permits employees to send and receive messages on their computer terminals. The idea seems to have pleased everyone. Messages are conveyed faster, and less paper is generated in the office. As office services manager, you met with a representative of the company that would install the system, shortly before the installation date. Your boss, Leonard Schwartz, expected a memo from you summarizing the installation process (see Model 6–1 on page 161). You also wrote a memo to all employees, giving them instructions about how to read electronic-mail messages (see Model 6–2, pp. 162–163).

This brief McDuff case study demonstrates two types of technical writing you will often face—process descriptions and instructions. Both are important patterns, but instructions will play the greater role in your career. As noted in one essay, "The field of technical writing has expanded greatly in the last ten years, and the new work is in writing *instructions*. . . . Many graduates of technical programs now find themselves writing *instructions* more often than they write descriptions and reports"[1] (italics mine). For this reason, instructions receive the most emphasis in this chapter.

Instructions and process descriptions share an important common bond. Both must accurately describe a series of steps leading toward a specific result. Yet they differ in purpose, audience, and format. This chapter (1) explores these similarities and differences, with specific reference to McDuff applications, (2) gives specific guidelines for developing both types, and (3) provides models to use in your own writing.

PROCESS DESCRIPTIONS VERSUS INSTRUCTIONS

In the McDuff example just given, the memo to your boss (Model 6–1, p. 161) explained the process by which the vendor installed electronic mail. The other memo (Model 6–2, pp. 162–163) gave users the directions needed to read on-line

---

[1]Reprinted by permission of the Modern Language Association of America from Janice C. Redish and David A. Schell, "Writing and Testing Usability Instructions," in *Technical Writing: Theory and Practice*, ed. Bertie E. Fearing and W. Keats Sparrow (New York: Modern Language Association of America, 1989), 63.

mail. In other words, you write a process description to help readers *understand* what has been, is being, or will be done, whereas you write instructions to show readers how to *perform* the process themselves.

Process descriptions are appropriate when the reader needs to be informed about the action but does not need to perform it. If you suspect a reader may in fact be a "user" (that is, someone who uses your document to perform the process), always write instructions. Figure 6–1 provides a list of contrasting features of process descriptions and instructions; the two subsections that follow give these features some realism by briefly describing some McDuff contexts.

Process Descriptions at McDuff

Process descriptions provide information for interested readers who do not need instructional details. At times, describing a process may be the sole purpose of your document, as in Model 6–1. More often, however, you use process description only as a pattern of organization within a document with a larger purpose. The following McDuff examples (1) show the supporting purpose of process descriptions and (2) reinforce the difference between process descriptions and instructions.

- **Accounting:** As an accountant at McDuff's corporate office, you have just finished auditing the firm's books. Now you must write a report to the vice president for business and marketing on the state of the firm's finances. Along with your findings, the vice president wants an overview of the procedure you followed to arrive at your conclusions.

PROCESS DESCRIPTIONS

Purpose:	Explain a sequence of steps in such a way that the reader understands a process
Format:	Use paragraph descriptions, listed steps, or some combination of the two
Style:	Use "objective" point of view ("2. The operator started the engine..."), as opposed to "command" point of view ("2. Start the engine...")

INSTRUCTIONS

Purpose:	Describe a sequence of steps in such a way that the reader can *perform* the sequence of steps
Format:	Employ numbered or bulleted lists, organized into subgroups of easily understandable units of information
Style:	Use "command" point of view ("3. Plug the phone jack into the recorder unit"), as opposed to "objective" point of view ("3. The phone jack was plugged into the recorder unit")

FIGURE 6–1
Process descriptions versus instructions

- **Maintenance:** As a maintenance supervisor at McDuff's Saudi Arabian branch, you traveled to a construction site to repair a machine that tests the strength of concrete. The procedure requires billing the client for additional charges. Along with your bill, you send an attachment that summarizes the procedure you followed.
- **Laboratory work:** As lab supervisor at the St. Louis office, you spent all day Saturday in the lab assembling a new gas chromatograph needed to analyze gases. To justify the overtime hours, you write a memo to your manager describing the assembly process.
- **Welding inspection:** As an NDT (nondestructive testing) expert at McDuff's San Francisco office, you were hired by a California state agency to x-ray all welds at a bridge damaged by an earthquake. The text of your report gives test results; the appendices describe the procedures you followed.
- **Marketing:** As McDuff's marketing manager, you have devised a new procedure for tracking contacts with prospective clients (from first sales call to getting the job). You must write a memo to McDuff's two vice presidents for operations, briefly describing the process. Their approval is needed before the new marketing technique can be introduced at the 15 branch offices.

In each case, you are writing for a reader who wants to know what has happened or will happen, but who does *not* need to perform the process.

Instructions at McDuff

Think of instructions this way: They must provide users with a road map to *do* the procedure, not just understand it. That is, someone must complete a task on the basis of words and pictures you provide. Clearly, instructions present you, the writer, with a much greater challenge *and* risk. The reader must be able to replicate the procedure without error and, most importantly, with full knowledge of any dangers. The McDuff situations that follow reflect this challenge. Note that they parallel the case studies presented for process descriptions.

- **Accounting:** As McDuff's lead accountant for the last 20 years, you have always been responsible for auditing the firm's books. Because you developed the procedure yourself over many years, there is no comprehensive set of instructions for completing it. Now you want to record the steps so that other company accountants besides you can perform them.
- **Maintenance:** As a maintenance supervisor at McDuff's branch in Saudi Arabia, you must repair a piece of equipment for testing concrete. You have never disassembled this particular machine, and there are no manufacturer's instructions available. Therefore, the job takes you three full days. To help other employees perform this task in the future, you write a set of detailed instructions for making the repair.
- **Laboratory work:** As lab supervisor for McDuff's St. Louis office, you have assembled one of the two new gas chromatographs just purchased by the company. You are supposed to send the other unit to the Tokyo branch, where it will be put together by Japanese technicians. Unfortunately, the manufacturer's in-

structions are poorly written, so you plan to rewrite them for the English-speaking technicians at the Tokyo office.

- **Welding inspection:** As McDuff's NDT (nondestructive testing) expert at the San Francisco office, you have seen a large increase in NDT projects. Given California's aging bridges and constant earthquake activity, you persuaded your branch manager to hire several NDT technicians. Now you must write a training manual that will instruct these new employees on methods for inspecting bridge welds.
- **Marketing:** As McDuff's marketing manager, you have suggested a new approach for tracking sales leads. Having had your proposal approved by the corporate staff, you now need to explain the marketing procedure to technical professionals at all 15 offices. Your written set of instructions must be understood by technical experts in many fields, who have little if any marketing experience.

In each case, your instructions must describe steps so thoroughly that the reader will be able to replicate the process, *without* having to speak in person with the writer of the instructions. The next two sections give rules for preparing both process descriptions and sets of instructions.

GUIDELINES AND MODELS FOR PROCESS DESCRIPTIONS

You have already learned that process descriptions are aimed at persons who need to understand the process, not perform it. Process descriptions often have these purposes:

- Describing an experiment
- Explaining how a machine works
- Recording steps in developing a new product
- Describing what happened during a field test

In each case, follow these guidelines for creating first-rate process descriptions:

■ *Process Guideline 1: Know Your Purpose and Audience*

Your intended purpose and expected audience influence every detail of your description. Here are some preliminary questions to answer before writing:

- Are you supposed to give just an overview or are details needed?
- Do readers understand the technical subject or are they laypersons?
- Do readers have mixed technical backgrounds?
- Does the process description supply supporting information (perhaps in an appendix) or is it the main part of the document?

Process descriptions are most challenging when directed to a mixed audience. In this case, write for the lowest common denominator—that is, for your least technical readers. It is better to write below the level of your most technical readers than to write above the level of your nontechnical readers.

For example, the process description in Model 6–3 on page 164 is directed to a mixed audience of city officials—some technical staff and some nontechnical political officials. It is contained in an appendix to a long McDuff report that recommends immediate cleanup of a toxic-waste dump. Note that the writer either uses nontechnical language or defines any technical terms used.

■ *Process Guideline 2: Follow the ABC Format*

In chapter 3 you learned about the ABC format (**A**bstract/**B**ody/**C**onclusion), which applies to all documents. The abstract gives a summary, the body supplies details, and the conclusion provides a wrap-up or leads to the next step in the communication process. Whether a process description forms all or part of a document, it usually subscribes to the following version of the three-part ABC plan:

- The **Abstract** component includes three background items:

  1. Purpose statement
  2. Overview or list of the main steps that follow
  3. List of equipment or materials used in the process

  Model 6–3 on page 164 includes all three, with a separate heading for equipment. First, the purpose statement places the description in the context of the entire document. Then the list of main steps gives readers a framework for interpreting details that follow. Finally, the list of equipment or materials provides a central reference point as readers work through all the steps.

- The **Body** component of the process description moves logically through the steps of the process. By definition, all process descriptions follow a chronological, or step-by-step, pattern of organization. These steps can be conveyed in two ways:

  1. **Paragraphs:** This approach weaves steps of the process into the fabric of typical paragraphs, with appropriate transitions between sentences. Use paragraphs when your readers would prefer a smooth explanation of the entire process, rather than emphasis on individual steps.
  2. **List of steps:** This approach includes a list of steps, usually with numbers or bullets. Much like instructions, a listing emphasizes the individual parts of the process. Readers prefer it when they will need to refer to specific steps later on.

Both paragraph and list formats have their places in process descriptions. In fact, most descriptions can be written in either format. See Figure 6–2 for a McDuff example showing both a paragraph and list description for the same process of laying a concrete patio. As a public-service gesture, McDuff, Inc., has written a pamphlet that briefly describes simple home improvements. It is intended for owners of small homes who complete renovations with little or no help from contractors. If home owners are interested in one of the projects, they can write for detailed instructions to an address listed in the pamphlet.

Building a concrete patio is one project covered; the process description contains a subsection about constructing the wooden form into which concrete is poured.

Steps of process are embedded in paragraph.

After brief lead-in, steps of process are placed in list format.

A. PARAGRAPH OPTION

The home owner should select rough-grade 2 x 4s for building the wooden form for the patio. The form is just a box, with an open top and with the ground for the bottom, into which concrete will be poured. First the four sides are nailed together, and then the form is leveled with a standard carpenter's level. Finally, 2 x 4 stakes are driven into the ground about every 2 or 3 ft on the outside of the form, to keep it in place during the pouring of the concrete.

B. LIST OPTION

Building a wooden form for a home concrete patio can be accomplished with some rough-grade 2 x 4s. This form is just a box with an open top and the ground for the bottom. Building involves three basic steps:

1. Nailing 2 x 4s into the intended shape of the patio
2. Leveling the box-shaped form with a standard carpenter's level
3. Nailing stakes (made from 2 x 4 lumber) every 2 or 3 ft at the outside edge of the form, to keep it in place during the pouring of the concrete

FIGURE 6–2
Two options for process description: (A) paragraph option; (B) list option

- The **Conclusion** component of a process description keeps the process from ending abruptly with the last step. Here you should help the reader put the steps together into a coherent whole. When the process description is part of a larger document, you can show how the process fits into a larger context (see Model 6–3, p. 164).

Process Guideline 3: Use an Objective Point of View

Process descriptions explain a process rather than direct how it is to be done. Thus they are written from an objective point of view—*not* from the personal "you" or "command" point of view common to instructions. Note the difference in these examples:

Process: The concrete is poured into the two-by-four frame.

or

The technician pours the concrete into the two-by-four frame.

Instructions: Pour the concrete into the two-by-four frame.

The process excerpts *explain* the step, whereas the instructions excerpt *gives a command* for completing the instructions.

Process Guideline 4: Choose the Right Amount of Detail

Only a thorough audience analysis will tell you how much detail to include. Model 6–3, (p. 164), for example, could have contained much more technical detail about the substeps for testing air quality at the site. The writer, however, decided that the city officials would not need more scientific and technical detail.

In supplying specifics, be sure to subdivide information for easy reading. In paragraph format, headings and subheadings can be used to make the process easier to grasp. In list format, an outline arrangement of points and subpoints may be appropriate. When such detail is necessary, remember this general rule of thumb: *place related steps into groups of from three to seven points*. Readers find it easier to remember several groupings with subpoints, as opposed to one long list. Following are two rough outlines for a process description. The second is preferred in that it groups the many steps into three easily grasped categories.

Employment Interview Process

1. Interviewer reviews job description
2. Interviewer analyzes candidate's application
3. Candidate and interviewer engage in "small talk"
4. Interviewer asks open-ended questions related to candidate's resume and completed application form
5. Interviewer expands topic to include matters of personal interest and the candidate's long-term career plans
6. Interviewer provides candidate with information about the position (salary, benefits, location, etc.)
7. Candidate is encouraged to ask questions about the position
8. Interviewer asks candidate about her or his general interest, at this point, in the position
9. Interviewer informs candidate about next step in hiring process

Employment Interview Process

■ **Preinterview Phase**

1. Interviewer reviews job description
2. Interviewer analyzes candidate's application

■ **Interview**

3. Candidate and interviewer engage in "small talk"
4. Interviewer asks open-ended questions related to candidate's resume and completed application form
5. Interviewer expands topic to include matters of personal interest and the candidate's long-term career plans
6. Interviewer provides candidate with information about the position (salary, benefits, location, etc.)
7. Candidate is encouraged to ask questions about the position

■ **Closure**

8. Interviewer asks candidate about his or her general interest, at this point, in the position
9. Interviewer informs candidate about next step in hiring process

■ *Process Guideline 5: Use Flowcharts for Complex Processes*

Some process descriptions contain steps that are occurring at the same time. In this case, you may want to supplement a paragraph or list description with a flowchart. Such charts use boxes, circles, and other geometric shapes to show progression and relationships among various steps.

Model 6–4 on page 165, for example, shows a flowchart and an accompanying process description at McDuff. Both denote services that McDuff's London branch provides for oil companies in the North Sea. The chart helps to demonstrate that the geophysical study (mapping by sonar equipment) and the engineering study (securing and testing of seafloor samples) take place at the same time. Such simultaneous steps are difficult to show in a list of sequential steps.

GUIDELINES AND MODELS FOR INSTRUCTIONS

Rules change considerably when moving from process descriptions to instructions. Although both patterns are organized by time, the similarity stops there. Instructions walk readers through the process so that they can *do* it, not just understand it. It is one thing to explain the process by which a word-processing program works; it is quite another to write a set of instructions for using that word-processing program. This section explores the challenge of writing instructions by giving you some basic writing and design guidelines.

These guidelines for instructions also apply to complete operating *manuals*, a document type that many technical professionals will help to write during their careers. Those manuals include the instructions themselves, as well as related information such as (1) features, (2) physical parts, and (3) troubleshooting tips. In other words, manuals are complete documents, whereas instructions can be part of a larger piece.

■ *Instructions Guideline 1: Select the Correct Technical Level*

This guideline is just another way of saying you need to know *exactly* who will be reading your instructions. Are your readers technicians, engineers, managers, general users, or some combination of these groups? Once you answer this question, select language that every reader can understand. If, for example, the instructions include technical terms or names of objects that may not be understood, use the techniques of definition and description discussed in the previous chapter.

■ *Instructions Guideline 2: Provide Introductory Information*

Like process descriptions, instructions follow the ABC format (**A**bstract/**B**ody/**C**onclusion) described in chapter 3. The introductory (or abstract) information should include (1) a purpose statement, (2) a summary of the main steps, and (3) a list or an illustration giving the equipment or materials needed (or a reference to an attachment with this information). These three items set the scene for the procedure itself.

Besides these three "musts," you should consider whether some additional items might help set the scene for your user:

- Pointers that will help with installation
- Definitions of terms
- Theory of how something works
- Notes, cautions, warnings, or dangers that apply to all steps

■ Instructions Guideline 3:
Use Numbered Lists in the Body

A simple format is crucial to the body of the instructions—that is, the steps themselves. Most users constantly go back and forth between these steps and the project to which they apply. Thus you should avoid paragraph format and instead use a simple numbering system. Model 6–5 on pages 166–167 shows a "before and after" example. The original version is written in paragraphs that are difficult to follow; the revised version includes nine separate, numbered steps.

■ Instructions Guideline 4:
Group Steps Under Task Headings

Readers prefer that you group together related steps under headings, rather than present an uninterrupted "laundry list" of steps. Model 6–6 on pages 168–170 shows how this technique has been used in a fairly long set of instructions for operating an answering machine. Given the number of steps in this case, the writer has used a separate numbering system within each grouping.

Groupings provide two main benefits. First, they divide fragmented information into manageable "chunks" that readers find easier to read. Second, they give readers a sense of accomplishment as they complete each task, on the way to finishing the whole activity.

■ Instructions Guideline 5: Place One Action in a Step

A common error is to "bury" several actions in a single step. This approach can confuse and irritate readers. Instead, break up complex steps into discrete units, as shown here:

- **Original:**

  Step 3: Fill in your name and address on the coupon, send it to the manufacturer within two weeks, return to the retail merchant when your letter of approval arrives from the manufacturer, and pick up your free toaster oven.

- **Revision:**

  Step 3: Fill in your name and address on the coupon.
  Step 4: Send the coupon to the manufacturer within two weeks.

Step 5: Show your retail merchant the letter of approval after it arrives from the manufacturer.

Step 6: Pick up your free toaster oven.

■ *Instructions Guideline 6:*
Lead Off Each Action Step With a Verb

Instructions should include the "command" form of a verb at the start of each step. This style best conveys a sense of action to your readers. Model 6–5 on pages 166–167 and Model 6–6 on pages 168–170 consistently use command verbs for all steps throughout the procedures.

■ *Instructions Guideline 7:*
Remove Extra Information From the Step

Sometimes you may want to follow the command sentence with an explanatory sentence or two. In this case, distinguish such helpful information from actions by giving it a label, such as "Note" or "Result" (for example, see Model 6–2, pp. 162–163).

■ *Instructions Guideline 8:*
Use Bullets or Letters for Emphasis

Sometimes you may need to highlight information, especially within a particular step. Avoid using numbers for this purpose, since you are already using them to signify steps. Bullets work best if there are just a few items; letters are best if there are many, especially if they are in a sequence. The revised version in Model 6–5 on pages 166–167 shows the appropriate use of letters, and Model 6–6 (pp. 168–170) shows the use of bullets.

In particular, consider using bullets at any point at which users have an *option* as to how they will respond. The following example uses bullets in this way; it also eliminates the problem of too many actions being embedded in one step.

Part of Procedure for Firing Clay in a Kiln

(*Note:* A pyrometric "cone" is a piece of test clay used in a kiln, an oven for baking pottery. The melting of the small cone helps the operator determine that the clay piece has completed the firing process.)

■ **Original:**

Step 6: Check the cone frequently as the kiln reaches its maximum temperature of 1850 degrees. If the cone retains its shape, continue firing the clay and checking the cone frequently. When the cone begins to bend, turn off the kiln. Then let the kiln cool overnight before opening it and removing the pottery.

■ **Revision:**

Step 6: Check the cone frequently as the kiln reaches its maximum temperature of 1850 degrees.

Step 7: Has the cone started to bend?
  - If *no,* continue firing the piece of pottery and checking the cone frequently to see if it has bent.
  - If *yes,* turn off the kiln.

Step 8: Let the kiln cool overnight after turning it off.

Step 9: Open the kiln and remove the pottery.

■ *Instructions Guideline 9: Emphasize Cautions, Warnings, and Dangers*

Instructions often require drawing attention to risks in using products and equipment. Your most important obligation is to highlight such information. Generally, the following three terms are used as "red flags" to the reader. The level of risk increases as you move from 1 to 3:

1. **Caution:** possibility of damage to equipment or materials
2. **Warning:** possibility of injury to people
3. **Danger:** probability of injury or death to people

If you are not certain that these distinctions will be understood by your readers, define the terms *caution, warning,* and *danger* in a prominent place before you begin your instructions.

As for placement of the actual cautions, warnings, or danger messages, here are your options:

- **Option 1:** *In a separate section right before the instructions begin.* This approach is most appropriate when you have a list of general warnings that apply to much of the procedure *or* when one special warning should be heeded throughout the instructions—for example, "DANGER: Keep main breaker on 'off' during entire installation procedure." Figure 6–3 shows such a warning at the start of instructions to install a security keypad.
- **Option 2:** *In the text of the instructions.* This approach works best if the caution, warning, or danger message applies to the step that immediately follows it. Thus users are warned about a problem *before* they read the step to which it applies. For an example, see the following:

  CAUTION: Use 220-grade sandpaper, to avoid scratching the surface of furniture.

In other words, give information about potential risks *before* the operator has the chance to make the mistake. Also, the caution, warning, or danger message can be made visually prominent by the following techniques:

- Underlining: <u>Warning</u>
- Bold: **Warning**
- Full Caps: WARNING
- Italics: *Warning*
- Oversized Print: Warning
- Boxing: | Warning |

FIGURE 6–3
Example of "danger" message

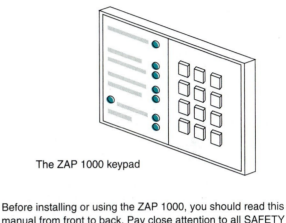

INTRODUCTION

You have purchased one of the most sophisticated security systems available for home, business, or industry use. The ZAP 1000 provides a multi-zoned blanket of protection for your family, your home, or your business. The ZAP 1000 will alert you to intrusions, fire, and smoke by sounding an alarm, calling the proper authorities, or both.

The ZAP 1000 keypad

Before installing or using the ZAP 1000, you should read this manual from front to back. Pay close attention to all SAFETY MESSAGES, such as the one below.

> **DANGER! You can be injured or killed by improper or careless use of this equipment. Consult a qualified electrician if you have any doubts about installation or use.**

- Color: Warning
- Combined Methods: *Warning*
                     WARNING
                     WARNING

Instructions Guideline 10: Keep a Simple Style

Perhaps more than any other type of technical writing, instructions must be easy to read. Readers expect a no-nonsense approach to writing that gives them required information without fanfare. Here are some techniques to use:

- Keep sentences short, with an average length of under 10 words.
- Use informal definitions (parenthetical, like this one) to define any terms not understood by all readers.

- Never use a long word when a short one will do.
- Be specific and avoid words with multiple interpretations (frequently, seldom, occasionally, etc.).

■ *Instructions Guideline 11: Use Graphics*

Illustrations are essential for instructions that involve equipment. Place an illustration next to every major step when (1) the instructions or equipment is quite complicated or (2) the audience may contain poor readers or people who are in a hurry. Such word-picture associations create a page design that is easy to follow.

In other cases, just one or two diagrams may suffice for the entire set of instructions. The one reference illustration in Model 6–6 (pp. 168–170) helps the user of a message recorder locate parts mentioned throughout the instructions.

Another useful graphic in instructions is the table. Sometimes within a step you need to show correspondence between related data. For example, the instructions that follow would benefit from a list.

- **Original:**

    Step 3:   Use pyrometric cones to determine when a kiln has reached the proper temperature to fire pottery. Common cone ratings are as follows: a Cone 018 corresponds to 1200°F; a Cone 07 corresponds to 1814°F; a Cone 06 corresponds to 1859°F; and a Cone 04 corresponds to 1940°F.

- **Revision:**

    Step 3:   Use pyrometric cones to determine when a kiln has reached the proper temperature for firing pottery. Common cone ratings are as follows:

Cone 018	1200°F
Cone 07	1814°F
Cone 06	1859°F
Cone 04	1940°F

■ *Instructions Guideline 12: Test Your Instructions*

Professional writers often test their instructions on potential users before completing the final draft. The most sophisticated technique for such testing involves a "usability laboratory." Here test subjects are asked to use the instructions or manual to perform the process, often while speaking their observations and frustrations (if any). The writers or lab personnel unobtrusively observe the process from behind a one-way mirror. Later they may review audiotaped or videotaped observations of the test subjects or they may interview these persons. This complex process helps writers to anticipate and then eliminate problems that users will confront when they follow written instructions.

Of course, you probably will not have access to a usability laboratory to test your instructions. However, you can adapt the following user-based approach to testing assignments in this class and projects in your career. Specifically, follow these four steps:

1. Team up with another class member (or a colleague on the job). This person should be unfamiliar with the process and should approximate the technical level of your intended audience.

2. Give this person a draft of your instructions and provide any equipment or materials necessary to complete the process. Of course, for the purposes of a class assignment, this approach would work only for a simple process with little equipment or few materials.

3. Observe your colleague following the instructions you have provided. You should record both your observations and any responses this person makes while moving through the steps.

4. Revise your instructions to solve problems your user encountered during the test.

CHAPTER SUMMARY

Both process descriptions and instructions share the same organization principle: time. That is, both relate a step-by-step description of events. Process descriptions address an audience that wants to be informed but does not need to perform the process itself. Instructions are geared specifically for persons who need to complete the procedure themselves.

In writing good process descriptions, follow these basic guidelines:

1. Know your purpose and audience.
2. Follow the ABC format.
3. Use an objective point of view.
4. Choose the right amount of detail.
5. Use flowcharts for complex processes.

For instructions, follow these twelve rules:

1. Select the correct technical level.
2. Provide introductory information.
3. Use numbered lists in the body.
4. Group similar steps under heads.
5. Place one action in a step.
6. Lead off each action step with a verb.
7. Remove extra information from the step.
8. Use bullets or letters for emphasis.
9. Emphasize cautions, warnings, and dangers.
10. Keep a simple style.
11. Use graphics.
12. Test your instructions.

ASSIGNMENTS

Part 1: Short Assignments

1. **Writing a Process Description—School-Related.** Your college or university has decided to evaluate the process by which students are advised about, and registered for, classes. As part of this evaluation, the registrar has asked a select group of students—you among them—to describe the actual process each of you went through individually during the last advising/registration cycle. These "case studies" collected from individual students, the customers, will be transmitted directly to a college-wide committee studying registration and advising problems.

   Your job is to give a detailed account of the process. Remain as objective as possible, without giving opinions. If you had problems during the process, the facts you relate will speak for themselves. Simply describe the process you personally experienced. Then let the committee members judge for themselves whether the steps you describe should or should not be part of the process.

2. **Writing Instructions—School-Related.** In either outline or final written form, provide a set of instructions for completing assignments in this class. Consider your audience to be another student who has been ill and missed much of the term. You have agreed to provide her with an overview that will help her to plan and then write any papers she has missed.

   Your instructions may include (1) highlights of the writing process from chapter 1 and (2) other assignment guidelines provided by your instructor in the syllabus or in class. Remember to present a generic procedure for all assignments in the class, not specific instructions for a particular assignment.

3. **Writing Instructions—McDuff Context.** As an employee at the corporate office of McDuff, you just received the job of writing a set of instructions for completing performance appraisal reviews (PARs). The instructions will be included in a memo that goes to all supervisors at all branches of the firm, along with related forms. To help you get started on the instructions, you have been given a narrative description of the process (see the following). Your task is to convert this narrative into a simple set of instructions to go into the memorandum to supervisors.

   PARs are conducted annually for each employee, during the anniversary month in which the employee was originally hired. Several days before the month in which the PARs are to be conducted, the corporate office will send each supervisor a list of employees in that supervisor's group who should receive PARs. The main part of the PAR process is an interview between the supervisor and the employee receiving the PAR. Before this interview takes place, however, the supervisor should give the employee a copy of the "McDuff PAR Discussion Guide," which offers suggestions for the topics and tone of a PAR interview. The supervisor completes a "PAR Report Form" after each interview and then sends a copy to corporate and to the employee, with the original staying in the personnel files of that respective supervisor's branch. If for any reason a PAR interview and report form are not completed in the required month, the supervisor must send a memo of explanation to the corporate Human Resources Department, with a copy to the supervisor's branch manager.

Part 2: Longer Assignments

These assignments test your ability to write and evaluate the two patterns covered in this chapter—process descriptions and instructions. Specifically, follow these guidelines:

- Write each exercise in the form of a letter report or memo report, as specified.
- Follow organization and design guidelines given in chapters 3 and 4, especially concerning the ABC format (**A**bstract/**B**ody/**C**onclusion) and the use of headings. Chapter 8 gives rules for short reports, but such detail is not necessary to complete the assignments here.
- Fill out a Planning Form (at the end of the book) for each assignment.

4. **Evaluating a Process Description.** Using a textbook in a technical subject area, find a description of a process. For example, a physics text might describe the process of waves developing and then breaking at a beach, an anatomy text might describe the process of blood circulating, or a criminal justice text might describe the process of a criminal investigation.

   Keeping in mind the author's purpose and audience, evaluate the effectiveness of the process description as presented in the textbook. Submit your evaluation in the form of a memo report to your instructor in this writing course, along with a copy of the textbook description.

   For the purposes of this assignment, assume that your writing instructor has been asked by the publisher of the text you have chosen to review the book as an example of good or bad technical writing. Thus your instructor would incorporate comments from your memo report into his or her comprehensive evaluation.

5. **Writing a Process Description—School-Related.** Conduct a brief research project in your campus library. Specifically, use company directories, annual reports, or other library sources to find information about a company or other organization that could hire students from your college.

   In a memo report to your instructor, (1) describe the process you followed in conducting the search and (2) provide an outline or paragraph summary of the information you found concerning the company or organization. Assume that your report will become part of a volume your college is assembling for juniors and seniors who are beginning their job search. These students will benefit both from information about the specific organization you chose and from a description of the process that you followed in getting the information—since they may want to conduct research on other companies.

6. **Writing a Process Description—McDuff Context.** As a project manager for McDuff's Atlanta office, you just found out that your office has been selected as one of the firms to help renovate Kiddieworld, a large amusement park in the Southeast. Before Kiddieworld officials sign the contract, however, they want you to report on the process McDuff uses to report and investigate accidents (since the project will involve some hazardous work). You found the following policy in your office manual, but you know it is not something you would want to send to a client. Take this stilted paragraph and convert it to a process description for your clients, in the form of a letter report. Remember: The readers will not be performing the process; they only want to understand it.

   Accident reporting and investigation are an important phase of operating McDuff, Inc. The main purpose of an accident investigation and report is to gain an objective insight into facts surrounding the accident in order to improve future accident control measures and activities and to activate the protection provided by our insurance policies. It is therefore imperative that all losses, no matter how minor, be reported as soon as possible and preferably within 48 hours to the proper personnel. Specifically, all accidents must be reported orally to the immediate supervisor. For minor accidents that do not involve major loss of equipment or hospitalization, that supervisor has the responsibility of filling out a McDuff accident report form and then sending the form to the safety personnel at the appropriate branch office, who later sends it to the safety manager at the corporate office. For serious accidents that involve major loss of equipment or hospitalization of any individuals involved, the supervisor must call or telex the safety personnel at the appropriate branch office, who then should call or telex the safety manager at the corporate office. (A list of pertinent telephone numbers should be kept at every job site.) These oral reports will be followed up with a written report.

7. **Evaluation of Instructions.** Find a set of operating or assembly instructions for a VCR, microwave oven, CD player, computer, timing light, or other electronic device. Evaluate all or part of the document according to the criteria for instructions in this chapter.

   Write a memo report on your findings and send it, along with a copy of the instructions, to Natalie Bern. As a technical writer at the company that produced the electronic device, Natalie wrote the set of instructions. In your position as Natalie's supervisor, you are responsible for evaluating her work. Use your memo report either to compliment her on the instructions or to suggest modifications.

8. **User Test of Instructions.** Find a relatively simple set of instructions. Then ask another person to follow the instructions from beginning to end. Observe the person's activity, keeping notes on any problems she or he encounters.

   Use your notes to summarize the effectiveness of the instructions. Present your summary as a memo report to Natalie Bern, using the same situational context as described in Assignment 7. That is, as Natalie's boss, you are to give her your evaluation of her efforts to produce the set of instructions.

9. **Writing Simple Instructions.** Choose a simple office procedure of 20 or fewer steps (for example, changing a printer ribbon, filling a mechanical pencil, adding dry ink to a copy machine, or adding paper to a laser printer). Then write a simple set of instructions for this process, in the form of a memo report. Your readers are assistants at the many offices of a large national firm. Consider them to be new employees who have no background or experience in office work and no education beyond high school. You are responsible for their training.

10. **Writing Complex Instructions, With Graphics—Group Project.** Complete this assignment as a group project (see the guidelines for group work in chapter 1). Choose a process connected with college life or courses—for example, completing a lab experiment, doing a field test, designing a model, writing a research paper, getting a parking sticker, paying fees, registering for classes, etc.

    Using memo report format, write a set of instructions for students who have never performed this task. Follow all the guidelines in this chapter. Include at least one illustration (along with warnings or cautions, if appropriate). If possible, conduct a user test before completing the final draft.

11. **Writing Instructions—McDuff Context.** McDuff does a good deal of environmental work around the country—cleaning up toxic-waste sites, building energy-efficient structures, removing asbestos from old buildings, and investigating construction sites to determine the most environmentally sound approach to design and construction. For business reasons—and also because of its sense of civic duty—the company encourages citizens to get directly involved in environmental action.

    As public relations manager for McDuff, you have just received an interesting assignment from the president, Rob McDuff. He wants you to prepare a set of instructions that will go out to citizen and school groups in the Baltimore-Washington area. In the form of a memo report, this document should give readers specific directions for recycling one or more types of waste. Your instructions should be directed toward a broad audience, of course. Moreover, they should give the kinds of details that allow someone to act without having to get more information.

    To get information for this report, you might consider (1) calling individuals in the waste-management department of your local government, (2) reading relevant articles from recent periodicals, or (3) checking an environmental science textbook at your college.

Mc Duff, Inc.

MEMORANDUM

DATE: May 29, 1993
TO: Leonard Schwartz
FROM: Your Name
SUBJECT: New Electronic-Mail System

Yesterday I met with Jane Ansel, the installation manager at BHG Electronics, about our new electronic-mail system. Ms. Ansel explained the process by which the system will be installed. As you requested, this memo summarizes what I learned about that process.

BHG technicians will be at our offices on June 18 to complete these five tasks:

1. Removing old cable from the building conduits
2. Laying cable to link remaining unconnected terminals with the central processing unit in the main frame
3. Installing software in the system that will give each terminal the capacity to operate the electronic-mail system
4. Testing each terminal to make sure that the system can operate from that location.
5. Instructing selected managers on the use of the system

As you and I have agreed, next week I will send a memo to all office employees who will have access to electronic mail. That memo will mention the installation date and summarize the procedures for reading mail. Shortly thereafter, I will send another memo instructing them about sending electronic mail.

Let me know, Leonard, if you have further suggestions about how I can help make our transition to electronic mail as smooth as possible.

MODEL 6–1
McDuff process description: electronic mail

Mc Duff, Inc.

MEMORANDUM

DATE: June 5, 1993
TO: All Employees with Computer Terminals
FROM: Your Name
SUBJECT: Basic Instructions for Reading Electronic Mail

Last month, you attended a brief seminar on the features of the new electronic-mail system. We have just learned that the system will be installed on June 18. This memo provides some basic instructions for reading mail sent to you on this system. Soon you will receive another set of instructions for sending electronic mail.

NOTE: In these instructions, the messages or prompts on your terminal screen appear in *italics*. Any key you push or response you type is shown in **bold** print.

1. Turn on the computer terminal.
   NOTE: The on/off button is on the right front corner of the unit.

2. Type in your terminal's number when the system requests it.
   terminal number: **23**

3. Respond with your initials when the system asks for "*login*."
   login: **wsp**

4. Give your password at the next system prompt.
   password: **Tex**
   RESULT: After the system has verified your password, it will respond with one of two messages: either *no mail* or *yes, you have mail*.

5. Respond to the message in one of these two ways:
   • If screen reads *no mail*, press "**e**" (for "**exit**") and begin another task on the terminal.
   • If screen reads *yes, you have mail*, press "**r**" (for "**read**") and continue with these instructions.

6. Read the first screen of your message.

7. Press the "**return**" key to discover whether there are additional screens with that message.

8. Press the "**backspace**" key if you wish to return to other screens in that message.

MODEL 6–2
McDuff instructions: electronic mail

Sidebar annotations (left margin):

Gives clear purpose.

Indicates what instructions *do* and *do not* cover.

Provides information to help reader understand instructions that follow.

Limits each step to *one* action.

Separates action from results.

Explains options clearly.

Memo to: Employees with Computer Terminals

9. Enter one of the abbreviations from the following list when you are finished with a message:

Uses list to
show choices.

    s = save message
    d = delete message
    p = print message

10. Press "**m**" (for "**move**") to move on to the next message.

11. Press "**e**" (for "**exit**") when the system indicates no more messages.

12. Turn off the machine or begin another activity with the terminal.

Restates impor-
tant date and
shows reader
how to get more
information.

    As noted earlier, the system will be installed on June 18. Feel free to call me if you have any questions about the installation or the instructions for reading messages.

MODEL 6–2, *continued*

APPENDIX A: ON-SITE MONITORING

The purpose of monitoring the air is to determine the level of protective equipment needed for each day's work. This appendix gives an overview of the process for monitoring on-site air quality each day. Besides describing the main parts of the process, it notes other relevant information to be recorded and the manner in which data will be logged.

EQUIPMENT
This process requires the following equipment:

- Organic vapor analyzers (OVA)
- Combustible-gas instruments
- Personal sampling devices

PROCESS
The project manager at the site is responsible for supervising the technician who performs the air-quality tests. At the start of every day, a technician uses an OVA to check the quality of air at selected locations around the site. Throughout the workday (at times specified by the project manager), the technician monitors the air with combustible-gas instruments and personal sampling devices. This monitoring takes place at these locations:

1. Around the perimeter of the site
2. Downwind of the site (to determine the extent of migration of vapors and gases)
3. Generally throughout the site
4. At active work locations within the site

Then at the end of every workday, the technician uses the OVA to monitor the site for organic vapors and gases.

CONCLUSION
Besides the air-quality data, the following information is collected by the technician at each sampling time: percent relative humidity, wind direction and speed, temperature, and atmospheric pressure. The project manager keeps records of air quality and weather conditions in dated entries in a bound log.

MODEL 6–3
Process description

Steps 1 and 2
are shown in top
center portion of
flowchart.

Steps 3 and 4
are shown in left
and right por-
tions of flow-
chart, respec-
tively.

Step 5 is shown
in bottom center
portion of flow-
chart.

Flowchart shows
relationship
among steps
occurring at the
same time.

COMBINED SITE INVESTIGATION

In helping to select the site for an offshore oil platform, McDuff recommends a combined site investigation. This approach achieves the best results by integrating sophisticated geophysical work with traditional engineering activities.

As the accompanying flowchart shows, a combined site investigation consists of these main steps:

1. Planning the program, with McDuff's scientists and engineers and the client's representatives
2. Reviewing existing data
3. Completing a high-resolution geophysical survey of the site, followed by a preliminary analysis of the data
4. Collecting, testing, and analyzing soil samples
5. Combining geophysical and engineering information into one final report for the client

The report from this combined study will show how geological conditions at the site may affect the planned offshore oil platform.

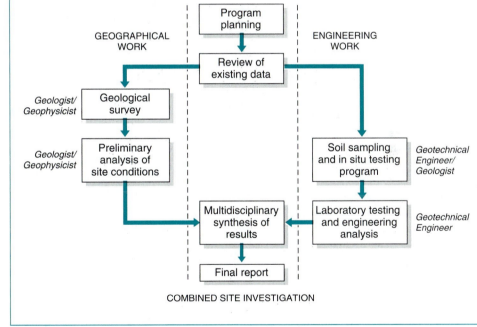

COMBINED SITE INVESTIGATION

MODEL 6–4
A McDuff process description with a flowchart (both are included in an appendix to a report to a client.)

MAKING TRAVEL ARRANGEMENTS
(Original Version)

When you're making travel arrangements, ask the person taking the trip to give you most of the details needed—dates, destinations, flight numbers, flight times, hotel requirements, rental car requirements, purpose of trip, and account number. Before proceeding, the first thing I do is confirm the flight information in the Official Airline Guide (OAG). You'll find the OAG on top of the credenza beside my typewriter. The next step is to call Turner Travel (566-0998). Although I've had great luck with all the people there, ask for Bonnie or Charlie—these two are most familiar with our firm. Turner Travel will handle reservations for flights, hotels, and rental cars. Remind them that we always use Avis midsize cars.

After you have confirmed the reservations information, fill out the McDuff travel form. Here's where you need to know the purpose of the trip and the traveler's McDuff account number. Blank forms are in the top drawer of my file cabinet in the folder labeled "Travel Forms—Blank." Once the form is complete, file the original in my "Travel Forms—Completed" folder, also in the top drawer of the file cabinet. Give the copy to the person taking the trip.

When you get the ticket in the mail from Turner Travel, check the flight information against the completed travel form. If everything checks out, give the ticket to the traveler. If there are errors, call Turner.

Also, when making any reservations for visitors to our office, call either the Warner Inn (566-7888) or the Hasker Hotel (567-9000). We have company accounts there, and they will bill us directly.

MODEL 6–5
Instructions for making travel arrangements. (McDuff's departed Baltimore travel coordinator left a narrative description of the procedure he followed [original version], which was then reformatted and edited [revised version].)

MAKING TRAVEL ARRANGEMENTS
(Revised Version)

Arranging Travel for Employees
To make travel arrangements for employees, follow these instructions:

Step	Action

1. Obtain the following information from the traveler:

   a. Dates
   b. Destinations
   c. Flight numbers
   d. Flight times
   e. Hotel requirements
   f. Rental car requirements
   g. Purpose of trip
   h. Account number

2. Confirm flight information in the Official Airline Guide (OAG).
   Note: The OAG is on the credenza beside my typewriter.

3. Call Turner Travel (566-0998) to make reservations.
   Note: Ask for Bonnie or Charlie.
   Note: For car rental, use Avis midsize cars.

4. Complete the McDuff travel form.
   Note: Blank forms are in the folder labeled "Travel Forms—Blank," in the top drawer of my file cabinet.

5. Make one copy of the completed travel form.

6. Place the original form in the folder labeled "Travel Forms—Completed," in the top drawer of my file cabinet.

7. Send the copy to the person taking the trip.

8. Check the ticket and the completed travel form after the ticket arrives from Turner Travel.

9. Do the ticket and the completed travel form agree?

   a. If *yes*, give the ticket to the traveler.
   b. If *no*, call Turner Travel.

Arrange Hotel Reservations for Visitors
   To make reservations for visitors, call the Warner Inn (566-7888) or the Hasker Hotel (567-9000). McDuff has company accounts there, and they will bill us.

Left margin annotations:

Action steps all begin with "command" form of verb.

Letters are used to show long list of subpoints, for easy reference.

Notes are used to provide reader with *extra* information, separate from action of steps.

Though closely related, Steps 5–7 are best separated—for convenient reference by reader.

As noted in Guideline 8, two subpoints can show reader the *options* that exist.

MODEL 6–5, *continued*

Mc Duff, Inc.

MEMORANDUM

DATE: June 23, 1993
TO: Employees Receiving New Message Recorders
FROM: Your Name, Purchasing Agent
SUBJECT: Instructions for New Message Recorders

INTRODUCTORY SUMMARY

We have just received the new phone message recorder you ordered. After processing, it will be delivered to your office within the week. The machine is one of the best on the market, but the instructions that accompany it are somewhat hard to follow. To help you begin using the recorder as soon as possible, I have simplified the instructions for setting up and operating the machine.

Listed here are the seven easy steps you need to follow. Also, I've attached an illustration that labels the machine's parts.

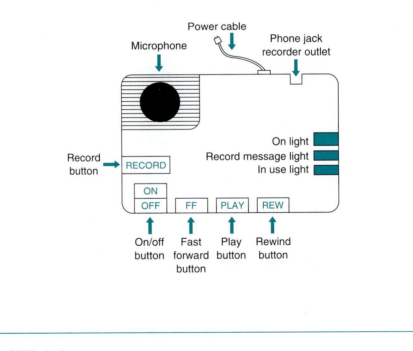

Abstract infor-
mation in ABC
format places
instructions in a
context.

Reader receives
purpose and
overview infor-
mation, along
with reference to
attachment.

MODEL 6–6
McDuff memo containing how-to instructions for a phone message recorder (rewritten
from manufacturer's difficult-to-follow instructions)

Memo to: Employees receiving new message recorders
Page 2

SETTING UP AND USING YOUR NEW RECORDER

If you devote about 15 minutes to these seven tasks, you can learn to operate your new message recorder.

1. Hooking Up Your Recorder
   a. Plug the recorder into any wall outlet using the *power cable.*
   b. Plug the *phone jack* into the *recorder outlet* located at the back of the unit.

2. Preparing Your Message
   a. Write down your:
      Greeting
      Name and department
      Time of return
   b. Write down the request you want to leave on the machine.
      NOTE: A sample request might be as follows: "Please leave your name, phone
                   number, and a brief message after you hear the tone."

3. Recording Your Message
   a. Press the *ON* button.
      RESULT: The red *on* light will come on.
   b. Press and hold down the *RECORD* button, and keep holding it down for the entire
      time you record.
      RESULT: While the button is held down, the red *record/message* light will come on.
   c. Record your message directly into the *microphone.*
   d. Release the *RECORD* button when you are finished recording.
   e. Do you need to record the message again?
      If *yes*, repeat steps a through d (your previous message will be erased each time you
      record).
      If *no*, go on to the next step.

4. Turning On and Testing Your Recorder
   a. Press the *ON* button.
      RESULT: The red *on* light will come on.
   b. Press the *PLAY* button.
      RESULT: The red *in use* light will come on.
   c. Call in your message from another phone to make sure the unit is recording properly.

5. Playing Back Messages
   a. Look at the red *record/message* light to see if it is blinking.
      NOTE: The number of times the light blinks in succession indicates the number of
      messages you have received.

MODEL 6–6, *continued*

continues

b. Press the *REW* (rewind) button.
NOTE: The tape will stop automatically when it is completely rewound.

> **CAUTION: Do not press *play* and *FF* (Fast Forward) at the same time! Doing so will break the tape. See the manufacturer's manual for process of replacing broken tape.**

Because of potential damage to equipment, "caution" appears *before* steps to which it applies.

c. Press *play* and listen to the messages.
d. Do you want to replay messages?
   If *yes*, repeat steps b and c.
   If *no*, go on to next step.
e. Do you want to skip ahead to other messages?
   If *yes*, push the *FF* (fast forward) button.
   If *no*, go on to next step.

6. **Erasing Received Messages**
   a. Press and continue holding down the *PLAY* and *REW* (rewind) buttons at the same time.
   b. Release the buttons when you hear a click.
      NOTE: The tape automatically stops when the messages are erased.

7. **Turning Off Your Recorder**
   a. Press the *OFF* button.
      RESULT: All red lights will go off.

CONCLUSION

Conclusion of ABC format wraps up memo by telling readers what to do if they encounter problems.

These new recorders are fully guaranteed for three years, so please report any problems right away. Paul Hansey (ext. 765) will be glad to help fix the machine or return the machine to the manufacturer for repair. In particular, you need to report these problems to Paul:

• Lost or incomplete messages
• Interference or noise on line
• Faulty equipment
• Inability to record

MODEL 6–6, *continued*

170

7 | *Letters and Memos*

Good letters and memos are related to good oral communication with customers and colleagues. Both forms of discourse benefit from natural, clear, and tactful language.

*M*arie Stargill, McDuff's fire-science expert, just returned from a seminar that emphasized new techniques for preventing injuries from job-site fires. Within 24 hours of her return, she has already done three things:

1. Written her manager an electronic mail (e-mail) message over the office computer network
2. Sent a letter to a McDuff client suggesting use of fire-retardant gloves she learned about at the seminar
3. Sent the conference director a letter of appreciation about the meeting

Like Marie, you will write many letters and memos in your career. In fact, you probably will write more of them than any other type of document.

Both letters and memos are short documents written to accomplish a limited purpose. They are alike in most respects except one: Letters are directed outside your organization, whereas memos are directed within your organization. (Longer, more complicated letters and memos—called "letter reports" and "memo reports"—are covered in chapter 8.) Here are some working definitions:

> **Letter:** document that conveys information to a member of one organization from someone outside that same organization. Also called "correspondence," letters usually cover one major point and go on one page. This chapter classifies letters into these four groups, according to type of message: (1) positive, (2) negative, (3) neutral, and (4) sales.

> **Memorandum:** document written from a member of an organization to one or more members of the same organization. Abbreviated "memo," it usually covers just one main point and no more than a few. Readers prefer one-page memos.

As with other forms of technical writing, your ability to write good memos and letters depends on a clear sense of purpose, thorough understanding of reader needs, and close attention to correct formats. This chapter prepares you for this challenge by presenting sections that cover (1) general rules that apply to both letters and memos and (2) specific formats for positive letters, negative letters, neutral letters, sales letters, and memoranda (printed memos and e-mail). Job letters and resumes are discussed in a separate chapter on the job search (chapter 14).

GENERAL GUIDELINES FOR LETTERS AND MEMOS

Letters convey your message to readers *outside* your organization, just as memos are an effective way to get things done *within* your own organization. By applying the guidelines in this chapter, you can master the craft of writing good letters and memos. You need to plan, draft, and revise each letter and memo as if your job depends on it—for it may.

Refer to Models 7–1 and 7–2 on pages 194 and 195–196 for McDuff examples that demonstrate the guidelines that follow. Later examples in this chapter show specific types of letters and additional memos.

■ *Letter/Memo Guideline 1: Know Your Purpose*

Before beginning your draft, write down your purpose in one clear sentence. This approach forces you to sift through details to find a main reason for writing every letter or memo. This ''purpose sentence'' often becomes one of the first sentences in the document. Here are some samples:

- **Letter purpose sentence:** ''As you requested yesterday, I'm sending samples of the new candy brands you are considering placing in McDuff's office vending machines.''
- **Memo purpose sentence:** ''This memo will explain McDuff's new policy for selecting rental cars on business trips.''

Some purpose statements are implied; others are stated. An implied purpose statement occurs in the second paragraph of Model 7–1 (p. 194). That paragraph shows that the writer wishes both to respond to requests for McDuff brochures and, just as important, to seek the professor's help in soliciting good graduates for McDuff's Atlanta office. In a sense, one purpose leads into the other. In Model 7–2 (pp. 195–196), you will find a more obvious purpose statement in the second sentence.

■ *Letter/Memo Guideline 2: Know Your Readers*

Whom are you trying to inform or influence? The answer to this question affects the vocabulary you choose, the arguments you make, and the tone you adopt. Pay particular attention when a letter or memo will be read by more than one person. If these readers are from different technical levels or different levels within an organization, the challenge increases. A complex audience compels you to either (1) reduce the level of technicality to that which can be understood by all readers or (2) write different parts of the document for different readers.

Model 7–1 (p. 194) is directed to a professor with whom the writer wants to develop a reciprocal relationship—that is, George Lux gives free guest lectures in civil-engineering classes, hoping the professor in turn will help him inform potential job applicants about McDuff. Model 7–2 (pp. 195–196), directed to an in-house technical audience, contains fairly general information about the new technical editor. This information would apply to, and be understood by, all readers.

■ *Letter/Memo Guideline 3: Follow Correct Format*

Most organizations adopt letter and memo formats that must be used uniformly by all employees. Here are the basic guidelines:

- **Letters:** There are three main letter formats—block, modified block, and simplified. Models 7–3, 7–4, and 7–5 on pages 197–199 show the basic page design of each; letter examples throughout the chapter use the three formats. As noted, you usually follow the preferred format of your own organization. Another, less common approach is to write letters in the style preferred by your reader—*if* you can discover that preference and *if* you think such a strategy might give you the persuasive edge.
- **Memos:** With minor variations, all memos look much the same. The obligatory ''date/to/from/subject'' information hangs at the top left margin, in whatever order your organization requires. These four lines allow you to dispense with lengthy introductory passages seen in more formal documents.

    Give the *subject line* special attention, for it telegraphs meaning to the audience immediately. In fact, readers use it to decide when, or if, they will read the complete memo. Be brief but also engage interest. For example, the subject line of the Model 7–2 memo could have been ''Editing.'' Yet that brevity would have sacrificed reader interest. The actual subject line, ''New Employee to Help with Technical Editing,'' conveys more information and shows readers that the contents of the memo will make their lives easier.

For both letters and memos, the recipient's name and the page number should appear on all pages after the first. (Some organizations prefer using both sender and recipient names, as in ''Jones to Bingham, 2.'')

■ *Letter/Memo Guideline 4: Follow the ABC Format for All Letters and Memos*

Letters and memos subscribe to the same three-part ABC (**A**bstract/**B**ody/**C**onclusion) format used throughout this book. This approach responds to each reader's need to know ''What does this document have to do with me?'' According to the ABC format, your letter or memo comprises these three main sections:

- **Abstract:** The abstract introduces the purpose and usually gives a summary of main points to follow. It includes one or two short paragraphs.
- **Body:** The body contains supporting details and thus makes up the largest part of a letter or memo. You can help your readers by using techniques like these:

Deductive patterns for paragraphs: In this general-to-specific plan, your first sentence should state the point that will help the reader understand the rest of the paragraph. This pattern avoids burying important points in the middle or end of the paragraph, where they might be missed. Fast readers tend to focus on paragraph beginnings and expect to find crucial information there. Note how most paragraphs in Model 7–2 (pp. 195–196) follow this format.

Personal names: If they know you, readers like to see their names in the body of the letter or memo. Your effort here shows concern for the reader's perspective, gives the letter a personal touch, and helps strengthen your personal relationship with the reader. (See the last paragraph in Model 7–1, page 194.) Of course, the same technique in direct mail can sometimes backfire, since it is an obvious ploy to create an artificially personal relationship.

Lists that break up the text: Listed points are a good strategy for highlighting details. Readers are especially attracted to groupings of three items, which create a certain rhythm, attract attention, and encourage recall. Use bullets, numbers, dashes, or other typographical techniques to signal the listed items. For example, the bulleted list in Model 7–1 (p. 194) draws attention to three important points about McDuff that the writer wants to emphasize.

Strongest points first or last: If your letter or memo presents support or makes an argument, include the most important points at the beginning and/or at the end—not in the middle. For example, Model 7–2 begins and ends with two crucial issues: the effect of poor editing on company productivity and the need to decide upon specific ways the new editor can improve McDuff writing.

Headings to divide information: Even one-page letters and memos sometimes benefit from the emphasis achieved by headings. The three headings in Model 7–2 (pp. 195–196) quickly steer the reader to main parts of the document.

■ **Conclusion:** Readers remember first what they read last. The final paragraph of your letter or memo should leave the reader with an important piece of information—for example, (1) a summary of the main idea or (2) a clear statement of what will happen next. The Model 7–1 (p. 194) letter makes an offer that will help to continue the reader's association with the university, while the Model 7–2 (pp. 195–196) memo gives readers a specific task to accomplish before the next meeting.

■ Letter/Memo Guideline 5: Use the 3Cs Strategy for Persuasive Messages

The ABC format provides a way to organize all letters and memos. Another pattern of organization for you to employ is the "3Cs strategy"—especially when your letter or memo has a persuasive objective. This strategy has three main goals:

- **Capture** interest with a good opener, which tells the reader what the letter or memo can do for him or her.
- **Convince** the reader with supporting points, all of which confirm the opening point that this document will make life easier.
- **Control** the closing, with a statement that puts you in the position of following up on the letter or memo *and* solidifies your relationship with the reader.

Although neither Model 7–1 nor Model 7–2 is overtly persuasive, each has an underlying persuasive purpose. Note how both employ the 3Cs strategy.

Letter/Memo Guideline 6: Stress the "You" Attitude

As noted earlier, using the reader's name in the body helps convey interest. But your efforts to see things from the reader's perspective must go deeper than a name reference. For example, you should perform these tasks:

- **Anticipate questions** your reader might raise and then answer these questions. You can even follow an actual question ("And how will our new testing lab help your firm?") with an answer ("Now McDuff's labs can process samples in 24 hours").
- **Replace the pronouns "I," "me," and "we" with "you" and "your."** Of course, you have to use first-person pronouns at certain points in a letter, but many pronouns should be second-person. The technique is quite simple. You can change almost any sentence from writer-focused prose ("We feel that this new service will . . .") to reader-focused prose ("You'll find that this new service will . . .").

Model 7–1 (p. 194) shows this "you" attitude by emphasizing what McDuff and the writer himself can do for the professor and his students. Model 7–2 (pp. 195–196) shows it by emphasizing that the new editor will make the readers' jobs easier.

Letter/Memo Guideline 7: Use Attachments for Details

Keep text brief by placing details in attachments, which readers can examine later, rather than bogging down the middle of the letter or memo. In this way, the supporting facts are available for future reference, without distracting readers from the main message. The memo in Model 7–2 (pp. 195–196) for example, includes a list of possible job tasks for the new McDuff editor. The listing would only clutter the body of the memo, especially since its purpose is to stimulate discussion at the next meeting.

Letter/Memo Guideline 8: Be Diplomatic

This guideline is pivotal. Without a tactful tone, all your planning and drafting will be wasted. Choose words that will persuade and cajole, not demand. Be especially careful of memos written to subordinates. If you sound too authoritarian, your message may be ignored—even if it is clear that what you are suggesting will help the readers.

For example, the letter in Model 7–1 (p. 194) would fail in its purpose if it sounded too pushy and one-sided about McDuff's interest in hiring graduates. Similarly, the editing memo in Model 7–2 (pp. 195–196) would be poorly received if it used stuffy, condescending wording such as "Be advised that starting next month, you are to make use of proofreading services provided in-house by. . . ."

■ *Letter/Memo Guideline 9: Edit Carefully*

Because letters and memos are short, editing errors may be obvious to readers. Take special care to avoid the following errors:

- **Mechanics:**
  - Misspelled words of any kind, but especially the reader's name
  - Wrong job title (call the reader's office to double-check, if necessary)
  - Old address (again, call the reader's office to check)

- **Grammar:**
  - Subject-verb and pronoun non-agreement
  - Misused commas

- **Style:**
  - Stuffy phrases such as "per your request" and "enclosed herewith"
  - Long sentences with more than one main and one dependent clause
  - Presumptuous phrases such as "Thanking you in advance for . . ."
  - Negative tone suggested by phrases such as "We cannot," "I won't," and "Please don't hesitate to"

The last point is crucial and gets more attention later in this chapter. Use the editing stage to rewrite any passage that could be phrased in a more positive tone. You must always keep the reader's goodwill, no matter what the message.

■ *Letter/Memo Guideline 10: Respond Quickly*

A letter or memo that comes too late will fail in its purpose, no matter how well written. Mail letters within 48 hours of your contact with, or request from, the reader. Send memos in plenty of time for your reader to make the appropriate adjustments in schedule, behavior, and so forth. This rule applies, for example, to letters and memos written in response to these situations:

- You want to write a follow-up letter after meeting or talking with a client.
- A customer requests information about a product or service.
- You discover that there will be a delay in your supply of a product or service to a customer.
- You select a candidate to interview for a position.
- You announce a change in company policy.
- You set the time for a company meeting.

The first sentence in the Model 7–1 (p. 194) letter, for example, shows that George Lux writes the day after his guest lecture. This responsiveness will help secure the goodwill of the professor.

SPECIFIC GUIDELINES FOR LETTERS AND MEMOS

Letters are to your clients and vendors what memos are to your colleagues. They relay information quickly and keep business flowing. This section gives you specific guidelines for these documents:

- Letters with a positive message
- Letters with a negative message
- Letters with a neutral message
- Letters with a sales message
- Memoranda

To be sure, many documents are hybrid forms that combine these patterns. As a technical sales expert for McDuff, for example, you may be writing to answer a customer question about a new piece of equipment just purchased from McDuff's Equipment Development group. Your main task is to solve a problem caused by a confusing passage in the owner's manual. At the same time, however, your concern for the customer's satisfaction can pave the way for purchase of a second machine later in the year. Thus the letter has both a positive message and a sales message. This example also points to a common thread that weaves all four letter types together: the need to maintain the reader's goodwill toward you and your organization.

The next five sections present (1) a pattern for each type of correspondence, based on the ABC (**A**bstract/**B**ody/**C**onclusion) format used throughout this text, and (2) one or more brief case studies in which the pattern might be used at McDuff.

Positive Letters

Everyone likes to give good news; fortunately, you will often be in the position of providing it when you write. Here are some sample situations:

- Replying to a question about products or services
- Acknowledging that an order has been received
- Recommending a colleague for a promotion or job
- Responding favorably to a routine request
- Responding favorably to a complaint or an adjustment
- Hiring an employee

The trick is to recognize the good-news potential of many situations. This section gives you an all-purpose format for positive letters, followed by a case study from McDuff.

ABC Format for Positive Letters.   All positive letters follow one overriding rule. You must always:

> State good news immediately!

Any delay gives readers the chance to wonder whether the news will be good or bad, thus causing momentary confusion. Here is a complete outline for positive letters that corresponds to the ABC format:

ABC Format: Positive Letter
Abstract
- Bridge between this letter and last communication with person
- Clear statement of good news you have to report

Body
- Supporting data for main point mentioned in abstract
- Clarification of any questions reader may have
- Qualification, if any, of the good news

Conclusion
- Statement of eagerness to continue relationship, complete project, etc.
- Clear statement, if appropriate, of what step should come next

---

McDuff Case Study for a Positive Letter.   As a project manager at McDuff's Houston office, Nancy Slade has agreed to complete a foundation investigation for a large church about 300 miles away. There are cracks in the basement floor slab and doors that do not close, so her crew will need a day to analyze the problem (observing the site, measuring walls, digging soil borings, taking samples, and so on). She took this small job on the condition that she could schedule it around several larger (and more profitable) projects in the same area during mid-August.

Yesterday Nancy received a letter from the minister (speaking for the church committee), who requested that McDuff change the date. He was just asked by the regional headquarters to host a three-day conference at the church during the same time that McDuff was originally scheduled to complete the project.

After checking her project schedule, Nancy determines that she can reschedule the church job. Model 7–6 on page 200 shows her response to the minister.

Negative Letters

It would be nice if all your letters could be as positive as those just described. Unfortunately, the real world does not work like that. You will have many opportunities to display both tact and clarity in relating negative information. Here are a few cases:

- Explaining delays in projects or delivery of services
- Declining invitations or requests
- Registering complaints about products or services
- Refusing to make adjustments based on complaints

- Denying credit
- Giving bad news about employment or performance
- Explaining changes from original orders

This section gives you a format to follow in writing sensitive letters with negative information. Then it provides one application at McDuff.

ABC Format for Negative Letters.   One main rule applies to all negative letters:

> ### Buffer the bad news, but still be clear.

Despite the bad news, you want to keep the reader's goodwill. Spend time at the beginning building your relationship with the reader by introducing less contro-versial information—*before* you zero in on the main message. Here is an overall pattern to apply in each negative letter:

ABC Format: Negative Letter
Abstract

- Bridge between your letter and previous communication
- General statement of purpose or appreciation—in an effort to find common bond or area of agreement

Body

- Strong emphasis on what *can* be done, when possible
- Buffered yet clear statement of what cannot be done, with clear statement of reasons for negative news
- Facts that support your views

Conclusion

- Closing remarks that express interest in continued association
- Statement, if appropriate, of what will happen next

McDuff Case Study for a Negative Letter.   Reread the letter situation described in the section on positive letters. Now assume that instead of being able to comply with the minister's request, the writer is unable to complete the work on another date without changing the fee. This change would be necessary because Nancy would have to send a new crew the 300 miles to the site, rather than using a crew already working on a nearby project.

Nancy knows the church is on a tight budget, but she also knows that McDuff would not be in business too long by working for free. Most important, since the church is asking for a change in the original agreement, she believes it is fair to request a change in the fee. Model 7–7 on page 201 is the letter she sends. Note her effort to buffer the negative news.

Neutral Letters

Some letters express neither positive nor negative news. They are simply the routine correspondence written every day to keep businesses and other organizations operating. Some situations follow:

- Requesting information about a product or service
- Inviting the reader to an event
- Responding to an invitation or routine request
- Placing orders

Use the following outline in writing your neutral letters. Also, refer to the McDuff examples that follow the outline.

ABC Format for Neutral Letters. Because the reader usually has no personal stake in the news, neutral letters require less emphasis on tone and tact than other types. Yet they still require careful planning. In particular, always abide by this main rule:

> Be absolutely clear about your inquiry or response.

Neutral letters operate a bit like good-news letters. You need to make your point early, without giving the reader time to wonder about your message. Neutral letters vary greatly in specific organization patterns. The "umbrella plan" suggested here emphasizes the main criterion of clarity.

ABC Format: Neutral Letters

Abstract

- Bridge or transition between letter and previous communication, if any
- Precise purpose of letter (request, invitation, response to invitation)

Body

- Details that support the purpose statement—for example,
    Description of item(s) requested
    Requirements related to the invitation
    Description of item(s) being sent

Conclusion

- Statement of appreciation
- Description of actions that should occur next

McDuff Case Studies for Neutral Letters. Letters with neutral messages get written by the hundreds each week at McDuff. Here are four situations that would require a neutral letter; items 2 and 4 provide the context for the examples in Model 7–8 on page 202 and Model 7–9 on page 203:

1. Zach Bowers, a lab assistant, writes a laboratory supply company for information about a new unbreakable beaker to use in testing.
2. Faron Abdullah, president of the Student Government Association at River College, asks representatives of McDuff's St. Louis office to attend a career fair.
3. Donna Martinich, a geologist, responds to the request of a past client for a copy of a report done three years ago.
4. Farah Linkletter, a supply assistant with McDuff's San Francisco office, orders three new transits, making sure to emphasize that one is not to include a field case.

Sales Letters

Upon hearing the term *sales letter*, some people have visions of direct-mail requests for magazine subscriptions, vacation land, or diet plans. In this text, however, sales letters mean something quite different. They name all your correspondence with a customer—from the first contact letter through the last thank-you note. This list gives you some idea of the possibilities for sales letters:

- Starting a relationship ("I'll be calling you. . . .")
- Following a phone call ("Good talking to you. . . . Can we meet to discuss your needs regarding. . . .")
- Following a meeting ("You mentioned that you could use more information . . . so here's a brochure on. . . .")
- Following completion of sale or project ("We enjoyed working with you on. . . .")
- Seeking repeat business ("I'd like to know how the new machinery has been working. . . .")

Notice that sales letters almost always work together with personal contacts, such as meetings and phone calls. Your goal is to build a continuing relationship with the customer. Consult the following outline when writing sales letters for any context; the McDuff example shows the outline in action.

ABC Format for Sales Letters. The one main rule that governs all sales letters is as follows:

> Help readers solve their problems.

Customers are interested in your product or service only insofar as it can assist them. You must engage the readers' interest by showing that you understand their needs and can help fulfill them. Here is a plan for writing a successful sales letter. Note reference to the 3Cs (Capture/Convince/Control) strategy mentioned earlier in the chapter:

ABC Format: Sales Letters
Abstract
(choose one or two to capture attention)

- Cite a surprising fact
- Announce a new product or service that client needs
- Ask a question
- Show understanding of client's problem
- Show potential for solving client's problem
- Present a testimonial
- Make a challenging claim
- Summarize results of a meeting
- Answer a question reader previously asked

Body
(choose one or two to convince the reader)

- Stress one main problem reader has concern about
- Stress one main selling point of your solution
- Emphasize what is unique about your solution
- Focus on value and quality, rather than price
- Put details in enclosures
- Briefly explain the value of any enclosures

Conclusion
(keep control of the next step in sales process)

- Leave the reader with one crucial point to remember
- Offer to call (first choice) or ask reader to call (last choice)

McDuff Case Study for a Sales Letter. McDuff provides customers with professional services and equipment, so sales letters have an important place in the firm. Barbara Feinstein is one employee who writes them almost every day. As a first-year employee with a degree in industrial hygiene, Barbara works in the newly formed asbestos-abatement group. Basically, she helps clients find out if there is any asbestos that needs to be removed from structures, recommends a plan for removal, and has the work done by another division of McDuff.

Here is one series of sales contacts that involves several letters. First, Barbara sent "cold call" sales letters to 100 schools and small businesses in the St. Louis area, suggesting that they might want to have their structures checked for unsafe levels of asbestos. The letter contained a reply card. After calling and then meeting with a number of the respondents, she sent individualized follow-up letters that answered questions that came up in discussions and provided additional information. After another series of phone calls and meetings with some of the potential customers, she negotiated contracts with five of the businesses and completed the projects. Then within a few months of completion, she sent a final letter proposing additional McDuff services and began the cycle again. Model 7–10 on page 204 provides a sample sales letter that Barbara used at the beginning of the cycle.

Memoranda

Memoranda (also called *memos*) may be the single most common type of writing in business today. You will write them to peers, subordinates, and superiors in your organization—from the first days of your career until you retire. Even if you work in an organization with "electronic mail"—that is, the capacity to send and receive messages by computer—you still will have to compose messages that convey your point with brevity, clarity, and tact. The medium may change, but not the message. This section covers both types of memoranda: (1) the traditional printed memo and (2) the less formal e-mail message.

Printed Memos.   Printed memos can contain all four types of messages discussed with respect to letters—positive, negative, neutral, and sales. Here are some situations that would require good memos in an organization.

Positive

- Announcing high bonuses for the fiscal year
- Commending an employee for performance on a project
- Informing employees about improved fringe benefits

Negative

- Reporting decreased quarterly revenues for the year
- Requesting closer attention to filling out time sheets
- Asking for volunteers to work on a holiday

Neutral

- Announcing a meeting
- Summarizing the results of a meeting with a client
- Explaining a new laboratory procedure

Sales

- Requesting funding for a training seminar
- Recommending another staff member for the proposals unit
- Suggesting changes in the performance evaluation system

Following is an ABC format for memos, along with several case studies from McDuff.

ABC Format for Printed Memoranda.   Abide by this one main rule in every memo-writing situation:

> Be clear, brief, and tactful.

Because many activities are competing for their time, readers expect information to be related as quickly and clearly as possible. Yet be sure not to sacrifice tact and sensitivity as you strive to achieve conciseness. This ABC format will help you accomplish both goals:

ABC Format: Memoranda
Abstract

- Clear statement of memo's purpose
- Outline of main parts of memo

Body

- Supporting points, with strong points at the beginning and/or end
- Frequent use of short paragraphs or listed items
- Absolute clarity about what memo has to do with reader
- Tactful presentation of any negative news
- Reference to attachments, when much detail is required

Conclusion

- Clear statement of what step should occur next
- Another effort to retain goodwill and cooperation of readers

---

McDuff Case Studies for Printed Memoranda. At McDuff, memoranda are written to and from employees at all levels. To reflect this diversity, this section describes two different contexts for writing memoranda and shows accompanying examples.

In the first context, the lead secretary at McDuff's Baltimore office has chaired an office committee to improve efficiency in using the centralized word-processing center. The committee was formed when the branch manager realized that many technical staff members had not been trained to use the center. This lack of training led to sloppy habits and loss of productivity. Rather than issue a "dictum" from his office, the branch manager established a small committee to review the problem and issue guidelines to the office staff. The memorandum in Model 7-11 on page 205 resulted from the committee's meetings.

In the second McDuff case, the St. Louis personnel director, Timothy Fu, must announce several changes in benefits to the entire staff. Some changes are good news in that they expand employee benefits, and others are bad news in that they further limit benefits. Timothy has the difficult task of imparting both types of information in one memo to a broad audience. The memorandum in Model 7-12 on page 206 is a result of his efforts.

Electronic Mail (E-mail). Though printed memoranda follow the ABC format, another form of in-house written communication is much less formal. Electronic mail—called *e-mail*, for short—allows you to communicate with others who are on your computer network. E-mail is fast, accessible, and informal. These features can increase your productivity on a project when you don't have time to write memos and when you cannot reach (or do not want to reach) someone by phone.

Chapter 1, which mentions e-mail in the context of group writing, shows how electronic mail helps you collaborate with others during the writing process—especially the planning stage. Interestingly, the e-mail medium has produced a casual writing style similar to that of handwritten notes. It even has its own set of abbreviations and shortcut language. Following is an e-mail message from one McDuff employee to another. Josh Bergen and Natalie Long are working together on a report wherein they must offer suggestions for designing an operator's control panel at a large dam. Josh has just learned about another control panel that McDuff designed and installed for a Russian nuclear power plant (see Project 6 in the color insert). Josh wrote this e-mail message to draw Natalie's attention to the related McDuff project:

DATE: September 15, 1993
TO: Natalie Long
FROM: Josh Bergen
SUBJECT: Zanger Dam Project

Natalie—I've got an idea that might save us A LOT of time on the Zanger Dam project. Check out the company project sheet on the Russian nuclear plant job done last year. Should we look at that project report in order to get started on the job?

Operators of hi-tech dams and nuke plants seem to face the same hassles—confusing displays, need to respond fast, distractions, etc. When either a dam or nuke operator messes up, there's often big-time trouble. IMO, we'd save our time—and our client's money—if we could go right to some of the technical experts used in the nuke job. At least as a starting place. Maybe we'd even make our deadline on this project ⟨grin⟩. That would be a change, considering the schedule delays this month on other jobs.

What do ya think about this idea?

This message displays some of the most common features of electronic mail. Although e-mail varies from company to company, group to group, and even person to person, the following suggestions are in wide use:

- Begin with standard "date/to/from/subject" information. The exact wording and order will depend on the particular e-mail system in your organization.
- Focus on one main subject in a message, and state it as briefly as possible. When e-mail messages get too long or complicated, they lose their usefulness and impact.
- Adopt a conversational style that resembles how you would talk to the recipient on the phone. Sentence fragments and slang are acceptable, as long as they contribute to your objectives and are in good taste.
- Use abbreviations if you know they'll be understood by the recipient. For example, the "IMO" used in the example message stands for "in my opinion" and is commonly used. Words in capitals are used with emphasis. Words within angle brackets—like "⟨grin⟩" in the example—are used to convey nonverbal communication and emotions.

While still in its infancy, e-mail will soon become a common form of written communication. The continuing computer revolution and the need for fast, informal communication action have ensured this. Our correspondents will be as varied as colleagues down the hall and clients on other continents.

CHAPTER SUMMARY

Letters and memoranda keep the machinery of business, industry, and government moving. Letters usually go to readers outside your organization, while memos go to readers inside. In both types of correspondence, abide by these rules:

1. Know your purpose.
2. Know your readers.
3. Follow correct format.
4. Follow the ABC format for all letters and memos.
5. Use the 3Cs strategy for persuasive messages.
6. Stress the "you" attitude.
7. Use attachments for details.
8. Be diplomatic.
9. Edit carefully.

Besides following these basics, you need to follow specific strategies for the four basic business letters and for memos. In letters with a positive message, the good news always goes first. In letters with a negative message, work on maintaining goodwill by placing a "buffer" before the bad news. Neutral letters, such as requests for information, should be absolutely clear in their message. Sales letters should show an interest in solving the readers' problems, more than an eagerness to sell a product or service. Memos should strive for brevity, clarity, and tact. E-mail messages give you the additional flexibility of adopting an informal, conversational style. Use e-mail when speed and informality are desired. Your relationship with both superiors and subordinates can depend in part on how well you write memoranda.

ASSIGNMENTS

Follow these general guidelines for all the assignments:

- Write brief responses (preferably one page or less).
- Print or design a letterhead when necessary.
- Use whatever letter format your instructor requires.
- Invent addresses when necessary.
- Invent any extra information you may need, but do not change the information presented here.

Part 1: Letters

1. **Positive Letter—Job Offer.** Assume that you are the personnel director for McDuff's San Francisco office. Yesterday, you and your hiring committee decided to offer a job to Ashley Tasker, one of ten recent graduates you interviewed for an entry-level position as a lab technician. Write Ashley an offer letter and indicate a starting date (in two weeks), a specific salary, and the need for her to sign and return an acceptance letter immedi-

ately. In the interview, you outlined the company's benefit plan, but you are enclosing with your letter a detailed description of fringe benefits (such as health insurance, long-term disability insurance, retirement plan, and vacation policy). Although Ashley is your first choice for this position, you will offer the job to another top candidate if Ashley is unable to start in two weeks at the salary you stated in the letter.

2. **Positive Letter—Recommendation.**

   Option A. Select one of the model resumes in chapter 14. Assume you have known the writer as his or her professor, supervisor, colleague, etc. Now write a letter of recommendation that highlights what you see as the person's high points. Keep in mind the job objective on the resume. Have a specific job and reader in mind for this exercise.

   Option B. Pair up with a class colleague and share information about your respective academic backgrounds, work experiences, and career goals. Now write a letter that recommends your counterpart for either (a) a scholarship based on grades, need, and/or some other criterion; (b) a specific job; (c) an internship in an organization related to the person's career interest; or (d) an academic award. Give yourself a simulated role that would give you firsthand knowledge of the person's background.

   Option C. Assume that a professor in a class you have already completed—and a class that you found valuable—has asked you to write a letter of recommendation for his or her promotion package at the college. You are assuming that this individual, in assembling his or her promotion materials, is permitted to include reference letters from a variety of sources such as fellow professors and former students. Your letter should only address what you know about the professor's abilities from your firsthand experience.

3. **Positive Letter—Favorable Response to Complaint.** The following letter was written in response to a complaint from the office manager at McDuff's Denver office. She wrote the manufacturer that the lunchroom toaster-oven broke down just three days after the warranty expired. Although she did not ask for a specific monetary adjustment, she did make clear her extreme dissatisfaction with the product. The manufacturer responded with the following letter. Be prepared to discuss what is right and what is wrong with the letter. Also, rewrite it using this chapter's guidelines.

   This letter is in response to your August 3 complaint about the Justrite toaster-oven you purchased about six months ago for your lunchroom at McDuff, Inc. We understand that the ''light-dark'' adjustment switch for the toaster device broke shortly after the warranty expired.

   Did you know that last year our toaster-oven was rated ''best in its class'' and ''most reliable'' by *Consumers Count* magazine? Indeed, we have received so few complaints about the product that a recent survey of selected purchasers revealed that 98.5% of first-time purchasers of our toaster-ovens are pleased that they chose our products and would buy another.

   Please double-check your toaster-oven to make sure that the switch is broken—it may just be temporarily stuck. We rarely have had customers make this specific complaint about our product. But if the switch is in need of repair, return the entire appliance to us, and we will have it repaired free of charge or have a new replacement sent to you. We stand behind our product, since the warranty period only recently expired.

   It is our sincere hope that you continue to be a satisfied customer of Justrite appliances.

4. **Negative Letter—Explanation of Project Delay.** You work for McDuff's Boston office. As project manager for the construction of a small strip shopping center, you have had delays about halfway through the project because of bad weather. Even worse, the

forecast is for another week of heavy rain. Yesterday, just when you thought nothing else could go wrong, you discovered that your concrete supplier, Atlas Concrete, has a truck drivers' strike in progress. Because you still need half the concrete for the project, you have started searching for another supplier.

Your client, an investor/developer named Tanya Lee located in a city about 200 miles away, probably will be upset by any delays in construction, whether or not they are within your control. Write her a letter in which you explain weather and concrete problems. Try to ease her concern, especially because you will want additional jobs from her in the future.

5. **Negative Letter—Declining a Request.** Assume that you work at the McDuff office that completed the Sentry Dam (see Project 1 in the color insert). Word of your good work has spread to the state director of dams. He has asked you, as manager of the Sentry project, to deliver a 20-minute speech on dam safety to the annual meeting of county engineers. Unfortunately, you have already agreed to be at a project site in another state on that day, and you cannot reschedule the site visit. Write the director of dams—who is both a former and, you hope, a future client—and decline the request. Though you know he expressly wanted you to speak, offer to send a substitute from your office.

6. **Negative Letter—Request for Prompt Payment.** Recently your McDuff office completed the project described in Project 4 included in the color insert. As indicated on the sheet, the city of Rondo was quite satisfied with your work. You billed the city within a week after the course and requested payment within 30 days of receipt of the bill, as you do for most clients. Forty-five days elapsed without payment being received, so you sent a second bill. Now it has been three months since completion of the project, and you still have not been paid. You suspect that your bill got lost in the paperwork at city hall, for there was a change of mayoral administrations shortly after you finished the project. Yet two phone calls to the city's business office have brought no satisfaction—two different assistants told you they could not find the bill and that you should rebill the city. You are steamed but would like to keep the client's goodwill, if possible. Write another letter requesting payment.

7. **Negative Letter—Change in Project Scope and Schedule.** As a marketing executive at the Cleveland office, you oversee many of the large accounts held by the office. One important account is a company that owns and operates a dozen radio and television stations throughout the Midwest. On one recent project, McDuff engineers and technicians did the foundation investigation for, and supervised construction of, a new transmitting tower for a television station in Toledo. First, your staff members completed a foundation investigation, at which time they examined the soils and rock below grade at the site. On the basis of what they learned, McDuff ordered the tower and the guy wires that connect it to the ground. Once the construction crew actually began excavating for the foundation, however, they found mud that could not support the foundation for the tower. While unfortunate, it sometimes happens that actual soil conditions cannot be predicted by the preliminary study. Because of this discovery of mud, the tower must be shifted to another location on the site. As a result, the precut guy wires will be the wrong length for the new site, requiring McDuff to order wire extenders. The extenders will arrive in two weeks, delaying placement of the tower by that much time. All other parts of the project are on schedule, so far.

Your client, Ms. Sharon West of Midwest Media Systems in Cleveland, doesn't understand much about soils and foundation work. But she does understand what construction delays mean to the profit margin of her firm's new television station as it attempts to compete with larger stations in Toledo. You need to console this important client, while informing her of this recent finding.

8. **Neutral Letter—Response to Request for Information.** As reservations clerk for the Best Central Inn in St. Louis, you just received a letter from Jerald Pelletier, an administrative secretary making arrangements for a meeting of McDuff managers from around the country. The group is considering holding its quarterly meeting in St. Louis in six months. Pelletier has asked you to send some brief information on hotel rates, conference facilities (meeting rooms), and availability. Send him some room rates for double and single rooms, and let him know that you have four conference rooms to rent out at $50 each per day. Also, tell him that at this time, the hotel rooms and conference rooms are available for the three days he mentioned.

9. **Neutral Letter—Request for Information.** As an official with the 1996 summer Olympic games in Atlanta, you are helping to select companies that will provide products and services related to Olympics construction. Recently a professional colleague of yours in Ohio remarked on the excellent work done by McDuff, Inc. McDuff monitored construction activities for the new Riverside Hospital, where your friend is executive administrator. (See Project 3 in the color insert.) Because so much Olympics construction is occurring at once, you may be interested in the services of a firm that will make sure no contractor takes shortcuts. Also, you've learned that McDuff has an Atlanta office. In your procurement role with the Olympics staff, write McDuff requesting information about the firm's construction-monitoring services.

10. **Sales Letter—to McDuff.** Select a product or service with which you are familiar because of home, work, or school experience. Now assume that you are responsible for marketing this product or service to Ms. Janis Black, purchasing agent for McDuff, Inc. In a phone conversation earlier today, she showed some initial interest in purchasing the product or service for her firm. In fact, you managed to set up a meeting for two weeks from today, after she returns from a business trip. Now you need to write a follow-up letter in which you summarize the phone conversation, confirm the meeting date and time, and offer some additional information about your product or service that will keep her interest.

   In selecting your product or service, you might want to review the information about McDuff, Inc., contained in chapter 2. Here are some sample products and services that Janis must routinely evaluate for use by the firm:

   - Cleaning supplies, such as lavatory soap and paper towels
   - Laboratory equipment, such as glass beakers and lab coats
   - Office materials, such as notepads and pencils
   - Office equipment, such as fax machines and pencil sharpeners
   - Leased products, such as automobiles for managers and salespeople

11. **Sales Letter—From McDuff.** For this assignment, choose one of the seven project sheets included with the color insert. Although you *will* be using the sheets as a source of information in writing your letter, you will *not* be sending the sheets as an attachment.
   Option A. Assume you are a McDuff employee writing to a new client. This client has shown initial interest in the service or product described in the project sheet you're using for this assignment. You've had one phone conversation with the potential client. Now you hope your letter will lay the foundation for a personal meeting. Write a sales letter that briefly describes (1) the service or product on the project sheet and/or (2) your success on the specific project.
   Option B. Assume you are a McDuff employee writing to the same client for whom you completed the project described on the sheet. You have heard there may be similar work available with this client, so you are seeking repeat business. Write your follow-up letter, referring to the success on the project and seeking information about possible future work.

Part 2: Memoranda

12. **Memo—Positive News.** As branch manager of McDuff's Atlanta office, you just learned from your accounting firm that last year's profits were even higher than previously expected. Apparently, several large construction jobs had not been counted in the first reporting of profits. You and your managers had already announced individual raises before you learned this good news. Now you want to write a memo that states that every branch employee will get a $500 across-the-board bonus, in addition to whatever individual raises have been announced for next year.

13. **Memo—Positive News.** Kevin Kehoe, an employee at the San Francisco office, is being considered for promotion to Manager of Technical Services. He has asked you to write a memo to the San Francisco branch manager on his behalf. Although you now work as a marketing expert at the corporate office, several years ago you worked directly for Kevin on the Ocean Exploration Program in California (see Project 2 in the color insert). Kevin has asked that your memo deal exclusively with his work on that program. Kevin was manager of the project; you believe that it was largely through his technical expertise, boundless energy, and organizational skills that the project was so successful. He developed the technical plan of work that led to the clear-cut set of findings. Write a memo that conveys this information to the branch manager considering Kevin for the promotion. Because the branch manager is new, he is not familiar with the project on which you and Kevin worked. Thus your memo may need to mention some details from Project 2 in the color insert.

14 **Memo—Negative News.** You are project manager of the construction-management group at McDuff's St. Paul office. The current policy in your office states that employees must pass a pre-employment drug screening before being hired. After that, there are no tests unless you or one of your job supervisors has reason to suspect that an employee is under the influence of drugs on the job.

    Lately a number of clients have strongly suggested that you should have a *random* drug-screening policy for all employees in the construction-management group. They argue that the on-the-job risk to life and property is great enough to justify this periodic testing, without warning. You have consulted your branch manager, who likes the idea. You have also talked with the company's attorney, who assures you that such random testing would be legal, given the character of the group's work. After considerable thought, you decide to implement the policy in three weeks. Write a memo to all employees of your group and relate this news.

15. **Memo—Negative News.** A year ago you introduced a pilot program to McDuff's London office, where you are branch manager. The program gave up to half the office employees the choice to work four 10-hour days each week, as opposed to five 8-hour days. You wanted to offer this flexibility to workers who, for whatever reason, desired longer weekends. As branch manager, you made it clear at the time that you would evaluate the program at the end of the one-year pilot.

    Having completed your review, you've decided that employees need to return to the old schedule. Your main reason is that having the office short of staff on Friday (when the four-day employees are gone) has proved awkward in dealing with current and prospective clients. On many occasions, clients have called to find that their company contact is not working that day. In addition, the secretarial and word-processing staff that does work on Friday cannot keep up with the end-of-week workload. Whereas you originally thought a split schedule in the office would work, now you know it causes more confusion than it's worth. People in the office are constantly forgetting who is working what schedule, though the schedule is published. So, your memo to office employees must inform them of your decision to discontinue the pilot program and to return to a five-day workweek for the whole office. The change will take place in one month.

16. **Memo—Neutral Message.** As mailroom supervisor at McDuff's Baltimore office, you have a number of changes to announce to employees of the corporate office. Write a memo that clearly relates the following information. Deliveries and pickups of mail, which currently are at 8:30 A.M. and 3:00 P.M., will change to 9:00 A.M. and 3:30 P.M., starting in two weeks. Also, there will be an additional pickup at noon on Monday, Wednesday, and Friday. The mailroom will start picking up mail to go out by Federal Express or any other one-day carrier, rather than the sender having to wait for the carrier's representative to come to the sender's office. The sender must call the mailroom to request the pickup; and the carrier must be told by the sender to go to the mailroom to pick up the package. The memo should also remind employees that the mail does not go out on federal holidays, even though the mailroom continues to pick up mail from the offices on those days.

17. **Memo—Persuasive Message.** For this assignment, choose either (1) a good reference book or textbook in your field of study or (2) an excellent periodical in your field. The book or periodical should be one that could be useful to someone working in a profession, preferably one that you may want to enter yourself.

    Now assume that you are an employee of an organization that would benefit by having this book or periodical in its staff library, customer waiting room, or perhaps as a reference book purchased for employees in your group. Write a one-page memo to your supervisor recommending the purchase. You might want to consider criteria such as those listed here:

    - Relevance of information in the source to the job
    - Level of material with respect to potential readers
    - Cost of book or periodical as compared with its value
    - Amount of probable use
    - Important features of the book or periodical (such as bibliographies or special sections)

18. **Memo—Neutral Message.** Assume you work in the public relations area at McDuff's corporate office in Baltimore. Recently your office began designing and producing company project sheets, each of which uses one page to describe a specific project completed by McDuff and provides an accompanying graphic on the page. The color insert in this textbook shows seven examples of these project sheets. Write a memo to all 15 branch managers. Give them the information that follows: Beginning next month, project sheets will be written for every McDuff job that grosses $10,000 or more. Your office will have each sheet finished within 30 days of project completion and will then send each office the sheets for projects that were coordinated by that office. Then, as time permits, the public relations office will go back to significant previous projects, like the ones in the color insert, to do additional project sheets on previous work.

19. **Memo—Persuasive Message.** Assume you work at a McDuff office and have no undergraduate degree. You are not yet sure what degree program you want to enter, but you have decided to take one night course each term. Your McDuff office has agreed to pay 100 percent of your college expenses on two conditions. First, before taking each course, you must write a memo of request to your supervisor, justifying the value of the class to your specific job or to your future work with the company. Clearly, your boss wants to know that the course has specific application or that it will form the foundation for later courses. Second, you must receive a *C* or better in every class for which you want reimbursement.

Write the persuasive memo just described. For the purposes of this assignment, choose one course that you actually have taken or are now taking. Yet in your simulated role for the assignment, write as if you have not taken the course.

20. **E-mail—Collaborative Project.** This assignment applies only if your campus offers students the use of electronic mail. Select one of the preceding memo assignments to complete as a group-writing project with two or three members of your class. Set up a plan of work that (1) involves the group in several face-to-face meetings and (2) requires each member of the group to send and receive at least one assignment-related e-mail message to and from every other member of the group.

Mc Duff, Inc.

12 Peachtree Street
Atlanta, GA 30056
(404) 555-7524

August 2, 1993

Professor Willard R. Burton, Ph.D.
Department of Civil Engineering
Southern University of Technology
Paris, GA 30007

Dear Professor Burton:

Expresses appreciation *and* provides lead-in to body.

Thanks very much for your hospitality during my visit to your class yesterday. I appreciated the interest your students showed in my presentation on stress fractures in highway bridges. Their questions were very perceptive.

Responds to question that arose at class presentation.

You may recall that several students requested some further information on McDuff, so I have enclosed a dozen brochures for any students who may be interested. As you know, job openings for civil-engineering graduates have increased markedly in the last five years. Some of the best opportunities lie in these three areas of the discipline:

Uses bulleted list to emphasize information of value to professor's students.

- Evaluation of environmental problems
- Renovation of the nation's infrastructure
- Management of construction projects

Adds unobtrusive reference to McDuff's needs.

These areas are three of McDuff's main interests. As a result, we are always searching for top-notch graduates from solid departments like yours.

Closes with offer to visit class again.

Again, I enjoyed my visit back to Southern last Friday, Professor Burton. Please call when you want additional guest lectures by me or other members of the McDuff staff.

Sincerely,

George F. Lux

George F. Lux, P.E.

Includes reference to enclosures.

Encl.: 12 McDuff brochures

MODEL 7–1
McDuff sample letter

Mc Duff, Inc.

MEMORANDUM

DATE: December 4, 1993
TO: Technical Staff
FROM: Ralph Simmons, Technical Manager RS
SUBJECT: New Employees to Help with Technical Editing

Uses informative
subject line.

Gives purpose
of memo and
highlights con-
tents.

Uses side head-
ings for easy
reading.

Shows that the
change arose
from *their* con-
cerns.

Adds evidence
from outside
observer.

Provides transi-
tion to next sec-
tion.

Gives important
information
about Ron in
first sentence.

Establishes his
credibility.

Refers to attach-
ment.

Focuses on
benefit of
change to
reader. Restates
next action to
occur.

Last week we hired an editor to help you produce top-quality reports, proposals, and other documents. This memorandum gives you some background on this change, highlights the credentials of our new editor, and explains what the change will mean to you.

BACKGROUND

At September's staff meeting, many technical staff members noted the excessive time spent editing and proofreading. For example, some of you said that this final stage of writing takes from 15-30 percent of the billable time on an average report. Most important, editing often ends up being done by project managers—the employees with the highest billable time.

Despite these editing efforts, many errors still show up in documents that go out the door. Last month I asked a professional association, the Engineers Professional Society (EPS), to evaluate McDuff–Boston documents for editorial correctness. (EPS performs this service for members on a confidential basis.) The resulting report showed that our final reports and proposals need considerable editing work. Given your comments at September's meeting and the results of the EPS peer review, I began searching for a solution.

SOLUTION: IN-HOUSE EDITOR

To come to grips with this editing problem, the office just hired Ron Perez, an experienced technical editor. He'll start work January 3. For the last six years, Ron has worked as an editor at Jones Technical Services, a Toronto firm that does work similar to ours. Before that he completed a master's degree in technical writing at Sage University in Buffalo.

At next week's staff meeting, we'll discuss the best way to use Ron's skills to help us out. For now, he will be getting to know our work by reviewing recent reports and proposals. Also, the attached list of possible activities can serve as a springboard for our discussion.

CONCLUSION

By working together with Ron, we'll be able to improve the editorial quality of our documents, free up more of your time for technical tasks, and save the client and ourselves some money.

I look forward to meeting with you next week to discuss the best use of Ron's services.

continues

MODEL 7–2
McDuff sample memo

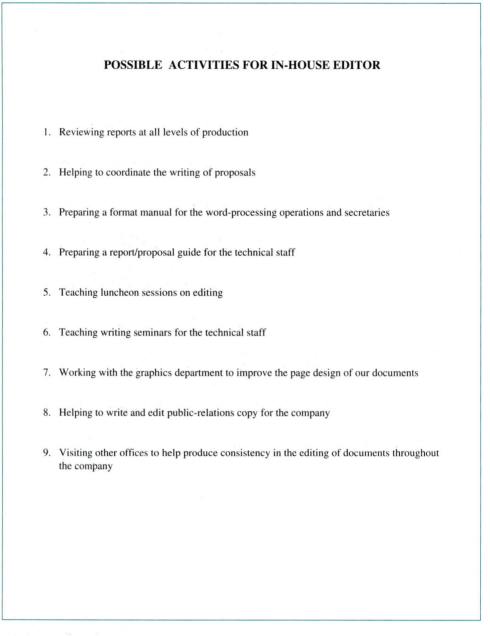

POSSIBLE ACTIVITIES FOR IN-HOUSE EDITOR

1. Reviewing reports at all levels of production

2. Helping to coordinate the writing of proposals

3. Preparing a format manual for the word-processing operations and secretaries

4. Preparing a report/proposal guide for the technical staff

5. Teaching luncheon sessions on editing

6. Teaching writing seminars for the technical staff

7. Working with the graphics department to improve the page design of our documents

8. Helping to write and edit public-relations copy for the company

9. Visiting other offices to help produce consistency in the editing of documents throughout the company

MODEL 7–2, *continued*

Letterhead of your organization

Two or more blank lines (adjust space to center letter on page)

Date of letter

Two or more blank lines (adjust space to center letter on page)

Address of reader

One blank line

Greeting

One blank line

Paragraph: single-spaced (indenting optional)

One blank line

Paragraph: single-spaced (indenting optional)

One blank line

Paragraph: single-spaced (indenting optional)

One blank line

Complimentary close

Three blank lines (for signature)

Typed name and title

One blank line

Typist's initials

Computer file # (if applicable)

One blank line

Enc.

MODEL 7-3
Block style for letters

Letterhead of your organization

Two or more blank lines (adjust space to center letter on page)

Date of letter

Two or more blank lines (adjust space to center letter on page)

Address of reader

One blank line

Greeting

One blank line

Paragraph: single-spaced, with first line indented 5 spaces

One blank line

Paragraph: single-spaced, with first line indented 5 spaces

One blank line

Paragraph: single-spaced, with first line indented 5 spaces

One blank line

Complimentary close

Three blank lines (for signature)

Typed name and title

One blank line

Typist's initials

Computer file # (if applicable)

One blank line

Enc.

MODEL 7–4
Modified block style (with indented paragraphs) for letters

Letterhead of organization

Two or more blank lines (adjust space to center letter on page)

Date of letter

Two or more blank lines (adjust space to center letter on page)

Address of reader

Three blank lines

Short subject line

Three blank lines

Paragraph: single-spaced, no indenting

One blank line

Paragraph: single-spaced, no indenting

One blank line

Paragraph: single-spaced, no indenting

Five blank lines (for signature)

Typed name and title

One blank line

Typist's initials

Computer file # (if applicable)

One blank line

Enc.

MODEL 7–5
Simplified style for letters

Mc Duff, Inc.

12 Post Street
Houston, Texas 77000
(713) 555-9781

July 23, 1993

The Reverend Mr. John C. Davidson
Maxwell Street Church
Canyon Valley, Texas 79195

Dear Reverend Davidson:

Mentions letter that prompted this response. Gives good news *immediately.*

Thanks for your letter asking to reschedule the church project from mid-August to another, more convenient time. Yes, we'll be able to do the project on one of two possible dates in September, as explained below.

Reminds reader of rationale for original schedule—*cost savings.* Offers two options—both save the church money.

As you know, McDuff originally planned to fit your foundation investigation between two other projects planned for the Canyon Valley area. In making every effort to lessen church costs, we would be saving money by having a crew already on site in your area—rather than having to charge you mobilization costs to and from Canyon Valley.

Shows McDuff's flexibility.

As it happens, we have just agreed to perform another large project in the Canyon Valley area beginning on September 18. We would be glad to schedule your project either before or after that job. Specifically, we could be at the church site for our one-day field investigation on either September 17 or September 25, whatever date you prefer.

Makes clear what should happen next.

Please call me by September 2 to let me know your scheduling preference for the project. In the meantime, have a productive and enjoyable conference at the church next month.

Sincerely,

Nancy Slade

Nancy Slade, P.E.
Project Manager

mh
File #34678

MODEL 7–6
Positive letter in block style

Mc Duff, Inc.

12 Post Street
Houston, Texas 77000
(415) 555-1381

July 23, 1993

The Reverend John C. Davidson
Maxwell Street Church
Canyon Valley, TX 79195

Dear Reverend Davidson:

Thanks for your letter asking to reschedule the foundation project at your church from mid-August to late August, because of the regional conference. I am sure you are proud that Maxwell was chosen as the site for the conference.

One of the reasons for our original schedule, as you may recall, was to save the travel costs for a project crew going back and forth between Houston and Canyon Valley. Because McDuff has several other jobs in the area, we had planned not to charge you for travel.

We can reschedule the project, as you request, to a more convenient date in late August, but the change will increase project costs from $1,500 to $1,800 to cover travel. At this point, we just don't have any other projects scheduled in your area in late August that would help defray the additional expenses. Given our low profit margin on such jobs, that additional $300 would make the difference between our firm making or losing money on the foundation investigation at your church.

I'll call you next week, Reverend Davidson, to select a new date that would be most suitable. McDuff welcomes its association with the Maxwell Street Church and looks forward to a successful project in late August.

Sincerely,

Nancy Slade

Nancy Slade, P.E.
Project Manager

mh
File #34678

Provides "bridge" and compliments Davidson on conference.

Reminds him about original agreement—in tactful manner.

Phrases negative message as *positively* as possible, giving rationale for necessary change.

Makes it clear what will happen next. Ends on *positive* note.

MODEL 7–7
Negative letter in modified block style (with indented paragraphs)

River
College

January 4, 1994

Mr. Timothy Fu, Personnel Director
McDuff, Inc.
127 Rainbow Lane
St. Louis, MO 63103

Dear Mr. Fu:

McDuff, Inc., has hired 35 graduates of River College since 1975. To help continue that tradition, we would like to invite you to the college's first Career Fair, to be held February 21, 1994, from 8 a.m. until noon.

Sponsored by the Student Government Association, the Career Fair gives juniors and seniors the opportunity to get to know more about a number of potential employers. We give special attention to organizations, like McDuff, that have already had success in hiring River College graduates. Indeed, we have already had a number of inquiries about whether your firm will be represented at the fair.

Participating in the Career Fair is simple. We will provide you with a booth where one or two McDuff representatives can talk with students that come by to ask about your firm's career opportunities. Feel free to bring along whatever brochures or other written information that would help our students learn more about McDuff's products and services.

I will call you next week, Mr. Fu, to give more details about the fair and offer a specific booth location. We at River College look forward to building on our already strong association with McDuff.

Sincerely,

Faron G. Abdulla

Faron G. Abdullah, President
Student Government Association

56 New Lane
Bolt, Missouri
65101
(314) 555-0272

MODEL 7–8
Neutral letter (invitation) in block style

Mc Duff, Inc.

345 Underwood Street
Belforth, California 90706
(713) 555-9781

April 2, 1994

Faraday Supply Company
34 State Street
San Francisco, CA 94987

ORDER FOR FIELD TRANSITS

Yesterday I called Ms. Gayle Nichols to ask what transits you had in current inventory. Having considered what you have in stock, I wish to order those listed below.

Please send us these items:

1. One Jordan #456 Transit, with special field case
2. One Smith-Beasley #101FR, with special field case
3. One Riggins #6NMG, without special field case

Note that we *do* want the special field cases with the Jordan and Smith-Beasley units, but do not want the case with the Riggins unit.

Please send the units and the bill to my attention. As always, we appreciate doing business with Faraday.

Farah Linkletter

Farah Linkletter
Supply Assistant

gh

MODEL 7–9
Neutral letter (placing order) in simplified style

Mc Duff, Inc.

127 Rainbow Lane
St. Louis, Missouri 63103
(314) 555-8175

August 21, 1994

Mr. James Swartz, Safety Director
Jessup County School System
1111 Clay Street
Smiley, MO 64607

NEW ASBESTOS-ABATEMENT SERVICE NOW AVAILABLE

We enjoyed working with you last year, James, to update your entire fire alarm system. Given the current concern in the country about another safety issue, asbestos, we wanted you to know that our staff now does abatement work.

As you know, many of the state's school systems were constructed during years when asbestos was used as a primary insulator. No one knew then, of course, that the material can cause illness and even premature death for those who work in buildings where asbestos was used in construction. Now we know that just a small portion of asbestos produces a major health hazard.

Fortunately, there's a way to tell whether you have a problem: the asbestos survey. This procedure, done by our certified asbestos-abatement professionals, results in a report that tells whether or not your buildings are affected. And if we find asbestos, we can remove it for you.

Jessup showed real foresight in modernizing its alarm system last year, James. Your desire for a thorough job on that project was matched, as you know, by the approach we take to our business. Now we'd like to help give you the peace of mind that will come from knowing that either (1) there is no asbestos problem in your 35 structures or (2) you have removed the material.

The enclosed brochure outlines our asbestos services. I'll give you a call in a few days to see whether McDuff can help you out.

Barbara Feinstein

Barbara H. Feinstein
Certified Industrial Hygienist

MODEL 7–10
Sales letter in simplified style

Mc Duff, Inc.

MEMORANDUM

DATE: August 1, 1994
TO: Technical Staff
FROM: Gini Preston, Chair, Word-Processing Committee *GP*
SUBJECT: Word-Processing Suggestions

The Word Processing Committee has met for 6 weeks to consider changes in McDuff's Word Processing Center. This memo highlights the recommendations that have been approved by management.

Please note these changes in your daily use of the company's Word Processing Center.

1. **Document status:** Documents will be designated either "rush" or "regular" status, depending on what you request. If at all possible, rush documents will be returned within 4 hours. Regular documents will be returned within one working day.
2. **Draft stages:** Both users and operators should make every effort to produce no more than three hard-copy drafts of any document. Typically, these would include:

   • **First typed draft** (typed from writer's handwritten or cut-and-paste copy)
   • **Second typed draft** (produced after user has made editing corrections on first-draft copy)
   • **Final typed draft** (produced after user makes final editing changes, after the proofreader makes a pass through the document, and after the operator incorporates final changes into the copy)

3. **New proofreader:** A company proofreader has been hired to improve the quality of our documents. This individual will have an office in the Word Processing Center and will review all documents produced by the word-processing operators.

These changes will all take effect August 15. Your efforts to implement them will help improve the efficiency of the center, the quality of your documents, and the productivity of the company.

Feel free to call me at ext. 567 if you have any questions.

MODEL 7–11
Memorandum: changes in procedures

Mc Duff, Inc.

MEMORANDUM

DATE: May 3, 1994
TO: All Employees of Cleveland Office
FROM: Timothy Fu, Personnel Director   *TF*
SUBJECT: New Cost Containment Measure for Health Care

The next fiscal year will bring several changes in the company fringe benefit plan. Later this month, you'll receive a complete report on all adjustments to go into effect July 1. For now, this memo will outline one major change in health care. Specifically, McDuff will adopt a cost-containment program called PAC—intended to help you and the company get more health care for the dollar.

WHAT IS PAC AND HOW DOES IT WORK?

Health costs have risen dramatically in the last 10 years. The immediate effect on McDuff has been major increases in insurance premiums. Both you and the company have shared this burden. This year McDuff will fight this inflationary trend by introducing a new cost-containment program called PAC—Pre-Admission Check.

Started by Healthco, our company medical supplier, PAC changes the procedure by which you and your dependents will be recommended for hospitalization. Except in emergencies, you or your physician will need to call the PAC hotline before admission to the hospital. The PAC medical staff will:

1. Review the length of stay recommended by your physician, to make sure it conforms to general practice
2. Request a second opinion if the PAC staff believes that such an opinion is warranted
3. Approve final plans for hospitalization

If your physician recommends that you stay in the hospital beyond the length originally planned, he or she will call PAC for authorization.

WILL PAC AFFECT THE LEVEL OR QUALITY OF HEALTH CARE?

No. PAC will in no way restrict your health care or increase your personal costs. Quite the contrary, it may reduce total costs considerably, leading to a stabilization of the employee contributions to premiums next year. The goal is to make sure physicians give careful scrutiny to the lengths of hospital stays, staying within the norms associated with a particular illness unless there is good reason to do otherwise.

Programs like PAC have worked well for many other firms around the country; there is a track record of lowering costs and working efficiently with physicians and hospitals. Also, you will be glad to know that Healthco has the firm support of its member physicians on this program.

WHAT WILL HAPPEN NEXT?

As mentioned earlier, this change goes into effect with the beginning of the new fiscal year on July 1. Soon you will receive a report about this and other changes in benefits. If you have any questions before that time, please call the Corporate Benefits Department at ext. 678.

MODEL 7–12
Memorandum: changes in benefits

8 | *Informal Reports*

Two McDuff scientists review the draft of a recommendation report that they have written for a client.

*A*lan Murphy, a salesperson for McDuff's St. Paul office, has a full day ahead. Besides having to make some sales calls in the morning, he must complete two short reports back in the office. The first is a short progress report to Brasstown Bearings, a company that recently hired McDuff to train its technical staff in effective sales techniques. (As noted in chapter 2, McDuff works in the field of training as well as in technical areas.) As manager of the project, Alan has overseen the efforts of three McDuff trainers for the last three weeks. According to the contract, he must send a progress report to Brasstown every three weeks during the project. Alan's second report is internal. His boss wants a short report recommending ways that McDuff can pursue more training projects like the Brasstown job.

Like Alan Murphy, you will spend much of your time writing informal reports in your career. Though short and easy to read like letters and memos, informal reports have more substance, are longer, and thus require more organization skills than correspondence. A working definition follows:

> **Informal report:** this document contains about two to five pages of text, not including attachments. It has more substance than a simple letter or memo but less than a formal report. It can be directed to readers either outside or inside your organization. If *outside*, it may be called a letter report. If *inside*, it may be called a memo report. In either case, its purpose can be informative (to clarify or explain) or persuasive (to convince) or both.

This chapter has three sections. The first one shows you when to use informal reports in your career by describing some McDuff cases. The second provides 10 main writing guidelines that apply to both letter and memo reports. The third focuses on specific suggestions for writing five common types of informal reports.

At the end of the chapter are examples with marginal annotations. They will give you specific, real-life applications of the chapter's writing guidelines. As such, the models will help you complete chapter assignments and do actual reports on the job. During your career, you will write many types of informal reports other than those presented here. If you grasp this chapter's principles, however, you can adapt to other formats.

WHEN TO USE INFORMAL REPORTS

As noted in the definition just given, informal reports are clearly distinguished from both formal reports and routine letters and memos. Early in your career, however, you may have trouble deciding exactly where to draw the line. To help you decide, the two sections that follow briefly describe situations in which informal reports would be appropriate at McDuff.

Letter Reports at McDuff

Written to people outside your organization, letter reports use the format of a business letter because of their brevity. Yet they include more detail than a simple business letter. Here are some sample projects at McDuff that would require letter reports:

- **Training recommendation:** McDuff's corporate training staff recommends changes in the training program of a large construction company. Courses that are recommended include technical writing, interpersonal communication, and quality management.
- **Seafloor study:** McDuff's Nairobi staff writes a preliminary report on the stability of the seafloor where an oil rig might be located off the coast of Africa. This preliminary study includes only a survey of information on file about the site. The final report, involving fieldwork, will be longer and more formal.
- **Marketing report:** A marketing specialist at the corporate office completes a study on "New Markets in the Year 2000." The report has been solicited by a professional marketing association to which McDuff belongs.
- **Asbestos project:** McDuff's Houston staff reports to a suburban school board about possible asbestos contamination of an old elementary school. After two days on site, the crew of two technicians determined that the structure had no asbestos in its walls, plumbing, floors, or storerooms.
- **Environmental study:** McDuff's San Francisco staff reports to the local Sierra Club chapter on possible environmental effects of an entertainment park, proposed for a rural area where eagles often nest. The project involved one site visit, interviews with a biologist, and some brief library research.
- **Equipment design project:** McDuff's equipment-development staff reports to a manufacturer on tentative designs for a computer-controlled device to cut plastic drainage pipe. The project involved several days' drafting work.

As these examples show, letter reports are the best format for projects with a limited scope. Also, this informal format is a good sales strategy when dealing with customers greatly concerned about the cost of your work. When reading letter reports, they realize—consciously or subconsciously—that these documents cost them less money than formal reports. Your use of letter reports for small jobs shows a sensitivity to their budget and may help gain their repeat work. See Model 8–1 on pages 230–231 for a letter report based on a small project at McDuff.

Memo Reports at McDuff

Memo reports are the informal reports that go back and forth among McDuff's own employees. Though in memorandum format, they include more technical detail and are longer than routine memos. These situations at McDuff show the varied contexts of memo reports:

- **Need for testing equipment:** Joan Watson, a lab technician in the Denver office, evaluates a new piece of chemical testing equipment for her department manager, Wes Powell. Powell discusses the report with his manager.
- **Personnel problem:** Werner Hoffman, a field engineer in the Munich office, writes to his project manager, Hans Schulman, about disciplinary problems with a field hand. Hoffman discusses the report with his manager and with the personnel manager.
- **Need for drafting tables:** Susan Gindle, an equipment-development technician in Baltimore, writes a report to the equipment-development manager, Ralph Peak. Gindle recommends the company purchase five drafting tables from Simulon, Inc., as opposed to similar tables from Sonet, Inc. Peak will discuss the report with the company's vice president for research and with the finance officer.
- **Progress in hiring minorities:** Scott Sampson, personnel manager, reports to Lynn Redmond, vice president of human resources, on the company's initial efforts to hire more minorities. Redmond will discuss Sampson's progress report with the company president and with all office managers.
- **Report on training session:** Pamela Martin, a field engineer in St. Louis, reports to her project manager, Mel Baron, on a one-week course she took in Omaha on new techniques for removing asbestos from buildings. Baron circulates the report to his office manager, Ramsey Pitt. Then Pitt sends copies to the manager of every company office, since asbestos projects are becoming common throughout the firm.

These five reports would require enough detail to justify writing memo reports, rather than simple memos. As for audience, each report would go directly to, or at least be discussed with, readers at *high* levels within the company. That means good memo reports can help advance your career. Model 8–2 on pages 232–233 provides an annotated example of a memo report about proposed computer software at McDuff.

GENERAL GUIDELINES FOR INFORMAL REPORTS

The following are 10 guidelines that focus mainly on report format.

■ *Informal Report Guideline 1:*
Plan Well Before You Write

Like other chapters in this book, this section emphasizes the importance of the planning process. Complete the Planning Form at the end of the book for each assignment in this chapter, as well as for informal reports you write in your career. Before you begin writing a draft, use the Planning Form to record specific information about these points:

- The document's purpose
- The variety of readers who will receive the document
- The needs and expectations of readers, particularly decision-makers
- An outline of the main points to be covered in the body

■ *Informal Report Guideline 2:*
Use Letter or Memo Format

Model 8–1 (pp. 230–231) shows that letter reports follow about the same format as typical business letters (see chapter 7). For example, both are produced on letterhead and both often include your last name, the reader's last name, and the page number on all pages after the first. Yet the format of letter reports differs from that of letters in these respects:

- The greeting is sometimes left out or replaced by an attention line, especially when your letter report will go to many readers in an organization.
- A report title often comes immediately after the inside address. It identifies the specific project covered in the report. You may have to use several lines because the project title should be described fully, in the same words that the reader would use.
- Spacing between lines might be single, one-and-one-half, or double, depending on the reader's preference.

Model 8–2 (pp. 232–233) shows the typical format for a memo report. Like most memos, it includes "date/to/from/subject" information at the top and has the writer's last name (optional), reader's last name, and page number on every page after the first. Also, both memos and memo reports have a subject line that should engage interest, give readers their first quick look at your topic, and be both specific and concise—for example, "Fracture Problems with Molds 43-D and 42-G" is preferable to "Problems with Molds."

There are, however, some format differences between memos and memo reports. Memo reports are longer and tend to contain more headings than routine memos. Also, spacing between lines varies from company to company, though one-and-one-half or double spacing is most common.

■ *Informal Report Guideline 3: Make Text Visually Appealing*

Your letter or memo report must compete with other documents for each reader's attention. Here are three visual devices that help get attention, maintain interest, and highlight important information:

- Bulleted points, for short lists like this one
- Numbered points, for lists that are longer or that include a list of ordered steps
- Frequent use of headings and subheadings

Headings are particularly useful in memo and letter reports. As Models 8—1 and 8—2 (pp. 230—233) show, they give readers much-needed visual breaks. Since informal reports have no table of contents, headings also help readers locate information quickly. (Chapter 4 gives more detail on headings and other features of page design.)

■ *Informal Report Guideline 4: Use the ABC Format for Organization*

Headings and lists attract attention, but these alone will not keep readers interested. You also need to organize information effectively. Most technical documents, including informal reports, follow what this book calls the ABC format. This approach to organization includes three parts: (1) **A**bstract, (2) **B**ody, and (3) **C**onclusion.

- **Abstract:** Start with a capsule version of the information most needed by decision-makers.
- **Body:** Give details in the body of the report, where technical readers are most likely to linger a while to examine supporting evidence.
- **Conclusion:** Reserve the end of the report for a description or list of findings, conclusions, or recommendations.

"Abstract," "body," and "conclusion" are only generic terms. They indicate the *types* of information included at the beginning, middle, and end of your reports—*not* necessarily the exact headings you will use. The next three guidelines give details on the ABC format as applied to memo and letter reports.

■ *Informal Report Guideline 5: Call the Abstract an Introductory Summary*

Abstracts should give readers a summary, the "big picture." This text suggests that in informal reports, you label this overview an "Introductory Summary," a term that gives the reader a good idea of what the section contains. (You do have the option of leaving off a heading label. In this case, your first few paragraphs would contain the introductory summary information, followed by the first body heading of the report.)

In letter reports, the introductory summary comes immediately after the title. In memo reports, it comes after the subject line. Note that informal reports do not require long, drawn-out beginnings; just one or two paragraphs in this first section give readers three essential pieces of information:

1. **Purpose** for the report—why are you writing it?
2. **Scope** statement—what range of information does the report contain?
3. **Summary** of essentials—what main information does the reader most want or need to know?

■ *Informal Report Guideline 6: Put Important Details in the Body*

The body section provides details needed to expand upon the outline presented in the introductory summary. If your report goes to a diverse audience, managers often read the quick overview in the introductory summary and then skip to conclusions and recommendations. Technical readers, on the other hand, may look first to the body section(s), where they expect to find supporting details presented in a logical fashion. In other words, here is your chance to make your case and to explain points thoroughly.

Yet the discussion section is no place to ramble. Details must be organized so well and put forth so logically that the reader feels compelled to read on. Here are three main suggestions for organization:

- **Use headings generously.** Each time you change a major or minor point, consider whether a heading change would help the reader. Informal reports should include at least one heading per page.
- **Precede subheadings with a lead-in passage.** Here you mention the subsections to follow, before you launch into the first subheading. (For example, "This section covers these three phases of the field study: clearing the site, collecting samples, and classifying samples.") This passage does for the entire section exactly what the introductory summary does for the entire report—it sets the scene for what is to come by providing a "road map."
- **Move from general to specific in paragraphs.** Start each paragraph with a topic sentence that includes your main point. Then give supporting details. This approach always keeps your most important information at the beginnings of paragraphs, where readers tend to focus first while reading.

Another important consideration in organizing the report discussion is the way you handle facts versus opinions.

■ *Informal Report Guideline 7: Separate Fact From Opinion*

Some informal reports contain strong points of view. Others contain only subtle statements of opinion, if any. In either case, you must avoid any confusion about what constitutes fact or opinion. The safest approach in the report discussion is to move logically from findings to your conclusions and, finally, to your recommendations. Since these terms are often confused, here are some working definitions:

- **Findings:** Facts you uncover (for example, you observed severe cracks in the foundations of two adjacent homes in a subdivision).

- **Conclusions:** Ideas or beliefs you develop based on your findings (for example, you conclude that foundation cracks occurred because the two homes were built on soft fill, where original soil had been replaced by construction scraps). Opinion is clearly a part of conclusions.
- **Recommendations:** Suggestions or action items based on your conclusions (for example, you recommend that the foundation slab be supported by adding concrete posts below it). Recommendations are almost exclusively made up of opinions.

■ Informal Report Guideline 8: Focus Attention in Your Conclusion

Letter and memo reports end with a section labeled "Conclusion," "Closing," "Conclusions," or "Conclusions and Recommendations." Choose the wording that best fits the content of your report. In all cases, this section gives details about your major findings, your conclusions, and, if called for, your recommendations. People often remember best what they read last, so think hard about what you place at the end of a report.

The precise amount of detail in your conclusion depends on which of these two options you choose for your particular report:

Option 1: If your major conclusions or recommendations have already been stated in the discussion, then you only need to restate them briefly to reinforce their importance (see Model 8–2, pp. 232–233).

Option 2: If the discussion leads up to, but has not covered, these conclusions or recommendations, then you may want to give more detail in this final section (see Model 8–1, pp. 230–231).

As in Models 8–1 and 8–2, lists often are mixed with paragraphs in the conclusion. Use such lists if you believe they will help readers remember your main points.

■ Informal Report Guideline 9: Use Attachments for Less Important Details

The trend today is to avoid lengthy text in informal reports. Yet technical detail is often needed for support. One solution to this dilemma is to replace as much report text as possible with clearly labeled attachments that could include these items:

- **Tables and figures:** Illustrations in informal reports usually appear in attachments unless it is crucial that one is within the text. Memo and letter reports are so short that attached illustrations are easily accessible.
- **Costs:** It is best to list costs on a separate sheet. First, you do not want to bury important financial information within paragraphs. Second, readers often need to circulate cost information, and a separate cost attachment is easy to photocopy and send.

■ Informal Report Guideline 10: Edit Carefully

Many readers judge you on how well you edit a report. A few spelling errors or some careless punctuation makes you seem unprofessional. Your career and your firm's future can depend on your ability to write final drafts carefully. Chapter 15 and the Handbook at the end of this text give detailed information about editing. For now, remember these basic guidelines:

- Keep most sentences short and simple.
- Proofread several times for mechanical errors such as misspellings (particularly personal names).
- Triple-check all cost figures for accuracy.
- Make sure all attachments are included, are mentioned in the text, and are accurate.
- Check the format and wording of all headings and subheadings.
- Ask a colleague to check over the report.

These 10 guidelines help memo and letter reports accomplish their objectives. Remember—both your supervisors and your clients will judge you as much on communication skills as they do on technical ability. Consider each report to be part of your resume.

SPECIFIC GUIDELINES FOR FIVE INFORMAL REPORTS

Report types vary from company to company. The ones described here are only a sampling of what you will be asked to write on the job. These five types were chosen because they are common, they can be written as either memo reports or letter reports, and they incorporate the writing patterns described in chapters 5 and 6. If you master these five informal reports, you can probably handle other types that come your way.

The sections that follow include an ABC format for each report being discussed, some brief case studies from McDuff, and report models based on the cases. Remember to consult chapters 5 and 6 if you need to review general patterns of organization used in short and long reports, such as cause-effect and technical description.

Problem Analyses

Every organization faces both routine and complex problems. The routine ones often get handled without much paperwork; they are discussed and then solved. But other problems often need to be described in reports, particularly if they involve many people, are difficult to solve, or have been brewing for a long time. Use this working definition of a report that analyzes a problem:

> **Problem analysis:** this informal report presents readers with a detailed description of problems in areas such as personnel, equipment, products, services, and so forth. Its main goal is to provide *objective* information so that the readers can choose the next step. Any opinions must be well supported by facts.

Problem analyses, which can be either internal or external documents, should follow the pattern of organization described here.

ABC Format for Problem Analyses.   Like other informal reports, problem analyses fit the simple ABC (**A**bstract/**B**ody/**C**onclusion) format recommended throughout this text. The three sections contain some or all of the following information, depending on the specific report. Note that solutions to problems are not mentioned; this chapter deals separately with (1) problem analyses, whose main focus is problems, and (2) recommendation reports, whose main focus is solutions. Of course, be aware that during your career, you will be called on to write reports that combine both types.

ABC Format: Problem Analysis

Abstract

- Purpose of report
- Capsule summary of problems covered in report discussion

Body

- Background on source of problems
- Well-organized description of the problems observed
- Data that support your observations
- Consequences of the problems

Conclusion

- Brief restatement of main problems (unless report is so short that such restatement would seem repetitious)
- Degree of urgency required in handling problems
- Suggested next step

McDuff Case Study for a Problem Analysis.   Model 8–3 on pages 234–235 presents a sample problem analysis that follows this chapter's guidelines. Harold Marshal, a longtime McDuff employee, supervises all technical work aboard the *Seeker II,* a boat that McDuff leases during the summer. Staffed with several technicians and engineers, the boat is used to collect and test soil samples from the ocean floor. Different clients purchase these data, such as oil companies that need to place oil rigs safely and telecommunications companies that need to lay cable.

After a summer on the *Seeker II,* Harold has severe reservations about the safety and technical adequacy of the boat. Yet he knows that his supervisor, Jan Stillwright, will require detailed support of any complaints before she seriously considers negotiating a new boat contract next season. Given this critical audience, Harold focuses on specific problems that affect (1) the safety of the crew, (2) the accuracy of the technical work performed, and (3) the morale of the crew. He believes that this pragmatic approach, rather than an emotional appeal, will best persuade his boss that the problem is serious.

Recommendation Reports

Most problem analyses contain both facts and opinions. You as the writer must make special efforts to separate the two, for this reason: Most readers want the opportunity to draw their *own* conclusions about the problem. Also, support all opinions with facts. This guideline holds true especially with recommendation reports, for they include more personal views than other report types. Use the following working definition for recommendation reports:

> **Recommendation report:** this informal report presents readers with specific suggestions that affect personnel, equipment, procedures, products, services, and so on. Although the report's main purpose is to persuade, every recommendation must be supported by objective data.

Recommendation reports can be either internal or external documents. Both follow the ABC format suggested here.

ABC Format for Recommendation Reports.   Problem analyses and recommendation reports sometimes overlap in content. You may recommend solutions in a problem analysis, just as you may analyze problems in a recommendation report. The ABC format assumes that you want to mention the problem briefly before proceeding to discuss solutions.

ABC Format: Recommendation Report

Abstract

- Purpose of report
- Brief reference to problem to which recommendations respond
- Capsule summary of recommendations covered in report discussion

Body

- Details about problem, if necessary
- Well-organized description of recommendations
- Data that support your recommendations (with reference to attachments, if any)
- Main benefits of recommendations you put forth
- Any possible drawbacks

Conclusion

- Brief restatement of main recommendations (unless report is so short that restatement would seem repetitious)
- The main benefit of recommended change
- Your offer to help with next step

McDuff Case Study for a Recommendation Report.   Model 8–1 (pp. 230–231) shows a typical recommendation report written at McDuff. The reader is a client oil firm about to place an oil rig at an offshore site in the Gulf of Mexico. Given the potential for risk to human life and to the environment, Big Muddy Oil wants to take every precaution. Therefore, it has hired McDuff to determine whether the preferred site is safe.

This example presents an important problem you may face in writing recommendation reports. Occasionally you may be asked to deliver recommendations sooner than you would prefer if you were working under ideal circumstances. In such cases, assume the cautious approach taken by Bartley Hopkins, the McDuff writer of Model 8–1. That is, make sure to state that your report is preliminary and based on incomplete data. This approach is even more important in situations like this case study in which there is risk to human life. Take pains to qualify your recommendations so that they cannot possibly be misunderstood by your audience.

Equipment Evaluations

Every organization uses some kind of equipment, and someone has to help buy, maintain, or replace it. Because companies put so much money into this part of their business, evaluating equipment has become an important activity. Here is a working definition of evaluation reports:

> **Equipment evaluation:** this informal report provides objective data about how equipment has, or has not, functioned. The report may cover topics such as machinery, tools, vehicles, and office supplies.

Like a problem analysis, an equipment evaluation may just focus on problems. Or like a recommendation report, it may go on to suggest a change in equipment. Whatever its focus, an equipment evaluation *must* provide a well-documented review of the exact manner in which equipment has performed. Follow this ABC format in evaluating equipment.

ABC Format for Equipment Evaluations.   Equipment evaluations that are informal reports should include some or all of the points listed here. Remember that in this type of report, the discussion must include evaluation criteria most important to the *readers,* not you.

ABC Format: Equipment Evaluation
Abstract

- Purpose of report
- Capsule summary of what your report says about the equipment

Body

- Thorough description of the equipment being evaluated
- Well-organized critique, either analyzing the parts of one piece of equipment or contrasting several pieces of similar equipment according to selected criteria
- Additional supporting data, with reference to any attachments

Conclusion

- Brief restatement of major findings, conclusions, or recommendations

McDuff Case Study for an Equipment Evaluation.   Like other firms, McDuff relies on word processing for almost all internal and external documents. Model 8–2 (pp. 232–233) contains an evaluation of a new word-processing package used on a trial basis. Melanie Frank, office manager in San Francisco, conducted the trial in her office and wrote the report to the branch manager, Hank Worley. Note that she analyzes each of the software's five main features. Then she ends with a recommendation, much like a recommendation report.

Pay special attention to the tone and argumentative structure of this example. Frank shows restraint in her enthusiasm, knowing that facts will be more convincing than opinions. Indeed, every claim about Best Choice software is supported either by evidence from her trial or by a logical explanation. For example, her praise of the file-management feature is supported by the experience of a field engineer who used the system for three days. And her statement about the well-written user's guide is supported by the few calls made to the Best Choice support center during the trial.

Progress/Periodic Reports

Some short reports are intended to cover activities that occurred during a specific period of time. They can be directed inside or outside your organization and are defined in this way:

> **Progress report:** this informal report provides your manager or client with details about work on a specific project. Often you agree at the beginning of a project to submit a certain number of progress reports at certain intervals.

> **Periodic report:** this informal report, usually directed within your own organization, summarizes your work on diverse tasks over a specific time period. For example, as supervisor of company publications, you may be asked to submit periodic reports each month on McDuff's new brochures, public-relations releases, and product flyers.

Progress and periodic reports contain mostly objective data. Yet both of them, especially progress reports, sometimes may be written in a persuasive manner. (See chapter 5 for techniques of argumentation.) After all, you are trying to put forth the best case for the work you have completed. The next section provides an ABC format for these two report types.

ABC Format for Progress/Periodic Reports.   Whether internal or external, progress and periodic reports follow a basic ABC format and contain some or all of these parts:

ABC Format: Progress/Periodic Report
Abstract

- Purpose of report
- Capsule summary of main project(s)
- Main progress to date or since last report

Body

- Description of work completed since last report, organized either by task or by time or by both
- Clear reference to any dead ends that may have taken considerable time but yielded no results
- Explanation of delays or incomplete work
- Description of work remaining on project(s), organized either by task or by time or by both
- Reference to attachments that may contain more specific information

Conclusion

- Brief restatement of work since last reporting period
- Expression of confidence, or concern, about overall work on project(s)
- Indication of your willingness to make any adjustments the reader may want to suggest

---

McDuff Case Study for a Progress Report. As Model 8–4 on pages 236–237 indicates, Scott Sampson, McDuff's personnel manager, is in the midst of an internal project being conducted for Kerry Camp, vice president of domestic operations. Sampson's goal is to find ways to improve the company's training for technical employees. Having completed two of three phases, he is reporting his progress to Camp. Note that Sampson organizes the body sections by task. This arrangement helps to focus the reader's attention on the two main accomplishments—the successful phone interviews and the potentially useful survey. If, instead, Sampson had completed many smaller tasks, he may have wanted to organize the body of the report by time, not tasks.

   Also note that Sampson adopts a persuasive tone at the end of the report. That is, he uses his solid progress as a way to emphasize the importance of the project. In this sense, he is "selling" the project to his "internal customer," Kerry Camp, who ultimately will be in the position to make decisions about the future of technical training at McDuff.

McDuff Case Study for a Periodic Report. Model 8–5 on page 238 shows the rather routine nature of most periodic reports. Here Nancy Fairbanks is simply submitting her usual monthly report. The greatest challenge in such reports is to classify, divide, and label information in such a way that readers can quickly find what they need. Fairbanks selected the kind of substantive headings that help the reader locate information (for example, "Jones Fill Project" and "Performance Reviews").

Lab Reports

College students write lab reports for courses in science, engineering, psychology, and other subjects. Yet this report type also exists in technical organizations such as hospitals, engineering firms, and computer companies. Perhaps more than any other type of informal report, the lab report varies in format from organization to organization (and from instructor to instructor, in the case of college courses). This chapter will present a format to use when no other instructions have been given. A working definition follows:

> **Lab report:** this informal report describes work done in any laboratory—with emphasis on topics such as purpose of the work, procedures, equipment, problems, results, and implications. It may be directed to someone inside or outside your own organization. Also, it may stand on its own or it may become part of a larger report that uses the laboratory work as supporting detail.

The next section shows a typical ABC format for lab reports, with the types of information that might appear in the three main sections.

ABC Format for Lab Reports. Whether simple or complicated, lab reports usually contain some or all of these parts:

ABC Format: Lab Report
Abstract

- Purpose of report
- Capsule summary of results

Body

- Purpose or hypothesis of lab work
- Equipment needed
- Procedures or methods used in the lab test
- Unusual problems or occurrences
- Results of the test with reference to your expectations (results may appear in conclusion, instead)

Conclusion

- Statement or restatement of main results
- Implications of lab test for further work

McDuff Case Study for a Lab Report. Model 8–6 on pages 239–240 shows a McDuff lab report that is not part of a larger document. In this case, the client sent McDuff some soils taken from borings made into the earth. McDuff has analyzed the samples in its company laboratory and then drawn some conclusions about the kind of rock from which the samples were taken. The report writer, a geologist named Joseph Rappaport, uses the body of the report to provide background information, lab materials procedures, and problems encountered. Note that the report body uses process analysis, a main pattern of organization covered in chapter 6.

CHAPTER SUMMARY

This chapter deals exclusively with the short, informal reports you will write throughout your career. On the job, you will write them for readers inside your organization (as **memo reports**) and outside your organization (as **letter reports**). In both cases, follow these 10 basic guidelines:

1. Plan well before you write.
2. Use letter or memo format.
3. Make text visually appealing.
4. Use the ABC format for organizing information.
5. Start with an introductory summary.
6. Put detailed support in the body.
7. Separate fact from opinion.
8. Focus attention in your conclusion.
9. Use attachments for details.
10. Edit carefully.

Although letter and memo reports come in many varieties, this chapter conveys only five common types: problem analyses, recommendation reports, equipment evaluations, progress/periodic reports, and lab reports. Each follows its own type of three-part ABC format for organizing information.

ASSIGNMENTS

This chapter includes both short and long assignments. The short assignments in Part 1 are designed to be used for in-class exercises and short homework assignments. The assignments in Part 2 generally require more time to complete.

Part 1: Short Assignments

1. **Problem Analysis—Critiquing a Report.** Using the guidelines in this chapter, analyze the level of effectiveness of the following McDuff problem analysis.

April 16, 1994

Mr. Jay Henderson
Christ Church
10 Smith Dr.
Jar, Georgia 30060

PROBLEM ANALYSIS
NEW CHURCH BUILDING SITE

Introductory Summary

Last week your church hired our firm to study problems caused by the recent incorporation of the church's new building site into the city limits. Having reviewed the city's planning and zoning requirements, we have found some problems with your original site design—which initially was designed to meet the county's requirements only. My report focuses on problems with four areas on the site:

1. Landscaping screen
2. Church sign
3. Detention pond
4. Fire truck access

Attached to this report is a site plan to illustrate these problems as you review the report. The plan was drawn from an aerial viewpoint.

Landscaping Screen

The city zoning code requires a landscaping screen along the west property line, as shown on the attached site illustration sheet. The former design does not call for a screen in this area. The screen will act as a natural barrier between the church parking lot and the private residence adjoining the church property. The code requires that the trees for this screen be a minimum height of 8 feet with a height maturity level of at least 20 feet. The trees should be an aesthetically pleasing barrier for all parties, including the resident on the adjoining property.

Church Sign

After the site was incorporated into the city, the Department of Transportation decided to widen Woodstock Road and increase the setback to 50 feet, as illustrated on our site plan. With this change, the original location of the sign fell into the road setback. Its new location must be out of the setback and moved closer to the new church building.

Detention Pond

The city's civil engineers reviewed the original site drawing and found that the detention pond was too small. If the detention pond is not increased, rainwater may build up and overflow into the building, causing a considerable amount of flood damage to property in the building and to the building itself. There is a sufficient amount of land in the rear of the site to enlarge and deepen the pond to handle all expected rainfall.

Fire Truck Access

On the original site plan, the slope of the ground along the back side of the new building is so steep that a city fire truck would not be able to gain access to the rear of the building in the event of a fire. This area is shown on our site illustration around the north and east sides of the building. The zoning office enforces a code that is required by the fire marshal's office. This code states that all buildings within the city limits must provide a flat and unobstructed access path around the buildings. If the access is not provided, the safety of the church building and its members would be in jeopardy.

Conclusion

The just-stated problems are significant, yet they can be solved with minimal additional cost to the church. Once the problems are remedied and documented, the revised site plan must be approved by the zoning board before a building permit can be issued to the contractor.

I look forward to meeting with you and the church building committee next week to discuss any features of this study and its ramifications.

Sincerely,

Thomas K. Jones
Senior Landscape Engineer

Enc.

---

2. **Problem Analysis.** Divide into three- or four-person teams, as your instructor directs. In your group, share information about any problems that team members have encountered with services or facilities at the college or university you attend. Then select a problem substantive enough to be described in a short report. As a group, write a problem analysis in the format put forth in this chapter. Assume that your group represents a McDuff technical team that has been hired to investigate, and then write a series of reports on, problems at the school. Your report is one in the series. Select as your audience the appropriate administrators at the college or university.

3. **Recommendation Report—Critiquing a Report.** Using the guidelines in this chapter, analyze the level of effectiveness of the following McDuff recommendation report.

April 20, 1994

Kenman Aircraft Company
76 Jonesboro Road
Sinman, Colorado 87885

Attention: Mr. Ben Randall, Facilities Manager

EMERGENCY EXIT STUDY

Introductory Summary

As you requested, I have just completed a study of the emergency exits in your accounting office at the plant. My study indicates that you have two main problems: (1) easier access to exits is needed and (2) more exit signs and better visibility of these signs are needed. This report contains recommendations for rearranging the floor plan and improving signage.

Problems with Current Floor Plan

Two main problems cause the accounting office to fail to meet the county's guidelines for access to fire exits. First, the file cabinets on the north wall of the office are partially blocking the Reynolds Lane exit. Second, the office photocopier partially blocks the exit to the east hallway. In the first case, the file cabinets are so heavy that they could not be moved by one person. In the second case, the photocopier could be rolled out of the way only by a very strong individual. Obviously, both situations are unacceptable and violate the current code.

The other problem is signage. The Reynolds Lane exit has an exit sign, but it is not easily seen. The east hallway exit has no sign at all. In addition, the rest of the office lacks any maps that show people the location of the two fire exits.

Recommendations for Solving Exit Problem

Fortunately, the existing problems can be corrected with only minor cost to the company. The following recommendations should be implemented immediately upon your receipt of this report.

1. Move the file cabinets on the north wall to the east wall so that they no longer block the Reynolds Street exit.
2. Relocate the photocopier to the office supply room or the cubicle adjacent to it.
3. Remove the undersized exit sign from the Reynolds Street exit.
4. Purchase and install two county-approved exit signs above the two fire exits.
5. Draw up an emergency plan map and post a copy in every cubicle within the accounting office.

When you implement these recommendations, you will be in accordance with the county's current fire regulations.

Conclusion

I strongly suggest that my recommendations be put into action as soon as possible. By doing so you will greatly reduce the risk to your employees and your associated liability.

If you have any questions or need additional information, please call me at your convenience.

Sincerely,

Howard B. Manwell
Field Engineer

---

4. **Writing a Recommendation Report.** Divide into groups of three or four students, as your instructor directs. Consider your group to be a technical team from McDuff. Assume that the facilities director of your college or university has hired your team to recommend changes that would improve your classroom. Write a group report that includes the recommendations agreed to by your group. For example, you may want to consider structural changes of any kind, additions of equipment, changes in the type and arrangement of seating, and so forth.

5. **Writing an Equipment Evaluation.** Assume you are a supervisor at McDuff's Equipment Development shop, located in Baltimore. The Procurement Office routinely asks you to write evaluations of new pieces of equipment being used in the shop. Such evaluations help the director of procurement, Brenda Seymour, decide on future purchases.

   Write a brief memo to Seymour, evaluating the Brakoh cordless drills that your staff began using in the shop about a year ago. The Brakoh brand replaced a more expensive brand that the shop had used for the 10 previous years. In the last few months, your technicians have reported that the cheaper models have been falling apart after six or eight months of use. Information coming to you suggests that there are two main problems: (1) a grinding noise can be heard in the housing, resulting in the failure of the chuck (the piece that holds the drill bit) to rotate; and (2) drilling time between rechargings tends to decrease as the drill gets older. You believe that the manufacturer, in order to cut the cost of the drill, has substituted poorly made components in high-wear locations. For example, the gears responsible for turning the chuck are made of plastic. With a little wear, the gears tend to slip, which produces the grinding sound and the rotation failure. As for the recharging problem, the power cell just seems to hold less charge than the previous drill. In summary, the drill has broken down four times faster than the other model, causing many repair bills and a loss in productivity.

In writing your memo, remember that in this case your main job is provide information to the director of procurement, *not* to make recommendations one way or the other. After receiving your memo, she probably will complete a cost analysis to determine if the problems with the cheaper drill outweigh the advantage of the initial cost savings.

Part 2: Longer Assignments

While planning some of these assignments—especially assignment 6— you may need to review information in chapter 2 about McDuff, Inc. Also, for each assignment you should complete a copy of the Planning Form (included at the end of the book). These assignments can be completed as individual or group projects.

6. **Report Based on Color Insert.** The color insert included in this book contains summaries of seven McDuff projects, in various topic areas. These summaries were written for marketing purposes, after the jobs were completed.

    Using the information on one of these sheets, write a brief informal report that summarizes the project for the client. If necessary, add details that are not on the sheet. *Caution:* Remember that marketing sheets may *not* be organized as reports are organized. Consider your purpose and audience carefully before writing.

7. **Problem Analysis.** Assume you are a McDuff field engineer working at the construction site of a nuclear power plant in Jentsen, Missouri. For the past three weeks, your job has been to observe the construction of a water cooling tower, a large cylindrical structure. As consultants to the plant's construction firm, you and your McDuff crew were hired to make sure that work proceeds properly and on schedule. As the field engineer, you are supposed to report any problems in writing to your project manager, John Raines, back at your St. Louis office. Then he will contact the construction firm's office, if necessary.

    Write a short problem analysis in the form of a memo report to Raines. (Follow the guidelines in the "Problem Analyses" section of this chapter.) Take the following randomly organized information and present it in a clear, well-organized fashion. If you wish, add information of your own that might fit the context.

    - Three cement pourings for the tower wall were delayed an hour each on April 21 because of light rain.
    - Cement-truck drivers need to slow down while driving through the site. Other workers complain about the excessive dust raised by the trucks.
    - Mary Powell, a McDuff safety inspector on the crew, has cited 12 workers for not wearing their hard hats.
    - You just heard from one subcontractor, Allis Wire, Inc., that there will be a two-day delay in delivering some steel reinforcing wires that go into the concrete walls. That delay will throw off next week's schedule. Last Monday's hard rain and flooding kept everyone home that day.
    - It is probably time once again to get all the subcontractors together to discuss safety at the tower site. Recently two field hands had bad cuts from machinery.

- Although there have not been any major thefts at the site, some miscellaneous boards and masonry pieces are missing each day—probably because nearby residents (doing small home projects) think that whatever they find at the site has been discarded. Are additional "No trespassing" signs needed?
- Construction is only two days behind schedule, despite the problems that have occurred.

8. **Problem Analysis.** As a landscape engineer for McDuff, one of your jobs is to examine problems associated with the design of walkways, the location of trees and garden beds, the grading of land around buildings, and any other topographical features. Assume that you have been hired by a specific college, community, or company with which you are familiar. Your objective is to evaluate one or more landscaping problems at the site.

   Write an informal report that describes the problem(s) in detail (Follow the guidelines in the "Problem Analyses" section of this chapter.) Be specific about how the problem affects people—the employees, inhabitants, students, etc. Here are some sample problems that could be evaluated:

   - Poorly landscaped entrance to a major subdivision
   - Muddy, unpaved walkway between dormitories and academic buildings on a college campus
   - Unpaved parking lot far from main campus buildings
   - Soil runoff into the streets from several steep, muddy subdivision lots that have not yet been sold
   - City tennis courts with poor drainage
   - Lack of adequate flowers or bushes around a new office building
   - Need for a landscaped common area within a subdivision or campus
   - Need to save some large trees that may be doomed because of proposed construction

9. **Recommendation Report.** For this paper, choose a design problem at your college or company. Now put yourself in the position of a McDuff employee hired by your school or company to recommend solutions to the problem.

   Your ideas must be in the form of a report that gives one or more recommendations resulting from your study. (Consult writing guidelines in the "Recommendation Reports" section of this chapter.) Assume that the problem is well enough understood to require only a brief summary, before launching into your recommendations. Because this is a short report, it may not contain many technical details for implementing your recommendations. Also, you need to choose a limited-enough topic so that it can be covered in a short memo report. Here are some sample topics:

   - Poor ventilation in an office or a classroom, such as one with sealed windows
   - Inadequate space for quick exits during emergencies
   - Poor visibility in a large auditorium
   - Poor acoustics in a large classroom or training room
   - Lack of, or improper placement of, lighting
   - Energy inefficiency caused by structural flaws, such as poor insulation or high ceilings
   - Rooms or walkways that are not handicap-accessible
   - Failure to take advantage of solar heating
   - Inefficient heating or air-conditioning systems

10. **Recommendation Report.** This project will require some research. Assume that your college plans either to embark on a major recycling effort or to expand a recycling program that has already started. Put yourself in the role of a McDuff environmental scientist or technician who has been asked to recommend these recycling changes.

First, do some research about recycling programs that have worked in other organizations. A good place to start would be periodical indexes such as *Readers' Guide to Periodicals* or the *Environment Index,* which will lead you to some magazine articles of interest. Choose to discuss one or more recoverable resources such as paper, aluminum, cardboard, plastic, or glass bottles. Be specific about how your recommendations can be implemented by the organization or audience about which you are writing. (Consult the guidelines in the ''Recommendation Reports'' section of this chapter.)

11. **Equipment Evaluation.** For six months you have driven a new Ford 150 company truck at remote job sites. As lead field hand for McDuff's Boston office, you have been asked to write an evaluation of the vehicle for Brenda Seymour, director of procurement at the corporate office in Baltimore. Seymour will use your report to decide whether to recommend ordering five more F-150s for other offices. She has told you that you need to discuss only major positive or negative features, not every detail. If she needs more information after reading your report, she will let you know.

    Consider the list below to be your random notes. Use all this information to write a memo report that evaluates the truck. Make sure to follow the guidelines in this chapter.

    - My 150 has been very reliable—it never failed to start, even during subzero ice storms last winter.
    - The 302 engine, Ford's small V-8, has provided plenty of power to handle any hauling I have done. No need to order the more expensive and less fuel-efficient 350 V-8.
    - Have been to 18 job sites with the truck, from marshes in Maine to mountains in New Hampshire. Have put about 12,000 miles on it, on all kinds of roads and in all conditions.
    - Tires that came with the truck did not work well in muddy locations, even with four-wheel drive. Suggest we buy all-terrain tires for future vehicles. Continue to order four-wheel drive—it is necessary at over half our job sites.
    - The short bed (6 ft) did not provide enough hauling room, once I put my toolbox across the truck bed near the back window. Suggest company buy long-bed trucks with the added 2 ft of room.
    - Given what I know now, I give the truck a good to excellent rating.
    - Automatic transmission worked great. Am told by other owners that the automatic is better than the manual for construction jobs because the manual tends to burn out clutches, especially when the truck needs to be ''rocked'' back and forth to get out of mud holes. My automatic has taken a lot of abuse without problems.
    - Have had some problems with front-end handling on rough roads. Suggest that future trucks be ordered with special handling package, which includes two shock absorbers—not just one—on each front wheel.
    - Have had no major repairs, just the regular maintenance checks at the dealer.
    - There was one recall from the manufacturer concerning an exhaust pipe hanger that might bend, but the dealer fixed the problem in 20 minutes.
    - Really need to have another six months to see how well truck holds up.

12. **Equipment Evaluation.** McDuff, Inc., has decided to make a bulk purchase of 20 typewriters or word processors. (For the purposes of this assignment, choose a typewriter or word processor with which you are familiar.) The machinery will go in a new department being set up in several months.

    Assume that McDuff now uses five different types of machines (again, remember you can choose typewriters or word-processing systems). In the interests of a fair comparison/contrast, Brenda Seymour, director of McDuff's corporate Procurement Department, has asked you and several other employees to evaluate the effectiveness of your own system. Write Seymour a memo report that includes your evaluation. (Consult guidelines in the "Equipment Evaluations" section of this chapter.) She will use the data and opinions in all the equipment evaluations she receives to make her choice for the bulk purchase. Your criteria for evaluation might include topics such as one or more of these:

- Physical design of the equipment
- Ease with which system can be learned
- Quality of the written instructions
- Frequency and cost of maintenance
- Availability of appropriate software
- Length of coverage of warranty
- Nearness to a service center
- Reputation of the manufacturer

13. **Progress or Periodic Report.** Assume that you have worked as a field hand at McDuff's Atlanta office for 15 years. Because of your reliability, good judgment, and intelligence, the company is paying for your enrollment at a local college. Also, you get half time off, with pay. Because of its investment in you, McDuff expects you to report periodically on your college work. Choose one of these two options for this assignment:

    Progress report: Select a major project you are now completing in any college course. Following the guidelines in the "Progress/Periodic Reports" section of this chapter, write a progress report on this project. Direct the memo report to the Atlanta office's manager of engineering, Wade Simkins. Sample topics might include a major paper, laboratory experiment, field project, or design studio.

    Periodic report: Assume that McDuff requires that you submit periodic reports on your schooling every few weeks. Following the guidelines in the "Progress/Periodic Reports" section of this chapter, write a periodic report on your recent course work (completed or ongoing classes or both). Direct the memo report to the manager of engineering, Wade Simkins. Organize the report by class, and then give specific updates on each one.

14. **Lab Report.** For this assignment, you must be taking a lab course now or have taken such a course recently. As in assignment 13, assume you work as a field hand with McDuff's Atlanta office. The company is sponsoring your schooling and has requested that you report on a specific college lab.

    Following the guidelines in this chapter's "Lab Reports" section, write a report to Wade Simkins, manager of engineering. The quality of your report may affect whether or not McDuff continues to fund your schooling. Be specific about the goals, procedures, and results of your laboratory—just as you would in an actual college lab report.

Mc Duff, Inc.

12 Post Street
Houston, Texas 77000
(713) 555-9781

April 22, 1994

Big Muddy Oil Company, Inc.
12 Rankin St.
Abilene, TX 79224

ATTENTION: Mr. James Smith, Engineering Manager

SHARK PASS STUDY
BLOCK 15, AREA 43-B
GULF OF MEXICO

Introductory Summary

You recently asked our firm to complete a preliminary soils investigation at an offshore rig site. This report presents the tentative results of our study, including major conclusions and recommendations. A longer, formal report will follow at the end of the project.

On the basis of what we have learned so far, your firm can safely place an oil platform at the Shark Pass site. To limit the chance of a rig leg punching into the seafloor, however, we suggest you follow the recommendations in this report.

Work at the Project Site

On April 16 and 17, 1993, McDuff's engineers and technicians worked at the Block 15 site in the Shark Pass region of the gulf. Using McDuff's leased drill ship, *Seeker II*, as a base of operations, our crew performed these main tasks:

- Seismic survey of the project study area
- Two soil borings of 40 feet each

Both seismic data and soil samples were brought to our Houston office the next day for laboratory analysis.

Laboratory Analysis

On April 18 and 19, our lab staff examined the soil samples, completed bearing capacity tests, and evaluated seismic data. Here are the results of that analysis.

Soil Layers

Our initial evaluation of the soil samples reveals a 7-9 ft layer of weak clay starting a few feet below the seafloor. Other than that layer, the composition of the soils seems fairly typical of other sites nearby.

Includes specific title.

Uses *optional* heading for abstract part of ABC format.

Draws attention to *main point* of report—that the platform *can* be placed safely at site (based on preliminary evidence).

Gives on-site details of project—dates, location, tasks.

Uses *lead-in* to subsections that follow.

Highlights most important point about soil layer—that is, the *weak clay*.

MODEL 8–1
Letter report on a small project

Bearing Capacity

We used the most reliable procedure available, the XYZ method, to determine the soil's bearing capacity (that is, its ability to withstand the weight of a loaded oil rig). That method required that we apply the following formula:

$$Q = cNv + tY, \text{ where}$$

Q = ultimate bearing capacity
c = average cohesive shear strength
Nv = the dimensionless bearing capacity factor
t = footing displacement
Y = weight of the soil unit

The final bearing capacity figure will be submitted in the final report, after we repeat the tests.

Seafloor Surface

By pulling our underwater seismometer back and forth across the project site, we developed a seismic "map" of the seafloor surface. That map seems typical of the flat floor expected in that area of the gulf. The only exception is the presence of what appears to be a small sunken boat. This wreck, however, is not in the immediate area of the proposed platform site.

CONCLUSIONS AND RECOMMENDATIONS

Based on our analysis, we conclude that there is only a slight risk of instability at the site. Though unlikely, it is possible that a rig leg could punch through the seafloor, either during or after loading. We base this opinion on (1) the existence of the weak clay layer, noted earlier, and (2) the marginal bearing capacity.

Nevertheless, we believe you can still place your platform if you follow careful rig-loading procedures. Specifically, take these precautions to reduce your risk:

1. Load the rig in 10-ton increments, waiting one hour between loadings.
2. Allow the rig to stand 24 hours after the loading and before placement of workers on board.
3. Have a soils specialist observe the entire loading process, to assist with any emergency decisions if problems arise.

As noted at the outset, these conclusions and recommendations are based on preliminary data and analysis. We will complete our final study in three weeks and submit a formal report shortly thereafter.

McDuff, Inc., enjoyed working once again for Big Muddy Oil at its Gulf of Mexico lease holdings. I will phone you this week to see if you have any questions about our study. If you need information before then, please give me a call.

Sincerely,

Bartley Hopkins

Bartley Hopkins, Project Manager
McDuff, Inc.

hg

MODEL 8–1, *continued*

Notes *why* this method was chosen (that is, reliability).

Explains both *how* the mapping procedure was done and *what results* it produced.

Leads off section with major conclusion, for emphasis.

Restates points (made in body) that support conclusion.

Uses list to emphasize recommendations to *reduce risk.*

Again mentions tentative nature of information, to prevent misuse of report.

Maintains control and shows initiative by offering to *call* client.

Mc Duff, Inc.

MEMORANDUM

DATE: July 26, 1994
TO: Melanie Frank, Office Manager
FROM: Hank Worley, Project Manager *HW*
SUBJECT: Evaluation of Best Choice Software

INTRODUCTORY SUMMARY

When the office purchased one copy of Best Choice Software last month, you suggested I send you an evaluation after 30 days' use. Having now used Best Choice for a month, I have concluded that it meets all our performance expectations. This report includes a critique of the package's five main features, as they relate to the needs of my group.

HOW BEST CHOICE HAS HELPED US

Best Choice provides five primary features: word processing, file management, spreadsheet, graphics, and a user's guide. Here is my critique of all five.

Word Processing

The system contains an excellent word-processing package that the engineers as well as the secretaries have been able to learn easily. This package can handle both our routine correspondence and the lengthy reports that our group generates. Of particular help is the system's 90,000-word dictionary, which can be updated at any time. The spelling correction feature has already saved much effort that was previously devoted to mechanical editing.

File Management

The file-manager function allows the user to enter information and then to manipulate it quickly. During one 3-day site visit, for example, a field engineer recorded a series of problems observed in the field. Then she rearranged the data to highlight specific points I asked her to study, such as I-beam welds and concrete cracks.

Spreadsheet

Like the system's word-processing package, the spreadsheet is efficient and quickly learned. Because Best Choice is a multipurpose software package, spreadsheet data can be incorporated into letter or report format. In other words, spreadsheet information can be merged with our document format to create a final draft for submission to clients or supervisors, with a real savings in time. For example, the memo I sent you last week on budget projections for field equipment took me only an hour to complete; last quarter, the identical project took four hours.

MODEL 8–2
Memo report

Uses optional first heading for abstract section of ABC format. Gives background, main points, and scope statement.

Notes five main points to be covered.

Begins paragraph with most important point. Supports claim with evidence.

Uses specific example to document opinion.

Gives simple explanation of how spreadsheet works.

Graphics

The graphics package permits visuals to be drawn from the data contained in the spreadsheet. For example a pie chart that shows the breakdown of a project budget can be created easily by merging spreadsheet data with the graphics software. With visuals becoming such an important part of reports, we have used this feature of Best Choice quite frequently.

User's Guide

Eight employees in my group have now used the Best Choice user's guide. All have found it well laid out and thorough. Perhaps the best indication of this fact is that in 30 days of daily use, we have placed only three calls to the Best Choice customer-service number.

CONCLUSION

Best Choice seems to contain just the right combination of tools to help us do our job, both in the field and in the office. These are the system's main benefits:

- Versatility—it has diverse functions
- Simplicity—it is easy to master

The people in our group have been very pleased with the package during this 30 day trial. If you like, we would be glad to evaluate Best Choice for a longer period.

MODEL 8–2, *continued*

Shows relevance of graphics to current work.

Supplies strong supporting statistic.

Wraps up report by restating main points.

Offers follow-up effort.

233

Mc Duff, Inc.

MEMORANDUM

DATE: October 15, 1994
TO: Jan Stillwright, Vice President of Research and Training
FROM: Harold Marshal, Technical Supervisor   HW
SUBJECT: Boat Problems During Summer Season

INTRODUCTORY SUMMARY

We have just completed a 1-month project aboard the leased ship, *Seeker II*, in the Pacific Ocean. All work went just about as planned, with very few delays caused by weather or equipment failure.

However, there were some boat problems that need to be solved before we lease *Seeker II* again this season. This report highlights the problems so that they can be brought to the owner's attention. My comments focus on four areas of the boat: drill rig, engineering lab, main engine, and crew quarters.

DRILL RIG

Thus far the rig has operated without incident. Yet on one occasion, I noticed that the elevator for lifting pipe up the derrick swung too close to the derrick itself. A quick gust of wind or a sudden increase in sea height caused these shifts. If the elevator were to hit the derrick, causing the elevator door to open, pipe sections might fall to the deck below.

I believe the whole rig assembly needs to be checked over by someone knowledgeable about its design. Before we put men near that rig again, we need to know that their safety would not be jeopardized by the possibility of falling pipe.

ENGINEERING LAB

Quite frankly, it is a tribute to our technicians that they were able to complete all lab tests with *Seeker II's* limited facilities. Several weeks into the voyage, these four main problems surfaced:

1. Ceiling leaks
2. Poor water pressure in the cleanup sink
3. Leaks around the window near the electronics corner
4. Two broken outlet plugs

Although we were able to devise a solution to the window leaks, the other problems stayed with us for the entire trip.

MODEL 8–3
Problem analysis

Uses *simple language* to describe technical problems.

Closes section with opinion that flows from facts presented.

Gives lead-in to three sections that follow.

Describes three problem areas in great detail—knowing the owner will want facts to support complaints.

Briefly restates problem, with emphasis on *safety* and *profits*.

Ends with specific recommendation.

MAIN ENGINE

On this trip, we had three valve failures on three different cylinder heads. From our experience on other ships, it is very unusual to have one valve fail, let alone three. Fortunately for us, these failures occurred between projects, so we did not lose time on a job. And fortunately for the owner, the broken valve parts did not destroy the engine's expensive turbocharger.

Only an expert will be able to tell whether these engine problems were flukes or if the entire motor needs to be rebuilt. In my opinion, the most prudent course of action is to have the engine checked over carefully before the next voyage.

CREW QUARTERS

When 15 men live in one room for three months, it is important that basic facilities work. On *Seeker II* we experienced problems with the bedroom, bathroom, and laundry room that caused some tension.

Bedroom

Three of the top bunks had such poor springs that the occupants sank 6 to 12 in. toward the bottom bunks. More important, five of the bunks are not structurally sound enough to keep from swaying in medium to high seas. Finally, most of the locker handles are either broken or about to break.

Bathroom

Poor pressure in three of the commodes made them almost unusable during the last two weeks. Our amateur repairs did not solve the problem, so I think the plumbing leading to the holding tank might be defective.

Laundry Room

We discovered early that the filtering system could not screen the large amount of rust in the old 10,000-gallon tank. Consequently, undergarments and other white clothes turned a yellow-red color and were ruined.

CONCLUSION

As noted at the outset, none of these problems kept us from accomplishing the major goals of this voyage. But they did make the trip much more uncomfortable than it had to be. Moreover, in the case of the rig and engine problems, we were fortunate that injuries and downtime did not occur.

I strongly urge that the owner be asked to correct these deficiencies before we consider using *Seeker II* for additional projects this season.

MODEL 8–3, *continued*

Mc Duff, Inc.

MEMORANDUM

DATE: June 11, 1994
TO: Kerry Camp, Vice President of Domestic Operations
FROM: Scott Sampson, Manager of Personnel   SS
SUBJECT: Progress Report on Training Project

INTRODUCTORY SUMMARY

On May 21 you asked that I study ways that our firm can improve training for technical employees in all domestic offices. We agreed that the project would take about six or seven weeks and involve three phases:

Phase 1: Make phone inquiries to competing firms
Phase 2: Send a survey to our technical people
Phase 3: Interview a cross section of our technical employees

I have now completed Phase 1 and part of Phase 2. My observation thus far is that the project will offer many new directions to consider for our technical training program.

WORK COMPLETED

In the first week of the project, I had extensive phone conversations with people at three competing firms about their training programs. Then in the second week, I wrote and sent out a training survey to all technical employees in McDuff's domestic offices.

Phone Interviews

I contacted three firms for whom we have done similar favors in the past: Simkins Consultants, Judd & Associates, and ABG Engineering. Here is a summary of my conversations:

1. Simkins Consultants
   Talked with Harry Roland, Training Director, on May 23. Harry said that his firm has most success with internal training seminars. Each technical person completes several one-or two-day seminars every year. These courses are conducted by in-house experts or external consultants, depending on the specialty.

2. Judd & Associates
   Talked with Jan Tyler, Manager of Engineering, on May 24. Jan said that Judd, like Simkins, depends mostly on internal seminars. But Judd spreads these seminars over one or two weeks, rather than teaching intensive courses in one or two days. Judd also offers short "technical awareness" sessions at the lunch hour every two weeks. In-house technical experts give informal presentations on some aspect of their research or fieldwork.

MODEL 8–4
Progress report

3. ABG Engineering

Talked with Newt Mosely, Personnel Coordinator, on May 27. According to Newt, ABG's training program is much as it was two decades ago. Most technical people at high levels go to one seminar a year, usually sponsored by professional societies or local colleges. Other technical people get little training beyond what is provided on the job. In-house training has not worked well, mainly because of schedule conflicts with engineering jobs.

Internal Survey

After completing the phone interviews noted, I began the survey phase of the project. Last week, I finished writing the survey, had it reproduced, and sent it with cover memo to all 450 technical employees in domestic offices. The deadline for returning it to me is June 17.

WORK PLANNED

With phone interviews finished and the survey mailed, I foresee this schedule for completing the project:

June 17:	Surveys returned
June 18-21:	Surveys evaluated
June 24-28:	Trips taken to all domestic offices to interview a cross section of technical employees
July 3:	Submission of final project report to you

CONCLUSION

My interviews with competitors gave me a good feel for what technical training might be appropriate for our staff. Now I am hoping for a high-percentage return on the internal survey. That phase will prepare a good foundation for my on-site interviews later this month. I believe this major corporate effort will upgrade our technical training considerably.

I would be glad to hear any suggestions you may have about my work on the rest of the project. For example, please call if you have any particular questions you want asked during the on-site interviews (ext. 348).

Margin notes:

Gives important details about the survey.

Organizes section chronologically, making sure to stay within a six or seven week schedule.

Looks to future tasks.

Emphasizes major benefit, to "sell" the project internally.

Indicates flexibility and encourages response from reader.

MODEL 8–4, *continued*

Mc Duff, Inc.

MEMORANDUM

DATE: August 2, 1994
TO: Ralph Buzby, Manager of Engineering
FROM: Nancy Fairbanks, Project Manager NJ
SUBJECT: Activity Report for July 1994

July has been a busy month in our group. Besides starting and finishing many smaller jobs, we completed the Jones Fill project. Also, the John Lewis Dam borings began just a week ago. Finally, I did some marketing work and several performance reviews.

SMALL PROJECTS

Last month, my group completed nine small projects, each with a budget under $20,000 and each lasting only a few days. These jobs were in three main areas:

1. Surveying subdivisions--five jobs
2. Taking samples from toxic sites--two jobs
3. Doing nearby soil borings--two jobs

All nine were completed within budget. Eight of the nine projects were completed on time. The Campbell County survey, however, was delayed for a day because of storms on July 12.

JONES FILL PROJECT

Our written report on this 12-month job was finally submitted to Trunk Engineering, Inc., on July 23. The most recent delay was caused by Trunk's decision to change the scope of the project again. Tramell wanted another soil boring, which we completed on July 22.

JONES LEWIS DAM PROJECT

As you know, we had hoped to start work at the dam site last month. However, the client decided to make a lot of design changes that had to be approved by subcontractors. The final approval to start came just last week; thus our first day on site was July 29.

MARKETING

During July, my main marketing effort was to meet with some previous clients, acquainting them with some of our new services. I met with eight different clients at their offices, with two meetings occurring on each of these dates: July 15, 16, 22, and 23. There's a good possibility that several of these meetings will lead to additional waste-management work in the next few months.

PERFORMANCE REVIEWS

As we discussed last month, I fell behind on my staff's performance reviews in June. In July, I completed the three delayed reviews, as well as the four that were due in July. Copies of the paperwork were sent to your office and to the Personnel Department on July 19. This brings us up to date on all performance reviews.

CONCLUSION

July was a busy month in almost all phases of my job. Because of this pace, I haven't had time to work on the in-house training course you asked me to develop. In fact, I'm concerned that time I devote to that project will take me away from my ongoing client jobs. At our next meeting, perhaps we should brainstorm about some solutions to this problem.

MODEL 8–5
Periodic report

Mc Duff, Inc.

105 Halsey Street
Baltimore, Maryland 21212
(301) 555-7588

December 12, 1993

Mr. Andrew Hawkes
Monson Coal Company
2139 Lasiter Dr.
Baltimore, MD 21222

LABORATORY REPORT
BOREHOLE FOSSIL SAMPLES
BRAINTREE CREEK SITE, WEST VIRGINIA

INTRODUCTORY SUMMARY

Last week you sent us six fossilized samples from the Braintree Creek site. Having analyzed the samples in our lab, we believe they suggest the presence of coal-bearing rock. As you requested, this report will give an overview of the materials and procedures we used in this project, along with any problems we had.

As you know, our methodology in this kind of job is to identify microfossils in the samples, estimate the age of the rock by when the microfossils existed, and then make assumptions about whether the surrounding rock might contain coal.

LAB MATERIALS

Our lab analysis relies on only one piece of specialized equipment: a Piketon electron microscope. Besides the Piketon, we use a simple 400-power manual microscope. Other equipment is similar to that included in any basic geology lab, such as filtering screens and burners.

LAB PROCEDURE

Once we receive a sample, we first try to identify the exact kinds of microfossils that the rocks contain. Our specific lab procedure for your samples consisted of these two steps:

Step 1

We used a 400-power microscope to visually classify the microfossils that were present. Upon inspection of the samples, we concluded that there were two main types of microfossils: nannoplankton and foraminifera.

continues

MODEL 8–6
Lab report

Margin annotations:

Gives *overview* of results.

Outlines *procedure* to be detailed in following paragraph.

Describes main equipment, in layperson's language.

Breaks down procedure into easy-to-read "chunks."

Step 2

Next, we had to extract the microfossils from the core samples you provided. We used two different techniques:

Nannoplankton Extraction Technique
a. Selected a pebble-size piece of the sample
b. Thoroughly crushed the piece under water
c. Used a dropper to remove some of the material that floats to the surface (it contains the nannoplankton)
d. Dried the nannoplankton-water combination
e. Placed the nannoplankton on a slide

Foraminifera Extraction Technique
a. Boiled a small portion of the sample
b. Used a microscreen to remove clay and other unwanted material
c. Dried remaining material (foraminifera)
d. Placed foraminifera on slide

PROBLEMS ENCOUNTERED

The entire lab procedure went as planned. The only problem was minor and occurred when we removed one of the samples from the container in which it was shipped. As the bag was taken from the shipping box, it broke open. The sample shattered when it fell onto the lab table. Fortunately, we had an extra sample from the same location.

CONCLUSION

Judging by the types of fossils present in the sample, they come from rock of an age that might contain coal. This conclusion is based on limited testing, so we suggest you test more samples at the site. We would be gald to help you with this additional sampling and testing.

I will call you this week to discuss our study and any possible follow-up you may wish us to do.

Sincerely,

Joseph Rappaport

Joseph Rappaport
Senior Geologist

Provides smooth transitions.

Itemizes steps because of their importance in procedure.

Uses *parallel form* in describing this process.

Does not bury sampling error— gives it proper treatment.

Ends with wrap-up that reinforces main point of report.

Offers follow-up services.

MODEL 8–6, *continued*

240

9 | *Formal Reports*

An excellent formal report reflects well on you, the writer, and on your organization.

*T*oby West, McDuff's director of marketing, was given an interesting assignment two months ago. Rob McDuff asked him to take a long, hard look at the company's clients. Are they satisfied with the service they received? Do they routinely reward McDuff with additional work? Are there any features of the company, its employees, or its services that frustrate them? What do they want to see changed? In other words, Toby West was asked to step back from daily events and evaluate the company's level of service. He attacked the project in four stages:

1. He designed and sent out a survey to all recent and current clients.
2. He followed up on some of the returned surveys with phone and personal interviews.
3. He evaluated the data he collected.
4. He wrote a *formal report* on the results of his study. Besides going to all corporate and branch managers, West's report later served as a basis for some in-house training sessions called "Quality at McDuff."

Like Toby West, you will write a number of long, formal reports during your career. Most will be written collaboratively with colleagues; others will be just your creations. All of them will require major efforts at planning, organizing, drafting, and revising. Though informal reports are the most common report in business writing, formal reports become a larger part of your writing as you move along in your career. This text uses the following working definition:

> **Formal report:** this report covers complex projects and is directed to readers at different technical levels. Although not defined by length, a formal report usually contains at least 6 to 10 pages of text, not including appendices. It can be directed to readers either inside or outside your organization. Often bound, it usually includes these separate parts: cover/title page, letter/memo of transmittal, table of contents, list of illustrations, executive summary, introduction, discussion sections, and conclusions and recommendations. (Appendices often appear after the report text.)

To prepare you to write excellent formal reports, this chapter includes three main sections. The first section briefly describes four situations that would require formal reports—two are in-house (to readers *within* McDuff) and two are external (to readers *outside* McDuff). The second section provides guidelines for writing the main parts of a long report. The third section includes a complete long report from McDuff that follows this chapter's guidelines.

WHEN TO USE FORMAL REPORTS

Like most people, you probably associate formal reports with important projects. What else would justify all that time and effort? In comparison to informal reports, formal reports usually (1) cover more complicated projects and (2) are longer than their informal counterparts.

While complexity of subject matter and length are the main differences between formal and informal reports, sometimes there is another distinction. Formal reports may have a more diverse set of readers. In this case, readers who want just a quick overview can turn to the executive summary at the beginning or the conclusions and recommendations at the end. Technical readers who want to check your facts and figures can turn to discussion sections or appendices. And all readers can flip to the table of contents for a quick outline of what sections the report contains. You need to think about the needs of all these readers as you plan and write your formal reports.

The intended audience for formal reports can be inside or outside, though the latter is more common in formal reports. Here are four situations at McDuff for which formal reports would be appropriate:

- **Salary study and recommendations (internal):** Mary Kennelworth, a supervisor at McDuff's San Francisco office, has just completed a study of technicians' salaries among McDuff's competitors on the West Coast. What prompted the study was the problem she has had hiring technicians to assist environmental engineers and geologists. Lately some top applicants have been choosing other firms. Because the salary scales of her office are set by McDuff's corporate headquarters, she wants to give the main office some data showing that San Francisco starting salaries should be higher. Mary decides to submit a formal report, complete with data and recommendations for adjustments. Her main audience includes the branch manager and the vice president of human resources (the company's top decision-maker about salaries and other personnel matters).

- **Analysis of marketing problems (internal):** For several years, Jim Springer, engineering manager at McDuff's Houston office, has watched profits decline in onshore soils work. (In this type of work, engineers and technicians investigate the geological and surface features of a construction site and then recommend foundation designs and construction practices.) One problem has been the "soft" construction market in parts of Texas. But the slump in work has continued despite the recent surge in construction. In other words, some other company is getting the work. Jim and his staff have analyzed past marketing errors, with a view toward developing a new strategy for gaining new clients and winning back old ones. He plans to present the problem analysis and preliminary suggestions in a formal report. The main readers will be his manager, the corporate marketing manager in Baltimore, and managers at other domestic offices who have positions that correspond to his.

- **Waste-management survey (external):** For the last several years, the city of Belton, Georgia, has noticed increased fish kills on the Channel River, which flows through the city and serves as the city's main source of drinking water. Pollution has always been fairly well monitored on the river, so city officials are puzzled by the kills. McDuff's Atlanta office was hired to analyze the problem and present its opinion about the cause. A team of chemists, environmental engineers, and field technicians has just completed a study and will present its formal report. The audience will be quite diverse—from the technical experts in the city's water department to the members of a special citizens' panel representing the residents.

- **Collapse of oil rig (external):** A 10-year-old rig in the North Sea recently collapsed during a mild storm. Several rig workers died, and several million dollars' worth of equipment was lost. Also, the accident created an oil spill that destroyed a significant amount of fish and wildlife before it was finally contained. McDuff's London office was hired to examine the cause of the collapse of this structure, which supposedly was able to withstand hurricanes. After three months of on-site analysis and laboratory work, McDuff's experts are ready to submit their report. It will be read by corporate managers of the firm, agencies of the Norwegian government, and members of several major wildlife organizations. Also, it will be used as the basis for some articles in magazines and newspapers throughout the world.

As these four situations show, formal reports are among the most difficult on-the-job writing assignments you face in your career. You will write some yourself; you will write others as a member of a team of technical and professional people. In all cases, you need to (1) understand your purpose, (2) grasp the needs of your readers, and (3) design a report that responds to these needs. Guidelines in the next two sections will help you meet these goals.

STRATEGY FOR ORGANIZING FORMAL REPORTS

You will encounter different report formats in your career, depending on your profession and your specific employer. Whatever format you choose, however, there is a universal approach to good organization that always applies. This approach is based on these main principles, discussed in detail in chapter 3:

> **Principle 1:** Write different parts for different readers.
>
> **Principle 2:** Place important information first.
>
> **Principle 3:** Repeat key points when necessary.

These principles apply to formal reports even more than they do to short documents, for these reasons:

1. A formal report often has a very mixed audience—from laypersons to highly technical specialists to executives.
2. The majority of readers of long reports focus on specific sections that interest them most, reading selectively each time they pick up the report.
3. Few readers have time to wade through a lot of introductory information before reaching the main point. They will get easily frustrated if you do not place important information first.

This chapter responds to these facts about readers of formal reports by following the ABC format (for **A**bstract, **B**ody, **C**onclusion). As noted in chapter 3, the three main rules are that you should (1) start with an abstract for decision-makers, (2) put supporting details in the body, and (3) use the conclusion to produce action. This simple ABC format should be evident in *all* formal reports, despite their complexity. Here is how the particular sections of formal reports fit within the ABC format:

ABC Format: Formal Report
Abstract

- Cover/title page
- Letter or memo of transmittal
- Table of contents
- List of illustrations
- Executive summary
- Introduction

Body

- Discussion sections
- [Appendices—appear after text but support Body section]

Conclusion

- Conclusions
- Recommendations

Several features of this structure deserve special mention. First, note that the generic abstract section includes five different parts of the report that help give readers a capsule version of the entire report. As you will learn shortly, the executive summary is by far the most important section for providing this "big picture" of the report. Second, appendices are placed within the body part of the outline, even though sequentially they come at the end of the report. The reason for this outline placement is that both appendices and body sections provide supporting details for the report. Third, remember that the generic conclusion section in the ABC format can contain conclusions and/or recommendations, depending on the nature of the report.

Before moving to a discussion of the specific sections that make up the ABC format, take note of the use of headings in complex formal reports. (See Figure 4−7 in chapter 4.) Much like chapter titles, these headings are often centered, in full caps, in bold type, and oversized. Also, they usually begin a new page. In this way, each major section of the formal report seems to exist on its own. Then you have three remaining heading levels−B, C, and D−for the rest of the report.

GUIDELINES FOR THE EIGHT PARTS OF FORMAL REPORTS

What follows is a description of eight parts of the formal report:

1. Cover/title page
2. Letter of transmittal
3. Table of contents
4. List of illustrations
5. Executive summary
6. Introduction
7. Discussion sections
8. Conclusions and recommendations

What could be considered a ninth part, appendices, is mentioned in the context of the discussion section.

Because formal reports can cover such a broad range of material, the guidelines here are rather general. For specific application of these guidelines, see the formal report in Model 9−9 on pages 268−284.

Cover/Title Page

Formal reports are normally bound, usually with a cover used for all reports in the writer's organization. (Reports prepared for college courses, however, are often placed in a three-tab folder, the outside of which serves as the report cover.) Since the cover is the first item seen by the reader, it should be attractive and informative. Usually it contains the same four pieces of information mentioned in the following list with regard to the title page; sometimes it may have only one or two of these items.

Inside the cover is the title page, which should include these four pieces of information:

- Project title (exactly as it appears on the letter/memo of transmittal)
- Your client's name ("Prepared for . . .")
- Your name and/or the name of your organization ("Prepared by . . .")
- Date of submission

To make your title page or cover distinctive, you might want to place a simple illustration on it. Do not clutter the page, of course. Use a visual only if it reinforces a main point and if it can be simply and tastefully done. For example, assume that McDuff, Inc., has submitted a formal report to a coastal city in California, concluding that an industrial park can be built near the city's bird sanctuary without harming the habitat—if stringent guidelines are followed. The report writer decides to place the picture of a nesting bird on the title page, punctuating the report's point about the industrial park, as in Model 9–1 on page 260.

Letter/Memo of Transmittal

Letters or memos of transmittal are like an appetizer—they give the readers a taste of what is ahead. If your formal report is to readers outside your own organization, write a letter of transmittal. If it is to readers inside your organization, write a memo of transmittal. Models 9–2 and 9–3 on pages 261–262 show examples of both. Follow these guidelines for constructing this part of your reports:

■ Transmittal Guideline 1: Place Letter/Memo Immediately After Title Page

This placement means that the letter/memo is bound with the document, to keep it from becoming separated. Some organizations paper-clip this letter or memo to the front of the report, making it a cover letter or memo. In so doing, however, they risk having it become separated from the report.

■ Transmittal Guideline 2: Include a Major Point From Report

Remember that readers are heavily influenced by what they read first in reports. Therefore, take advantage of the position of this section by including a major finding, conclusion, or recommendation from the report—besides supplying necessary transmittal information.

■ Transmittal Guideline 3: Follow Letter and Memo Conventions

Like other letters and memos, letters and memos of transmittal should be easy to read, inviting readers into the rest of the report. Keep introductory and concluding paragraphs relatively short—no more than three to five lines each. Also, write in a conversational style free of technical jargon and stuffy phrases such as "per your request'' or ''enclosed herewith.'' See chapter 7 for more details and options concerning letter/memo format. For now, here are some highlights about the mechanics of format:

Letters and Memos

- Use single spacing and ragged-edge copy, even if the rest of the report is doubled-spaced and right-justified.
- Use only one page.

Letters

- Include company project number with the letter date.
- Correctly spell the reader's name.
- Be sure the inside address includes the mailing address to appear on the envelope.
- Use the reader's last name (''Dear Mr. Jamison:'') in the salutation because of the formality of the report—unless your close association with the reader would make it more appropriate to use first names (''Dear Bill:'').
- Usually include a project title, as with letter reports. It is treated like a main heading. Use concise wording that matches wording on the title page.
- Use ''Sincerely'' as your closing.
- Include a line to indicate those who will receive copies of the report (''cc'' for carbon copy, ''pc'' for photocopy, or just ''c'' for copy).

Memos

- Give a clear description of the project in the subject line of the memo, including a project number if there is one.
- Include a distribution list to indicate those who will receive copies.

Table of Contents

Your contents page acts as an outline. Many readers go there right away to grasp the structure of the report, and then return again and again to locate report sections of most interest to them. Guidelines follow for assembling this important component of your report. See Model 9–4 on page 263 for an example.

■ Table of Contents Guideline 1: Make It Very Readable

The table of contents must be pleasing to the eye so that readers can find sections quickly and see their relationship to each other. Be sure to:

- Space items well on the page
- Use indenting to draw attention to subheadings
- Include page numbers for every heading and subheading, unless there are many headings in a relatively short report, in which case you can delete page numbers for all of the lowest-level headings listed in the table of contents

■ *Table of Contents Guideline 2: Use the Contents Page to Reveal Report Emphases*

Choose the wording of headings and subheadings with care. Be specific yet concise so that each heading listed in the table of contents gives the reader a good indication of what the section contains.

Readers associate the importance of report sections with the number of headings and subheadings listed in the table of contents. If, for example, a discussion section called "Description of the Problem" contains many more heading breakdowns than other sections, you are telling the reader that the section is more important. When possible, it is best to have about the same number of breakdowns for report sections of about the same importance. In short, the table of contents should be balanced.

■ *Table of Contents Guideline 3: Consider Leaving Out Low-Level Headings*

In very long reports, you may want to unclutter the table of contents by removing low-level headings. As always, the needs of the readers are the most important criterion to use in making this decision. If you think readers need access to all levels of headings on the contents page, keep them there. If you think they would prefer a simple contents page instead of a comprehensive one, delete all the lowest-level headings from the table of contents (see Model 9–5 on page 264 compared to Model 9–4 on page 263).

■ *Table of Contents Guideline 4: List Appendices*

Appendices include items such as tables of data or descriptions of procedures that are inserted at the end of the report. Typically, they are listed at the end of the table of contents. Often no page numbers are given, since many appendices contain "off-the-shelf" material and are thus individually paged (for example, Appendix A might be paged A-1, A-2, A-3, etc.). Tabs on the edges of pages can help the reader locate these sections.

■ *Table of Contents Guideline 5: Use Parallel Form in All Entries*

All headings in one section, and sometimes even all headings and subheadings in the report, have parallel grammatical form. Readers find mixed forms distracting. For example, "Subgrade Preparation" and "Fill Placement" are parallel, in that they are both the same type of phrase. However, if you were to switch the wording of the first item to "Preparing the Subgrade" or "How to Prepare the Subgrade," parallel structure would be lost.

■ *Table of Contents Guideline 6: Proofread Carefully*

The table of contents is one of the last report sections to be assembled; thus it often contains errors. Wrong page numbers and incorrect heading wording are two common mistakes. Another is the failure to show the correct relationship of head-ings and subheadings. Obviously, errors in the table of contents can confuse the reader and prove embarrassing to the writer. Proofread the section carefully.

List of Illustrations

Illustrations within the body of the report are usually listed on a separate page right after the table of contents. When there are few illustrations, another option is to list them at the bottom of the table of contents page rather than on a separate page. In either case, this list should include the *number, title,* and *page number* of every table and figure within the body of the report. If there are many illustrations, separate the list into tables and figures. See the example in Model 9–6 on page 265. (For more information on illustrations, see chapter 11.)

Executive Summary

No formal report would be complete without an executive summary. This short section provides decision-makers with a capsule version of the report. Consider it a stand-alone section that should be free of technical jargon. See Model 9–7 on page 266 for an example. Follow these basic guidelines in preparing this important section of your formal reports:

■ Executive Summary Guideline 1: Put It on One Page

The best reason to hold the summary to one page is that most readers expect and prefer this length. It is a comfort to know that somewhere within a long report there is one page to which one can turn for an easy-to-read overview. Moreover, a one-page length permits easy distribution at meetings. When the executive sum-mary begins to crowd your page, it is acceptable to switch to single spacing if such a change helps keep the summary on one page—even though the rest of the report may be space-and-a-half or double-spaced.

Some extremely long formal reports may require that you write an executive summary of several pages or longer. In this case, you still need to provide the reader with a section that summarizes the report in less than a page. The answer to this dilemma is to write a brief "abstract," which is placed right before the executive summary. Consider the abstract to be a condensed version of the executive sum-mary, directed to the highest-level decision-makers. (See chapter 13 for further discussion of abstracts.)

■ Executive Summary Guideline 2: Avoid Technical Jargon

Include only that level of technical language the decision-makers will comprehend. It makes no sense to talk over the heads of the most important readers.

■ *Executive Summary Guideline 3: Include Only Important Conclusions and Recommendations*

The executive summary mentions only major points of the report. An exhaustive list of findings, conclusions, and recommendations can come later at the end of the report. If you have trouble deciding what is most important, put yourself in the position of the readers. What information is most essential for them? If you want to leave them with one, two, or three points about the report, what would these points be? *That* is the information that belongs in the executive summary.

■ *Executive Summary Guideline 4: Avoid References to the Report Body*

Avoid the tendency to say that the report provides additional information. It is understood that the executive summary is only a generalized account of the report's contents. References to later sections do not provide the busy reader with further understanding.

An exception is those instances when you are discussing issues that involve danger or liability. Here you may need to add qualifiers in your summary—for example, "As noted in this report, further study will be necessary." Such statements protect you and the client in the event the executive summary is removed from the report and used as a separate stand-alone document.

■ *Executive Summary Guideline 5: Use Paragraph Format*

Whereas lists are often appropriate for body sections of a report, they can give executive summaries a fragmented effect. Instead, the best summaries create unity with a series of fairly short paragraphs that flow together well. Within a paragraph, there can be a short listing of a few points for emphasis (see Model 9–7 on page 266), but the listing should not be the main structural element of the summary.

Occasionally, you may be convinced that the paragraph approach is not desirable. For example, a project may involve a series of isolated topics that would not mesh into unified paragraphs. In this case, use a modified list. Start the summary with a brief introductory paragraph, followed by a numbered list of three to nine points. Each numbered point should include a brief explanation (for example, "3. *Sewer Construction:* We believe that seepage influx can be controlled by. . . . 4. *Geologic Fault Evaluation:* We found no evidence of surficial. . . .").

■ *Executive Summary Guideline 6: Write the Executive Summary Last*

Only after finishing the report do you have the perspective to write a summary. Approach the task this way. First, sit back and review the report from beginning to end. Then ask yourself, "What do my readers really need to know if they had only three to five minutes to read?" The answer to that question becomes the core of your executive summary.

Introduction

View this section as your chance to prepare both technical and nontechnical readers for the discussion ahead. You do not need to summarize the report, for your executive summary has accomplished that goal. Instead, give information on the report's purpose, scope, and format, as well as a project description. Follow these basic guidelines, which are reflected in the Model 9–8 example on page 267:

■ *Introduction Guideline 1: State Your Purpose and Lead-in to Subsections*

The purpose statement for the document should appear immediately after the main introduction heading ("Smith Technologies hired McDuff Engineers to design a foundation for the new Hilltop Building in Franklin, Maine"). Follow it with a sentence that mentions the introduction subdivisions to follow ("This introduction provides a description of the project site and explains the scope of activities we conducted").

■ *Introduction Guideline 2: Include a Project Description*

Here you need to be precise about the project. Depending on the type of project, you may be describing a physical setting, a set of problems that prompted the report study, or some other data. The information may have been provided to you or you may have collected it yourself. Accuracy in this section will help prevent any later misunderstandings between you and the reader. (When the project description would be too long for the introduction, sometimes it is placed in the body of the report.)

■ *Introduction Guideline 3: Include Scope Information*

This section must outline the precise objectives of the study. Include all necessary details, using bulleted or numbered lists when useful. Your listing or description should parallel the order of the information presented in the body of the report. Like the project description, this subsection must be accurate in every detail. Careful, thorough writing here can prevent later misunderstanding about the tasks you were hired to perform.

■ *Introduction Guideline 4: Consider Including Information on Report Format*

Often the scope section lists information as it is presented in the report. If this is not the case, end the introduction with a short subsection on the report format. Here you can give readers a brief preview of the main sections that follow. In effect, the section acts as a condensed table of contents and may list the report's major sections and appendices.

Discussion Sections

Discussion sections compose the longest part of formal reports. In general, they are written for the most technically oriented members of your audience. You can focus on facts and opinions, demonstrating the technical expertise that the reader expects from you. General guidelines for writing the report discussion are listed here. For a complete example of the discussion component, see the formal-report example in Model 9–9 on pages 268–284.

■ Discussion Guideline 1: Move From Facts to Opinions

As you have learned, the ABC format requires that you start your formal report with a summary of the most important information. That is, you skip right to the essential conclusions and recommendations the reader needs and wants to know. Once into the discussion section, however, you back up and adopt a strategy that parallels the stages of the technical project itself. You begin with hard data and move toward conclusions and recommendations (that is, those parts that involve more opinion).

One way to view the discussion is that it should follow the order of a typical technical project, which usually involves these stages:

First, you collect data (samples, interviews, records, etc.).

Second, you subject these data to verification or testing (lab tests or computer analyses, for example).

Third, you analyze all the information, using your professional experience and skills to form conclusions (or convictions based on the data).

Fourth, you develop recommendations that flow directly from the conclusions you have formed.

Thus, the body of your report gives technical readers the same movement from fact toward opinion that you experience during the project itself. There are two reasons for this approach, one ethical and the other practical. First, as a professional, you are obligated to draw clear distinctions between what you have observed and what you have concluded or recommended. Second, your reports will usually be more persuasive if you give readers the chance to draw conclusions for themselves. If you move carefully through the four-stage process just described, readers will be more likely to reach the same conclusions that you have drawn.

■ Discussion Guideline 2: Use Frequent Headings and Subheadings

Headings give readers handles by which to grasp the content of your report. They are especially needed in the report body, which presents technical details. Your readers will view headings, collectively, as a sort of outline by which they can make their way easily through the report.

■ *Discussion Guideline 3: Use Listings to Break Up Long Paragraphs*

Long paragraphs full of technical details irritate readers. Use paragraphs for brief explanations, not for descriptions of processes or other details that could be listed.

■ *Discussion Guideline 4: Use Illustrations for Clarification and Persuasion*

A simple table or figure can sometimes be just the right complement to a technical discussion in the text. Incorporate illustrations into the report body to make technical information accessible and easier to digest.

■ *Discussion Guideline 5: Place Excessive Detail in Appendices*

Today's trend is to place cumbersome detail in appendices that are attached to formal reports, rather than weighing down the discussion with this detail. In other words, you give readers access to supporting information without cluttering up the text of the formal report. Of course, you need to refer to appendices in the body of the report and label appendices clearly so that readers can locate them easily.

Tabbed sheets are a good way to make all report sections, including appendices, very accessible to the reader. Consider starting each section with a tabbed sheet so that the reader can "thumb" to it easily.

Conclusions and Recommendations

This final section of the report should give readers a place to turn to for a *comprehensive* description—sometimes in the form of a *listing*—of all conclusions and recommendations. The points may or may not have been mentioned in the body of the report, depending on the length and complexity of the document. Conclusions, on the one hand, are convictions or beliefs based on the findings of your study. Recommendations, on the other hand, are actions you are suggesting based on your conclusions. For example, your conclusion may be that there is a dangerous level of toxic chemicals in a town's water supply. Your recommendation may be that the toxic site near the reservoir should be immediately cleaned.

What distinguishes this last section of the report text from the executive summary is the level of detail and the audience. The conclusions and recommendations section provides an *exhaustive* list of conclusions and recommendations for technical and management readers. The executive summary provides a *selected* list or description of the most important conclusions and recommendations for decision-makers, who may not have technical knowledge.

In other words, view the conclusions and recommendations section as an expanded version of the executive summary. It usually assumes one of these three headings, depending of course on the content:

1. Conclusions
2. Recommendations
3. Conclusions and Recommendations

Another option for reports that contain many conclusions and recommendations is to separate this last section into two sections: (1) Conclusions and (2) Recommendations.

FORMAL REPORT EXAMPLE

Model 9–9 (pp. 268–284) provides a long, formal technical report from McDuff, Inc. It contains the main sections discussed previously, with the exception of the list of illustrations. (Because the report's figures are in the appendices, they are listed at the end of the table of contents.) Marginal annotations indicate how the model reflects proper use of this chapter's guidelines for format and organization.

The report results from a study that McDuff completed for the City Transit Authority (CTA) of Alba, Texas. The CTA wants to install part of its light-rail transit system along a right-of-way now owned by the Southern Pacific Railroad. Before it begins the project, however, it needs to know what environmental hazards exist along the corridor. McDuff completes what it calls a "Phase 1 Environmental Site Assessment."

Members of the CTA come from both technical and nontechnical backgrounds. Some are full-time professionals hired by the city, whereas others are part-time, nonpaid citizens appointed by the mayor. The paid professionals include engineers, environmental specialists, accountants, city planners, managers, lawyers, real estate experts, and public-relations specialists. The part-time appointees include citizens who work in a variety of blue-collar and white-collar professions or who are homemakers.

CHAPTER SUMMARY

In your career you will write formal reports for large, complex projects—either inside or outside your organization. In both cases, you will be sending the report to people with different technical backgrounds. This complex audience will respond best to reports that subscribe to the ABC format, for it organizes information so that different readers can read different sections of the report. Although long-report formats vary according to company and profession, most will have these eight basic parts: cover/title page, letter/memo of transmittal, table of contents, list of illustrations, executive summary, introduction, discussion sections, and conclusions and recommendations. Follow the specific guidelines in this chapter for these sections. The annotated model can serve as your reference.

ASSIGNMENTS

Part 1 assignments ask you to evaluate a whole report, to write an individual section, or to evaluate an individual section. Part 2 assignments ask you to write complete formal reports. Remember to submit Planning Forms with the Part 2 assignments.

Part 1: Short Assignments

These assignments can be completed by individuals or by teams. If you are instructed to use teams, first review guidelines in chapter 1 on group writing.

1. **Evaluation—A Formal Report.** Use Model 9–9 on pages 268–284, the complete formal report example in this chapter, for this assignment. The audience for the report is described in the chapter section entitled "Formal Report Example." Although the writer directed the report to a mixed technical and nontechnical audience, some sections clearly are more technical than others.

   - Evaluate the likely audience for each section of the report.
   - Discuss ways that the writer did, or did not, address the needs of specific audience types.
   - Offer suggestions for improving the manner in which the report meets the needs of its intended audience.

2. **Evaluation—A Formal Report.** Locate a formal report written by a private firm or government agency *OR* use a long report provided by your instructor. Determine the degree to which the example follows the guidelines in this chapter. Depending on the instructions given by your teacher, choose between these options:

   - Present your findings orally *or* in writing.
   - Select part of the report *or* all of the report.

3. **Executive Summary.** Choose one of the seven project sheets included in the color insert. Write a brief executive summary for the project. If necessary, provide additional information or transitional wording not included on the sheet, but do not change the nature of the information already provided.

4. **Evaluation—An Introduction.** Review the chapter guidelines for writing an effective introduction to a formal report. Then evaluate the degree to which the following example follows or does not follow the guidelines presented.

INTRODUCTION

McDuff, Inc., has completed a three-week study of the manufacturing and servicing processes at King Radio Company. As requested, we have developed a blueprint for ways in which Computer-Aided Testing (CAT) can be used to improve the company's productivity and quality.

Project Description

M. Dan Mahoney familiarized our project team with the problems that prompted this study of computer-aided testing. According to Mr. Mahoney, the main areas of concern are as follows:

- Too many units on the production line are failing postproduction testing and thus returning to the repair line.
- Production bottlenecks are occurring throughout the plant because of the testing difficulties.
- Technicians in the servicing center are having trouble repairing faulty units because of their complexity.
- Customers' complaints have been increasing, both for new units under warranty and repaired units.

Scope

From May 3–5, 1993, McDuff, Inc., had a three-person team of experts working at the King Radio Company plant. This team interviewed many personnel, observed all the production processes, and acquired data needed to develop recommendations.

Upon returning to the McDuff office, team members met to share their observations and develop the master plan included in this report.

Report Format

This report is largely organized around the two ways that CAT can improve operation at the King Radio plant. Based on the detailed examination of the plant's problems in this regard, the report covers two areas for improvement and ends with a section that lists main conclusions and recommendations. The main report sections are as follows:

- Production and Servicing Problems at King Radio
- CAT and the Manufacturing Process
- CAT and the Servicing Process
- Major Conclusions and Recommendations

The report ends with two appendices. Appendix A offers detailed information on several pieces of equipment we recommend that you purchase. Appendix B provides three recent articles from the journal *CAT Today.* All three deal with the application of CAT to production and service problems similar to those you are experiencing.

Part 2: Longer Assignments

This section contains assignments for writing entire formal reports. Remember to complete the Planning Form for each assignment.

These assignments can be written by individual writers or by group-writing teams. If your instructor has made this a team assignment, review the chapter 1 guidelines on group writing.

5. **Research-Based Formal Report.** Complete the following procedure for writing a research-based report:

- Conduct either a computer-assisted search or a traditional library search on a general topic in a field that interests you. Do some preliminary reading to screen possible specific topics.
- Choose three to five specific topics that would require further research and for which you can locate information.
- Work with your instructor to select the one topic that would best fit this assignment, given your interests and the criteria set forth here.
- Develop a simulated context for the report topic, whereby you select a *purpose* for the report, a specific *audience* to whom it could be addressed (as if it were a "real" report), and a specific *role* for you as a writer.

  For example, assume you have selected "earth-sheltered homes" as your topic. You might be writing a report to the manager of a local design firm on the features and construction techniques of such structures. As a newly hired engineer or designer, you are presenting information so that your manager can decide whether the firm might want to begin building and marketing such homes. This report might present only data, or it could present data and recommendations.
- Write the report according to the format guidelines in this chapter and in consideration of the specific context you have chosen.
- Document your sources appropriately (see chapter 13).

6. **Work-Based Formal Report.** This assignment is based on the work experience that you may have had in the past or that you may be experiencing now.

   ▪ Choose 5 or 10 report topics that are based on your current or past work experience. For example, you could choose "warehouse design" if you stock parts, "check-out procedure" if you work behind the counter at a video-rental store, "report-production procedures" if you work as a secretary at an engineering firm, etc. In other words, find a subject that you know about or about which you can find more information, especially through interviews.
   ▪ Work with your instructor to select the one topic that holds out the best possibilities for a successful report, on the basis of the criteria given here.
   ▪ Develop a context for the report in which you give yourself a *role* in the company where you work(ed). This role should be one in which you would actually write a formal in-house or external report about the topic you have chosen, but the role does not have to be the exact one you had or have. Then select a precise *purpose* for which you might be writing the report and finally a set of *readers* that might read such a report within or outside the organization. Your report can be a presentation of data and conclusions *or* a presentation of data, conclusions, and recommendations.
   ▪ Follow the guidelines included in this chapter for format and organization.

7. **School-Based Formal Report.** This assignment can be completed as an individual project or as a group project. As an individual project, it will rely on observations you have made during the time you spent at a high school, college, or university—either the one where you are taking this course or another you attended previously. As a group project, it will rely on either (1) group members from diverse majors using their varied backgrounds to examine a common campus problem, or (2) group members majoring in the same field or working in the same department exploring a problem they have in common.
   Whether you write an individual or a group report, follow this general procedure:

   ▪ Assemble a list of 5 or 10 problems that you have observed at your school. These problems might concern (1) the physical campus (as in poor design of parking lots or inadequate lab space), (2) the curriculum (as in the need to update certain courses), (3) extracurricular activities (as in the need for more cultural or athletic events), or (4) difficulties with campus support services (as in red tape during registration).
   ▪ Work with your instructor to choose the one topic for which you can find the most information and for which you can develop the context described here.
   ▪ Collect information in whatever ways seem useful—for example, site observations, surveys to students, follow-up phone calls, or interviews.
   ▪ Submit progress reports at intervals requested by your instructor. (Consult guidelines in the "Progress/Periodic Reports" section of chapter 8.)
   ▪ Consider your *role* to be the one that you, in fact, have—a student or a group of students at the school. Then select as your *reader(s)* the school officials who would actually be in charge of solving the problem you have identified. (You may or may not end up sending the report. Follow the advice of your instructor in this matter.) The *purpose* of this report will be to explain, in great detail, all aspects of the problem *and* to form conclusions as to its cause. If it seems appropriate, you may take one further step to suggest recommendations for a solution—*if* your research has taken you this far. In any case, detail and also tact are important criteria.

8. **McDuff-Based Formal Report.** For this assignment you will place yourself into a role of your choosing at McDuff, Inc. Use the following procedure, which may be modified by your instructor:

1. Review the section at the beginning of this chapter that lists McDuff cases for formal reports to get a sense of when formal reports are used at companies like McDuff.

2. Review the McDuff information in chapter 2, especially with regard to the kinds of jobs people hold at the company and the kinds of projects that are undertaken.

3. Choose a specific job that *you* could assume at McDuff—based on your academic background, your work experience, or your career interests.

4. Choose a specific project that (a) could conceivably be completed at McDuff by someone in the role you have chosen, (b) would result in a formal report directed either inside or outside the company, and (c) would be addressed to a complex audience at two or three of the levels indicated on the Planning Form at the end of the book. Be sure you have access to information that will be used in this simulated report—for example, from work experience, from academic study, or from your interviews of individuals already in the field. (For this assignment, you may want to talk with a professional already in this field, such as a recent graduate in your major.)

5. Prepare a copy of the Planning Form at the end of the book for your instructor's approval—*before* proceeding further with the project.

6. Complete the formal report, following the guidelines in this chapter.

Oceanside's New Industrial Park

Prepared for: City Council
                      Oceanside, California

Prepared by: McDuff, Inc.
                      San Francisco, California

Date: March 3, 1994

MODEL 9–1
Title page with illustration

Mc Duff, Inc.

12 Post Street
Houston, Texas 77000
(713) 555-9781

Report #82-651

July 19, 1994

Belton Oil Corporation
P.O. Box 301
Huff, Texas 77704

Attention: Mr. Paul A. Jones

GEOTECHNICAL INVESTIGATION
DREDGE DISPOSAL AREA F
BELTON OIL REFINERY
HUFF, TEXAS

This is the second volume of a three-volume report on our geotechnical investigation concerning dredge materials at your Huff refinery. This study was authorized by Term Contract No. 604 and Term Contract Release No. 20-6 dated May 6, 1994.

This report includes our findings and recommendations for Dredge Disposal Area F. Preliminary results were discussed with Mr. Jones on July 16, 1994. We consider the soil conditions at the site suitable for limited dike enlargements. However, we recommend that an embankment test section be constructed and monitored before dike design is finalized.

We appreciate the opportunity to work with you on this project. We look forward to assisting you with the final design and providing materials-testing services.

Sincerely,

George Fursten

George H. Fursten
Geotechnical/Environmental Engineer

GHF/dnn

MODEL 9–2
Letter of transmittal

Mc Duff, Inc.

MEMORANDUM

DATE: March 18, 1994
TO: Lynn Redmond, Vice President of Human Resources
FROM: Abe Andrews, Personnel Assistant   *AA*
SUBJECT: Report on Flextime Pilot Program at Boston Office

   As you requested, I have examined the results of the six-month pilot program to introduce flextime to the Boston office. This report presents my data and conclusions about the use of flexible work schedules.

   To determine the results of the pilot program, I asked all employees to complete a written survey. Then I followed up by interviewing every fifth person off an alphabetical list of office personnel. Overall, it appears that flextime has met with clear approval by employees at all levels. Productivity has increased and morale has soared. This report uses the survey and interview data to suggest why these results have occurred and where we might go from here.

   I enjoyed working on this personnel study because of its potential impact on the way McDuff conducts business. Please give me a call if you would like additional details about the study.

MODEL 9–3
Memo of transmittal

TABLE OF CONTENTS

MODEL 9–4
Table of contents (level C headings included)

TABLE OF CONTENTS

MODEL 9–5
Table of contents with limited heading levels

LIST OF ILLUSTRATIONS

MODEL 9–6
List of illustrations—formal report

EXECUTIVE SUMMARY

Quarterly monitoring of groundwater showed the presence of nickel in Well M–17 at the Hennessey Electric facility in Jones, Georgia. Nickel was not detected in any other wells on the site. Hennessey then retained McDuff's environmental group to determine the source of the nickel.

The project consisted of four main parts. First, we collected and tested 20 soil samples within a 50-yard radius of the well. Second, we collected groundwater samples from the well itself. Third, we removed the stainless steel well screen and casing and submitted them for metallurgical analysis. Finally, we installed a replacement screen and casing built with Teflon.

The findings from this project are as follows:

- The soil samples contained *no* nickel.
- We found *significant* corrosion and pitting in the stainless steel screen and casing that we removed.
- We detected *no* nickel in water samples retrieved from the well after replacement of the screen and casing.

Our study concluded that the source of the nickel in the groundwater was corrosion of the stainless steel casing and screen.

Gives brief background of project.

Keeps verbs in *active voice,* for clarity and brevity.

Uses short list to emphasize major findings.

Emphasizes major conclusion in separate paragraph (note that this major point *could* have been placed after first paragraph, for a different effect).

MODEL 9–7
Executive summary—formal report

Gives purpose
of report and
overview of in-
troduction (as
lead-in).

Describes the
task he was
given.

Denotes the ma-
jor activities that
were accom-
plished.

Provides reader
with a preview
of main sections
to follow (as a
sort of "mini"
table of
contents).

INTRODUCTION

This document examines the need for a McDuff, Inc., <u>Human Resources Manual</u>. Such a manual would apply to all U.S. offices of the firm. As background for your reading of this report, I have included (1) a brief description of the project, (2) the scope of my activities during the study, and (3) an overview of the report format.

Project Description

Three months ago, Rob McDuff met with the senior staff to discuss diverse human resources issues, such as performance appraisals and fringe benefits. After several meetings, the group agreed that the company greatly needed a manual to give guidance to managers and their employees. Shortly thereafter, I was asked to study and then report on three main topics: (1) the points that should be included in a manual, (2) the schedule for completing the document, and (3) the number of employees that should be involved in writing and reviewing policies.

Scope of Activities

This project involved seeking information from any McDuff employees and completing some outside research. Specifically, the project scope involved:

- Sending a survey to employees at every level at every domestic office
- Tabulating the results of the survey
- Interviewing some of the survey respondents
- Completing library research on the topic of human resource manuals
- Developing conclusions and recommendations that were based on the research completed

Report Format

To fulfill the report's purpose of examining the need for a McDuff <u>Human Resources Manual</u>, this report includes these main sections:

Section 1: Research Methods
Section 2: Findings of the Survey and Interviews
Section 3: Findings of the Library Research
Section 4: Conclusions and Recommendations

Appendices at the end of the text contain the survey form, interview questions, sample survey responses, and several journal articles of most use in my research.

MODEL 9-8
Introduction—formal report

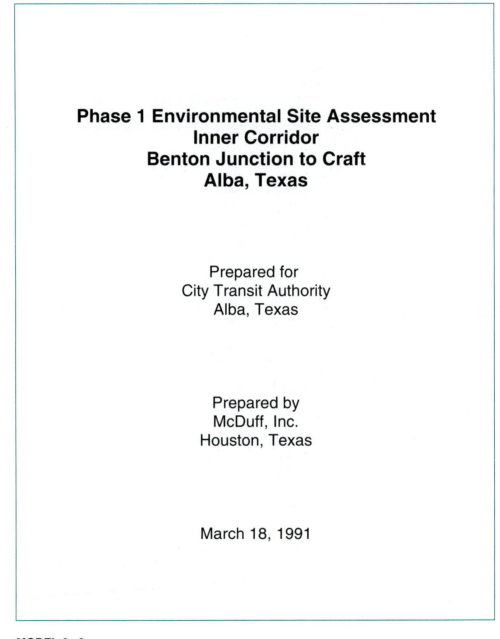

Phase 1 Environmental Site Assessment
Inner Corridor
Benton Junction to Craft
Alba, Texas

Prepared for
City Transit Authority
Alba, Texas

Prepared by
McDuff, Inc.
Houston, Texas

March 18, 1991

MODEL 9–9
Formal report

Mc Duff, Inc.

12 Post Street
Houston, Texas 77000
(713) 555-9781

March 18, 1991

City Transit Authority
500 State Street
Alba, Texas 75001

ATTENTION: Mr. Ralph Letson
                      Manager of Engineering Support Services

Lists project title as it appears on title page.

Gives brief project description and authorization information.

Provides single most important finding or conclusion.

PHASE I ENVIRONMENTAL SITE ASSESSMENT
INNER CORRIDOR
BENTON JUNCTION TO CRAFT
ALBA, TEXAS

We have now completed our Phase I environmental site assessment for the proposed City Transit Authority (CTA) Inner Corridor. The project was authorized on January 6, 1991. We performed the work in accordance with our revised proposal letter, McDuff Proposal No. 88-P34, dated December 16, 1990, and under the provisions of CTA Contract #601.

The enclosed report presents a preliminary evaluation of possible environmental risks at sites located within 300 ft of the proposed CTA Inner Corridor. The corridor is within the Southern Gulf Railroad right-of-way near West Drive. We have considered about 2.75 miles of the alignment located between Benton Junction and Craft.

In this Phase 1 environmental site assessment, we identified 24 sites along the alignment that may have *some* potential for environmental risks, 13 of which are considered to have a *high* potential for such risks. We have outlined an additional scope of work (Phase 2) to evaluate the possibility that these 13 sites have affected the proposed CTA project.

Thank you for the opportunity to work with you on this Phase 1 environmental site assessment. If we may be of further assistance, please call us.

Sincerely,

Marie S. Harris

Marie S. Harris, P.E.

MODEL 9–9, *continued*

continues

TABLE OF CONTENTS

Uses white space and indenting to accent organization of report.

MODEL 9–9, *continued*

270

EXECUTIVE SUMMARY

The City Transit Authority (CTA) hired McDuff, Inc., to perform a preliminary environmental site assessment of a 2.75-mile segment of the Inner Corridor alignment. The section of alignment addressed in this study is along the Southern Gulf Railroad right-of-way between Benton Junction and Craft. For this study, we have considered a 300-ft wide band along both the north and south boundaries of the right-of-way.

McDuff has identified 24 sites with *some* potential for environmental risks that could affect the CTA alignment. Of the 24 sites, we believe that 13 have a *high* potential for environmental risk. These professional opinions are based on a review of environmental site listings, a review of historical aerial photographs, and fieldwork at the site.

McDuff recommends that a Phase 2 sampling and analytical testing program be performed near these high-risk sites. This program will help assess the possible impact of environmental problems on the proposed CTA alignment.

1

MODEL 9–9, *continued*

continues

INTRODUCTION

States purpose of study.

McDuff, Inc., performed a Phase 1 environmental site assessment for a 2.75-mile segment of the Inner Corridor. This report includes a review of potential environmental risks from (1) environmental sites, (2) historical land use, and (3) current land-use conditions.

Project Description

Gives a brief project description.

The proposed CTA Inner Corridor is located along the Southern Gulf Railroad right-of-way near West Drive in Alba, Texas. The portion of the Inner Corridor alignment addressed in this study is located between Benton Junction and Craft, as shown in Figure 1. The alignment is about 2.75 miles long.

Scope

The purpose of the Phase 1 environmental site assessment was to evaluate the potential for selected environmental risks. The project sites were located within 300 ft north and 300 ft south along the alignment. Selected environmental risks considered for this project include:

- Nearness to sites and hazardous waste spills documented by the regulatory agencies

Uses bulleted list to emphasize scope of activities.

- Current and former land-use activities associated with waste disposal or industrial operations

To accomplish this purpose, our study included four main tasks:

- A search for listed and potential environmental sites within 300 ft north and south of the 2.75-mile alignment
- A review of historical aerial photographs to evaluate former land-use activities associated with waste disposal, oil and gas explorations, or industrial operations
- A site reconnaissance to evaluate current land use and to evaluate any anomalies observed during the aerial photograph review
- A report to present our preliminary evaluation for sites located within 300 ft north and south of the project alignment

Report Format

This proposal includes these four main sections:

Leads into rest of report with list of main sections to follow.

1. **Technical Approach:** a complete discussion of our activities in completing the project
2. **Results:** a presentation of the data accumulated on the sites, particularly with regard to previous and current use
3. **Site Risk Evaluation:** a description of the criteria for evaluating risk
4. **Conclusions and Recommendations:** a summary of the potential effects of the result of this study, along with recommendations for further study

2

MODEL 9–9, *continued*

TECHNICAL APPROACH

The technical approach used for this Phase 1 study consisted of an environmental site search, a review of aerial photograhs, and a site reconnaissance. A
description of each task follows:

Environmental Site Search

McDuff reviewed possible environmental hazards located within 300 ft north
and 300 ft south of the alignment. We collected information from the following
selected sources:

- Texas Water Commission (TWC)
- Region VI U.S. Environmental Protection Agency (EPA)
- Texas Department of Health (TDH)
- Geomap Company (GC)
- Alba Fire Department (AFD)

Texas Water Commission (TWC) Listing. The TWC listing includes the
generator's list, the underground storage tank (UST) registration list, and the Texas
Registry List.

The generator's list typically includes sites that generate, transport, store, or
dispose of potential industrial solid waste materials. A TWC generator's listing does
not imply that a listed site has documented hazardous waste problems. The TWC
generator's file dated May 17, 1988, was used in our site search (TWC Industrial
Solid Waste System, General Information Report).

The UST registration list typically includes underground storage tanks that have
been registered with the TWC. A UST listing does not imply that the underground
storage tank leaks. The UST registration list, dated October 1985, was used in our
search.

The Texas Registry List includes a listing of sites published in the January 22,
1991, Texas Registry and includes "hazardous waste facilities or areas which may
constitute an imminent and substantial endangerment to public health and safety or
to the environment." The Texas Registry is more commonly known as the Texas
Superfund sites.

Region VI EPA Listing. The Region VI EPA listing consists of two lists:
RCRA and CERCLIS. (These acronyms refer to two legislative acts that help
identify, control, and clean up environmental waste.) The RCRA list is composed of
sites that generate, transport, store, or dispose of regulated hazardous materials.
The RCRA notifiers listing, dated May 31, 1989, was used in our search. A RCRA
listing does not imply that the site has documented hazardous waste problems.

3

MODEL 9–9, *continued*

continues

The CERCLIS list is composed of sites that have documented on-site hazardous waste problems. The CERCLIS listing, dated September 29, 1988, was used in our search. The CERCLIS listing also includes National Priorities List (NPL) sites, or Superfund sites.

Texas Department of Health (TDH) Listing. The TDH listing includes sites that have been permitted or for which applications have been submitted to receive municipal solid wastes. The TDH listing, dated September 30, 1990, was used in our search.

Geomap Company Map. Geomap Company provides a map that locates documented oil and gas wells and provides the well name and operator. This map is constructed at a 1 in.-4000 ft scale. The Geomap, dated April 15, 1988, was used in our search for documented oil and gas wells.

Alba Fire Department Listing. The Alba Fire Department-Hazardous Materials Division provides a listing of documented hazardous materials spills. The approximate spill location, spill type, and spill date are provided. This listing is updated daily by the Hazardous Materials Division. Spill records from 1980 through January 11, 1991, were included in this search.

Aerial Photograph Review

McDuff reviewed photographs for surficial anomalies indicative of possible fill areas, oil and gas exploration activities, and industrial development. Photographic coverage throughout the proposed alignment was obtained from the U.S.D.A. Agricultural Stabilization and Conservation Service (ASCS) for the following years:

Year Scale	Photograph Number	Approximate Scale
1953	BQY-16M-33R	1 in.–200 ft.
1953	BQY-16M-33L	1 in.–200 ft.
1957	BQY-6T-32R	1 in.–200 ft.
1957	BQY-6T-32L	1 in.–200 ft.
1957	BQY-6T-57R	1 in.–200 ft.
1964	BQY-3FF-71R	1 in.–200 ft.
1964	BQY-3FF-71L	1 in.–200 ft.
1964	BQY-3FF-14R	1 in.–200 ft.
1973	48201-173-172X	1 in.–200 ft.
1981	48201-281-78X	1 in.–200 ft.
1981	48201-281-96X	1 in.–200 ft.

Incorporates simple, informal table into text of report.

4

MODEL 9–9, *continued*

Uses *action* verbs and maintains parallel structure in listing.

Site Reconnaissance

We performed site visits on January 13, 1991, and February 8, 1991. They consisted of traveling along the alignment from Benton Junction to Craft. The purposes of the site visits were as follows:

- Observe sites that were identified during the environmental site search
- Observe currently operating businesses along the alignment and identify those businesses associated with potential environmental risks
- Observe anomalies identified from the aerial photograph review
- Document the existing condition of the alignment with photographs

Potential Environmental Sites. A reconnaissance of the sites identified during the environmental site search was performed from adjacent properties. Access to these sites was not attempted. The approximate distance from the proposed CTA alignment was also checked.

Current Businesses. The site reconnaissance also included observations of the types of currently operating businesses located within about 300 ft north and south of the CTA alignment. Observations were made from adjacent properties. These observations were aimed at identifying nonlisted sites that may have the potential for high environmental risk operations.

Aerial Photograph Anomalies. The site reconnaissance was also aimed at field observation of anomalous areas disclosed from the aerial photograph review.

Site Photographs. Photographs were taken during the February 8, 1991, site reconnaissance to document the existing condition of the alignment. After completion of the environmental site risk evaluation, a site reconnaissance was made on March 15, 1991, to photograph the high-risk sites.

5

MODEL 9–9, *continued*

continues

States main
point up front in
section.

Uses expanded
listing to orga-
nize information
from sites.

RESULTS

We have based this finding on our environmental site search, aerial photograph review, and site reconnaissance. We consider 24 sites located within about 300 ft north and 300 ft south of the alignment to have potential for selected environmental risks.

A list of the potential sites is presented in Appendix A. The following section presents the results of our Phase 1 study and provides a preliminary evaluation of environmental sites, historical land use, and current land use.

Environmental Sites

Eleven sites were documented as having registered underground storage tank (UST) facilities, as indicated on Appendix A. None of the registered tank facilities is listed within the TWC active leaking underground storage tank data base, dated December 15, 1990. Seven AFD spill sites were documented and are indicated on Appendix A. No registered TDH facilities were documented within the alignment area. The Geomap Company did not document any oil and gas wells located within the project alignment. Seven sites were identified within the TWC generator's listing and the Region VI EPA RCRA listing, as indicated on Appendix A.

Regulatory agency site files for the TWC and RCRA sites were reviewed for possible compliance violations and enforcement actions. Available files were received from the TWC district office in Deer Park, Texas, and the EPA Region VI headquarters in Dallas, Texas. What follows is a summary of the pertinent findings for each site for which files were available:

• **J.S. Jones Company**—EPA files show that this facility generates hazardous wastes. The materials are ink wastes. A Generators Biennial Hazardous Waste Report dated 1988 indicated that the firm had generated about 55 gal. of ink wastes since 1982. No violations were noted.

• **XYZ Type Compositors**—EPA file information shows that this facility was not registered by the EPA as a hazardous-waste generator. All records were transmitted to the Texas Department of Health. No violations were noted.

• **Jack Parch Ford**—EPA file information shows that this facility generates less than 1000 kg per month of ignitable wastes. No violations were noted.

• **Manoplate**—EPA file information shows that this facility is a small-quantity generator that produces nonspecific ignitable, corrosive, and toxic wastes. No violations were noted. TWC file information indicates that the facility engages in printing, developing, and color manufacturing. A letter to Manoplate from the TWC, dated June 4, 1987, cited violations for not registering all generated hazardous wastes. TWC lists the facility as a small-quantity generator.

6

MODEL 9–9, *continued*

- **Paste-Randall Printing**—EPA file information indicates that this facility generates hazardous wastes. Typical wastes include formic acid, trichlorethylene sludge, cooling oil, and naptha. No violations were noted. TWC file information included diazo coating wastes and indications of an aboveground and a belowground storage tank. No violations were noted.

- **Tansley Palms Chrysler/Plymouth**—EPA file information indicates that this facility generates small quantities of ignitable wastes. No violations were noted.

- **Buck Service Station**—EPA file information indicates that this facility generates small quantities of ignitable, corrosive, and toxic wastes. A letter from Buck Oil to the EPA, dated February 23, 1988, stated that after evaluating the station, Buck finds that it is not a hazardous-waste generator and has cancelled the generator number. No ackowledgment by the EPA was noted. No violations were noted.

Historical Land Use

McDuff reviewed aerial photographs to evaluate historical land use throughout the alignment. Aerial photographs indicated that the Southern Gulf Railroad right-of-way was established before 1953. We observed that land along the alignment is used for a mixture of undeveloped, commercial, industrial, and residential purposes.

We observed no significant surficial anomalies that would indicate possible waste ponds, dumps, or oil and gas well sites. However, we observed an electrical substation located next to the alignment at Wakeforest on the 1957 photographs.

Our review of historical aerial photographs suggests that most of the adjacent land use to the north included commercial and industrial operations. Sites located south of the alignment generally were used for residential purposes. However, in the vicinity of Narden Speedway to Gilbert, the land use immediately south of the alignment is typically more commercial and industrial. Most residential land use in this area is located south of the 300-ft area included in this study.

Current Land Use

We conducted a site reconnaissance on January 13, 1991, to assess current land use. The findings from our site reconnaissance indicate that three sites, in addition to the environmental site listings, could have a potential for environmental risks. These sites include the A&L Substation, Quansi-Cola Bottling Co., and SDS Electronics. The sites were previously identified during the aerial photo review and relisted on Appendix A because of the industrial/commercial nature of the site operations. The A&L Substation was noted on the 1957 aerial photographs.

During the site reconnaissance, we noted no evidence of significant waste or remnant product spillage at the documented AFD sites, at the regulatory agency sites, or at other industrialized areas along the alignment. We did not observe other significant industrial or commercial areas identified on the 1981 aerial photographs.

7

MODEL 9–9, *continued*

continues

SITE RISK EVALUATION

After identifying the 24 potential environmental sites, we screened each site into two environmental risk classifications—high and low. This section outlines the screening criteria and discusses the ranking of sites.

Environmental Screening Criteria

Criteria used in the environmental site screening included:
- Presence of registered underground storage tanks (UST)
- Presence of unregistered UST
- Presence of potential PCB-containing electrical equipment
- Documented EPA CERCLIS sites
- Regulatory compliance violations and enforcement actions
- Presence of oil and gas wells
- Presence of hazardous material spill remnants
- Presence of industrial and selected commercial operations

A high environmental risk potential was assigned to each site meeting one or more of the screening criteria.

Several screening criteria deserve special mention. First, note that we considered the possible presence of leaking undergound storage tanks (USTSs) and polychlorinated byphenyls (PCBs). Either one can cause soil contamination. Second, note that we also considered any listing on EPA's CERCLIS list, for such a listing indicates that a site has documented hazardous waste problems. However, our study did not evaluate the type or extent of any actual soil contamination or migration of hazardous waste.

Site Risk Ranking

On the basis of the results of our environmental site screening, a total of 13 out of the 24 environmental sites were classified as having a high potential for preexisting environmental liabilities. A summary of the high-risk environmental sites is presented in Appendix B. A site reconnaissance was made on March 15, 1991, to obtain photographic documentation of the 13 high-risk sites. (Photographs of each high-risk site are available upon request.)

We observed no indications of on-site underground storage tanks or waste handling practices at SDS Electronics and Quansi-Cola Bottling Co.; therefore, these sites were not included as high-risk sites. Documented underground storage tanks were present at 11 of the 13 high-risk sites. The remaining two high-risk sites have the potential for PCB contamination.

8

MODEL 9–9, *continued*

CONCLUSIONS AND RECOMMENDATIONS

What follows are the conclusions of our current study and some recommendations for further work at some of the sites.

Conclusions

High-risk environmental sites located within about 300 ft north and south of the
proposed alignment may affect the condition of the property being considered by
CTA. We consider the 13 high-risk sites identified during our preliminary study to
have a greater likelihood of environmental impact than other properties located
within the 600-ft study band.

Potential environmental impacts from the high-risk sites may result from
transporting contaminants from adjacent properties onto the proposed CTA
alignment. Contaminant transport may occur by surface water runoff, groundwater
migration, airborne particulate emissions, and physical soil/waste removal.

Because of the nature of our preliminary study, we have identified sites that
may affect the CTA alignment. Further evaluation of high-risk sites is necessary to
assess whether environmental impacts are *probable*.

Recommendations for Phase 2 Evaluation

We recommend a Phase 2 study to evaluate the potential for environmental
risks along the CTA alignment. The Phase 2 study would consist of these tasks:

- A field exploration along the alignment with approximate boring locations at the
projections of the high-risk sites on the alignment

- A laboratory test program to analyze selected samples for possible hydrocarbon
and/or PCB contamination

- The preparation of a report documenting our findings and our opinion of probable
environmental damage

General Scope of Work. Underground fuel storage tanks were present at 11
of the 13 high-risk sites. Therefore, we consider hydrocarbons the primary
contaminant of concern. Possible PCB compounds may be present at the remaining
two high-risk sites. The general scope of work for each high-risk site located next to
the alignment right-of-way is presented as follows.

The field exploration will consist of drilling soil borings and collecting soil
samples to evaluate the potential presence of shallow subsurface and surface
hydrocarbon contaminants. Borings will be drilled within the alignment right-of-way
near the high-risk sites. Boring depths will extend to the first water-bearing zone
where hydrocarbons, if present, are typically encountered.

9

MODEL 9–9, *continued*

continues

We will sample the individual borings at 2-ft intervals throughout the borehole depth. The soil sampler will be cleaned between each sampling interval with a trisodium phosphate (TSP) wash followed by a deionized-water rinse, to reduce the potential for cross-contamination. Upon completion of soil sampling activities, the boreholes will be backfilled with a bentonite-cement grout. All field activities will be conducted in accordance with an OSHA site safety plan using personnel trained for hazardous-waste site work, in accordance with 29 CFR Part 1910.120.

The laboratory test program will consist of screening the recovered samples to select representative soil samples for analytical testing. Each recovered sample will be screened for volatile organic response using a portable photoionization detector (PID). On the basis of the field observations and the PID headspace screening results, we will select representative soil samples from each borehole for testing. The laboratory analysis will be concerned with (1) total petroleum hydrocarbons (TPH), in accordance with EPA Method 418.1; (2) benzene, toluene, ethylbenzene, and total xylenes (BTEX), in accordance with EPA Method 5030; and/or (3) polychlorinated biphenyls (PCBs) in accordance with EPA Method 608.

We will prepare an engineering report documenting the data and findings of our field exploration and laboratory testing. The report will describe the work tasks and procedures we used to complete these tasks. We will also provide our opinions about the potential for on-site environmental damage from the 13 high-risk sites.

Scope Quantities. We propose to drill 42 borings to evaluate the 13 high-risk sites. These borings will be located along the alignment at the approximate projection of the high-risk environmental sites. For estimating purposes, we assumed that most boring depths will be 25 to 30 ft. We based this assumption on limited subsurface information from our previous geotechnical report of the System Connector (McDuff File No: 88–139; dated May 27, 1990). The boring depths may be adjusted depending on the actual depths of target soil strata. The proposed Phase 2 work scope is presented in Appendix C. Supporting price estimates are presented in Appendix D.

10

MODEL 9–9, *continued*

APPENDIX A: POTENTIAL ENVIRONMENTAL SITES

Facility Name	Location	File
1. J.S. Jones Company	304 South St.	RCRA/TWC
2. XYZ Type Compositors	102 South St.	RCRA/TWC
3. Jack Parch Ford	304 South Street	RCRA/TWC/UST
4. Jack Parch Ford Paint and Body Shop	26 Mill Street	Recon/UST
5. Manoplate	15 South St.	RCRA/TWC
6. Paste-Randall Printing	32 Greenbriar	RCRA/TWC/UST
7. Tansley Palms Chrysler/Plymouth	35 Kirb Drive	RCRA/TWC/UST
8. Buck Service Station	39 South St.	RCRA/TWC/UST
9. Tex Service Station	67 South St.	UST
10. Barnes Building Material	45 Greenbriar	UST
11. Monser Auto Rental	988 Wakeforest	UST
12. Alba Van Lines	488 South St.	UST
13. Phil's Gas Station	766 Anster Place	UST
14. Argo 44 Service Station	66 South St.	UST
15. A&L Substation	877 Wakeforest	Recon
16. SDS Electronics	430 South St.	Recon
17. Quansi-Cola Bottling	26 Bisonet	Recon
18. Transformer explosion	Wakeforest at Gren	AFD
19. Diesel spill	South St. at Bender Ave.	AFD
20. Diesel spill	South St. at Yazo St.	AFD
21. Gasoline spill	229 South St.	AFD
22. Hydrochloric acid spill	34 South St.	AFD
23. Methane	South St. at Kirb	AFD
24. Diesel spill	South St. at Bisonet	AFD

11

MODEL 9-9, *continued*

continues

APPENDIX B: POTENTIAL HIGH-RISK ENVIRONMENTAL SITES

Facility (number and name from Appendix A)	Files
3. Jack Parch Ford	RCRA/TWC/UST
4. Jack Parch Ford Paint and Body Shop	Recon/UST
6. Paste-Randall Printing	RCRA/TWC/UST
7. Tansley Palms Chrysler/Plymouth	RCRA/TWC/UST
8. Buck Service Station	RCRA/TWC/UST
9. Tex Service Station	UST
10. Barnes Building Material	UST
11. Monser Auto Rental	UST
12. Alba Van Lines	UST
13. Phil's Gas Station	UST
14. Argo 44 Service Station	UST
15. A&L Substation	Recon
18. Transformer explosion	AFD

12

MODEL 9–9, *continued*

	No. of Borings	Total Footage	No. of Headspace Readings	Laboratory Tests		
				TPH	BTEX	PCB
3. Jack Parch Ford	4	120	60	4	4	--
4. Jack Parch Ford Paint and Body Shop	3	90	45	3	3	--
6. Paste-Randall Printing	3	90	45	3	3	--
7. Tansley Palms Chrysler/Plymouth	4	120	60	4	4	--
8. Buck Service Station	3	90	45	3	3	--
9. Tex Service Station	3	90	45	3	3	--
10. Barnes Building Material	3	90	45	3	3	--
11. Monser Auto Rental	2	60	30	2	2	--
12. Alba Van Lines	3	90	45	3	3	--
13. Phil's Gas Station	3	90	45	3	3	--
14. Argo 44 Service Station	3	90	45	3	3	--
15. A&L Substation	4	40	20	--	--	10
18. Transformer explosion	4	40	20	--	--	10
	42	1100	550	34	34	20

MODEL 9–9, *continued*

continues

APPENDIX D: PHASE 2 COST ESTIMATES

Task 1 – Soil Borings	Item	Units	Rate	Total
Field Coordination	Project Engineer	30	43.58	1,307
Borehole Staking	Senior Soil Tech.	25	26.80	670
Utility Clearances	Senior Soil Tech.	20	26.80	536
Drilling & Sampling	Contract Driller	140	94.00	13,160
Borehole Logging	Senior Soil Tech.	140	26.80	3,752
Site Safety Supplies	Disposables	expense	cost	700
Decontamination Supplies	Disposables	expense	cost	1,100
Borehole Backfill	Grouting	1,100	1.50	1,650
			Total Task 1:	$22,875

Task 2 – Laboratory Testing	Item	Units	Rate	Total
Sample Screening	Staff Engineer	40	30.17	1,207
Sample Delivery	Staff Engineer	35	30.17	1,056
TPH Tests	Soil	34	40	1,360
BTEX Tests	Soil	34	75	2,550
PCB Tests	Soil	20	70	1,400
			Total Task 2:	$ 7,573

Task 3 – Evaluation and Report	Item	Units	Rate	Total
Project Administration	Principal	30	76.64	2,299
Project Management	Project Manager	75	57.83	4,337
Report Preparation	Project Engineer	90	43.58	3,922
	Staff Engineer	70	34.75	2,433
	Draftsman	30	15.51	465
	Typist	50	19.94	997
Miscellaneous	Reproduction	expense	cost	350
			Total Task 3:	$14,803
			Total Phase 2 Budget:	$45,251

14

MODEL 9–9, *continued*

10 | *Proposals and Feasibility Studies*

A McDuff project manager discusses proposed changes to a client's retail outlet.

G ini Harris, a structural technician at McDuff's Houston office, has written many reports in her two years at the company. Last month, however, she had the chance to write her first major proposal. Anchor Productions, a Hollywood film company, asked McDuff for a proposal to build a full-scale replica of the Alamo. In an upcoming film about the Texas battle, most scenes will be filmed in this replica to be built on leased land west of San Antonio. To write her proposal for Anchor, Gini had to seek advice from a variety of technical specialists, as well as architectural historians.

As it happens, Gini wrote a successful proposal that got the job. The work went so well that she asked her supervisor, Ken Blair, if McDuff could start a technical group to work just on historical restoration/replication projects around the country. Intrigued by the idea, Blair asked Harris to prepare a *feasibility study* that would focus on start-up costs, involvement of McDuff offices, hiring needs, and potential profit during the first five years. If the feasibility study showed promise, Blair would present the idea to McDuff's corporate management.

Although you might not write proposals to build an Alamo, you will write proposals and feasibility studies in your own field during your career. These modes of writing are crucial to most organizations. Indeed, many companies rely on them, especially proposals, for their very survival. Proposals and feasibility studies are defined as follows:

Proposals: documents written to convince your readers to adopt an idea, a product, or a service. They can be directed to colleagues inside your own organization (*in-house proposals*), to clients outside your organization (*sales proposals*), or to organizations that fund research and other activities (*grant proposals*).

In all three cases, proposals can be presented in either a short, simple format (*informal proposal*) or a longer, more complicated format (*formal proposal*). Also, proposals can be either requested by the reader (*solicited*) or submitted without a request (*unsolicited*).

Feasibility studies: documents written to show the *practicality* of a proposed policy, product, service, or other change within an organization. Often prompted by ideas suggested in a *proposal,* they examine details such as costs, alternatives, and likely effects. Though they must reflect the objectivity of a report, most feasibility studies also try to convince readers either (1) to adopt or reject the one idea discussed or (2) to adopt one of several alternatives presented in the study.

Feasibility studies can be *in-house* (written to decision-makers in your own organization) or *external* (requested by clients from outside your organization).

There are four main sections in this chapter. The first gives specific situations in which you might write proposals and feasibility studies at McDuff. The second and third sections discuss informal and formal proposal formats, while also denoting differences between in-house and sales versions of these formats. The fourth section covers feasibility studies. All chapter guidelines are followed by annotated examples to use as models for your own writing.

PROPOSALS AND FEASIBILITY STUDIES AT MCDUFF

As the Alamo example shows, proposals and feasibility studies often work together. The proposal may suggest a topic upon which a feasibility study is then written. The flowchart in Figure 10–1 shows another possible communication cycle that would involve both a proposal and a feasibility study. Note that the diagram includes the term *RFP,* which stands for "request for proposal":

Request for proposal (RFP): documents sometimes sent out by organizations that want to receive proposals for a product or service. The RFP gives guidelines on (1) what the proposal should cover, (2) when it should be submitted, and (3) to whom it should be sent. As writer, you should follow the RFP religiously in planning and drafting your proposal.

RFPs generally are *not* used in these situations:
- When the proposal is solicited from within your own organization
- When the proposal is requested less formally, as through a letter, phone call, or memo
- When the proposal is unsolicited, meaning that you are writing it without a request from the person who will read it

The sections that follow describe additional situations in which proposals and feasibility studies would be written at McDuff. Reading through these brief cases will show you the varied contexts for persuasive writing within just one company.

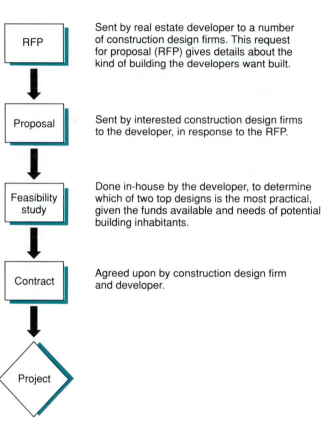

FIGURE 10–1
Flowchart showing the main documents involved in one possible construction project

RFP — Sent by real estate developer to a number of construction design firms. This request for proposal (RFP) gives details about the kind of building the developers want built.

Proposal — Sent by interested construction design firms to the developer, in response to the RFP.

Feasibility study — Done in-house by the developer, to determine which of two top designs is the most practical, given the funds available and needs of potential building inhabitants.

Contract — Agreed upon by construction design firm and developer.

Project

McDuff Proposals

Like many organizations, McDuff depends on (1) in-house proposals to breathe new life into its internal operations, (2) sales proposals to request work from clients, and (3) occasional grant proposals to seek research funds from outside organizations. Proposals are a main activity in healthy, growing organizations like McDuff.

Of the five cases described here, the first and second are internal, requiring in-house proposals; the third and fourth are external, requiring sales proposals; and the fifth is external, requiring a grant proposal.

- **In-house proposal for computer drafting equipment:** Meg Stevens, a drafter at the Denver office, writes an in-house proposal to the drafting manager, Elvin Lipkowsky, in which she proposes that the company purchase a new computer-drafting system. Her proposal includes a schedule whereby the department can shift entirely to computer drafting in the next five years.
- **In-house proposal for retaining legal counsel:** Jake Washington, an employment specialist in the Human Resources Department in the Baltimore office, writes an in-house proposal to Lynn Redmond, vice president of human resources. In it he proposes that the company retain legal counsel on a half-time

basis (20 hours a week). In his position at McDuff, Jake uses outside legal advice in dealing with new hiring laws, unemployment compensation cases, affirmative action regulations, and occasional lawsuits by employees who have been fired. He is proposing that the firm retain regular half-time counsel, rather than dealing with different lawyers as is done now.

■ **Sales proposal for asbestos removal:** Jane Wiltshire, asbestos department manager at McDuff's St. Paul office, regularly talks with owners of buildings that may contain asbestos. After an initial discussion with the head minister of First Street Church, she writes an informal sales proposal in which she offers McDuff's services in performing an asbestos survey of the church building. Specifically, she explains how McDuff will examine the structure for possible asbestos, gives a schedule for completing the survey and writing the final report, and proposes a lump-sum price for the project.

■ **Sales proposal for work on wind turbine project:** A utility company in California plans to build 10 giant turbines in a desert valley in the southern part of the state. The "free" power that is generated will help offset the large increases in fuel costs for the company's other plants. Although the firm has selected a turbine design and purchased the units, it needs to decide where to place them and what kind of foundations to use. Thus it sent out a request for proposal to companies that have experience with foundation and environmental engineering. Louis Bergen, engineering manager at McDuff's San Francisco office, writes a proposal that offers to test the soils at the site, pinpoint the best locations for the heavy turbines, and design the most effective foundations.

■ **Grant proposal for new equipment design:** Oilarus, Ltd., a British oil company, sometimes gives research and development funds to small companies. Such funding usually goes toward development of new technology or products in the field of petroleum engineering. Angela Issam, who works in McDuff's Equipment Development group, decides to apply for some of the funding. Her proposed project, if successful, would provide a new piece of oil drilling safety equipment that would reduce the chance of offshore oil spills at production sites.

McDuff Feasibility Studies

The next two examples show that feasibility studies often flow from proposals. They can be internal, to managers who need facts before making a decision, or they can be external, to clients who request a service.

■ **In-house feasibility study about legal counsel:** Lynn Redmond, the vice president of human resources, recently received Jake Washington's proposal that McDuff retain half-time legal counsel (see second case in previous section). The idea interests her, but she is not convinced of its practicality. She calls a meeting with Jake Washington and Scott Sampson, the personnel manager, who also uses legal counsel on a part-time basis. Because Lynn will have to sell her boss on this idea, she asks Jake and Scott to write a feasibility study on it.

This study, unlike the proposal, must include a detailed comparison of the present mode of operating versus the proposed strategy of retaining a lawyer for 20 hours per week on a regular basis. The study must examine criteria such as current costs versus projected costs, current level of satisfaction versus projected level of satisfaction, and current level of services provided versus projected level of services provided. Lynn asks that the report include a clear recommendation, made on the basis of the data.

- **External feasibility study on plant site:** Tarnak, Inc., a large furniture manufacturing company in North Carolina, has decided to build a new plant in northern Georgia. After getting proposals from many cities that want the plant, the company has narrowed its choices to three spots that are about equal in cost of living, access to workers, standard of living, construction costs, transportation facilities, and access to raw materials. Yet the firm has not studied any site with respect to waste management.

  Specifically, Tarnak needs to know which of the three cities is best prepared to handle the solid and chemical wastes from the plant in a safe and economical manner. It hires McDuff's Atlanta office to write a feasibility study. The first goal of the study is to determine what sites, if any, meet Tarnak's criteria for waste management. If the first objective yields more than one site, McDuff then must compare the sites and recommend the best one.

GUIDELINES FOR INFORMAL PROPOSALS

Like informal reports, informal proposals are short documents that cover projects with a limited scope. But how short is short? And just what does "limited" mean? Following are some specific guidelines that you can use if your employer or client has not provided others:

Use Informal Proposals When:

- The text of the proposal (excluding attachments) is less than 10 pages
- The size of the proposed project is such that a long, formal proposal would appear to be inappropriate
- The client has expressed a preference for a leaner, less formal document

Use Formal Proposals When:

- The text of the proposal (excluding attachments) is 10 pages or more
- The size and importance of the project is such that a formal proposal would be appropriate
- The client has expressed a preference for a more formal document

These two formats can be used for proposals that are either *in-house* (to readers within your own organization) or *external* (to readers outside your organization). The rest of this section provides writing guidelines and an annotated model for informal proposals, the type of persuasive writing you will do most often in your career.

Informal proposals have two formats: (1) memos (for in-house proposals) and (2) letters (for external proposals). The guidelines recommended apply to both. With some variations, they are essentially the guidelines suggested in chapter 8 for informal reports. The formats are much the same, though the content and tone are different. Reports explain, whereas proposals persuade.

■ Informal Proposal Guideline 1: Plan Well Before You Write

Complete the Planning Form at the end of the book for all proposal assignments in class, as well as for proposals you write on the job. Carefully consider your purpose, audience, and organization. Two factors make this task especially difficult in sales-proposal writing:

1. You may know nothing more about the client than what is written on the RFP.
2. Proposals almost always are on a tight schedule that limits your planning time.

Despite these limitations, try to find out exactly who will be making the decision about your proposal. Many clients will tell you if you give them a call. In fact, they may be pleased that you care enough about the project to target the audience. Once you identify the decision-makers, spend time brainstorming about their needs *before* you begin writing. Proposals that betray an ignorance of client needs often do so because the writer began writing too soon about the product or service.

■ Informal Proposal Guideline 2: Use Letter or Memo Format

Letter proposals, such as the example shown in Model 10−1 on pages 311−313, basically follow the format of routine business letters (see chapter 3). This casual style gives readers the immediate impression that your document will be "approachable"—that is, easy to get through and limited in scope. Memo proposals, such as the example shown in Model 10−2 on pages 314−315, follow the format of the internal memorandum (see chapter 7). Here are a few highlights:

- Line spacing is usually single, but it may be 1½ or double, depending on the reader's or company's preference.
- Your last name, the recipient's last name, and the page number appear on sheets after the first.
- Most readers prefer an uneven or "ragged-edge" right margin, as opposed to an even or "right-justified" margin.

Your subject line in a memo proposal (if you choose to use one) gives readers the first impression of the proposal's purpose. Make sure to choose concise yet accurate wording. Furthermore, the wording must match that which you have used in the proposal text. See Model 10−2 on pages xxx–xxx for wording that gives the appropriate information and tries to engage the reader's interest.

■ Informal Proposal Guideline 3: Make Text Visually Appealing

The page design of informal proposals must draw readers into the document. Never forget that you are trying to *sell* a product, a service, or an idea. If the layout is unappealing, then you will lose readers before they even get to your message. Also, remember that your proposal may be competing with others. Put yourself in the place of the reader who is wondering which one to pick up first. How the text looks on the page can make a big difference. Here are a few techniques to follow to help make your proposal visually appealing:

- Use lists (with bullets or numbered points) to highlight main ideas.
- Follow your readers' preferences as to font size, type, line spacing, and so forth. Proposals written in the preferred format of the reader will gain a competitive edge.
- Use headings and subheadings to break up blocks of text.

These and other techniques help to reveal the proposal's structure and lead readers through the informal proposal. Given that there is no table of contents, you need to take advantage of such strategies.

■ *Informal Proposal Guideline 4: Use the ABC Format for Organization*

With its "hook" to gain the reader's attention, this structure makes good sense in proposal writing. Here are the main parts:

- **Abstract:** Gives the summary or "big picture" for those who will make decisions about your proposal. Usually includes some kind of "hook" or "grabber"—a point that will interest the audience in reading further.
- **Body:** Gives the details about exactly what you are proposing to do.
- **Conclusion:** Drives home the main benefit and makes clear the next step.

Note that beginning and ending sections are easy to read and stress just a few points. They provide a short "buffer" on both ends of the longer, more technical body section in the middle. The next three guidelines give more specific advice for writing the main parts of an informal proposal.

■ *Informal Proposal Guideline 5: Use the Heading "Introductory Summary" for the Generic Abstract Section*

Here you capture the client's attention with a capsule summary of the entire proposal. This one-paragraph or two-paragraph starting section permits space only for what the reader *really* needs to know at the outset, such as the following:

- Purpose of proposal
- Reader's main need
- Main benefit you offer
- Overview of proposal sections to follow

As Models 10−1 and 10−2 on pages 311−315 show, the introductory summary appears immediately after the subject line of the memo proposal or salutation of the letter proposal. As with informal reports, you have the option of actually labeling the section "Introductory Summary" or simply leaving off the heading. In either case, keep this overview very brief. Answer the one question clients are thinking: "Why should we hire this firm instead of another?" If you find yourself starting to give too much detail, move background information into the first section of the discussion.

■ *Informal Proposal Guideline 6: Put Important Details in the Body*

The discussion of your proposal should address these basic questions:

1. What problem are you trying to solve, and why?
2. What are the technical details of your approach?
3. Who will do the work, and with what?
4. When will it be done?
5. How much will it cost?

Discussion formats vary from proposal to proposal, but here are some sections commonly used to respond to these questions:

1. Description of problem or project and its significance. Give a precise technical description, along with any assumptions that you have made on the basis of previous contact with the reader. Explain the importance or significance of the problem, especially to the reader of the proposal.

2. Proposed solution or approach. Describe the specific tasks you propose in a manner that is clear and well organized. If you are presenting several options, discuss each one separately—making it as easy as possible for the reader to compare and contrast information.

3. Personnel. If the proposal involves people performing tasks, it may be appropriate to explain qualifications of participants.

4. Schedule. Even the simplest proposals usually require some sort of information about the schedule for delivering goods, performing tasks, and so on. Be both clear and realistic in this portion of the proposal. Use graphics when appropriate (see chapter 11 for guidelines on Gantt and milestone charts).

5. Costs. Place complete cost information in the body of the proposal unless you have a table that would be more appropriately placed in an attachment. Above all, do not bury dollar figures in paragraph format. Instead, highlight these figures with indented or bulleted lists, or at least place them at the beginnings of paragraphs. Because your reader will be looking for cost data, it is to your advantage to make that information easy to find.

■ *Informal Proposal Guideline 7: Give Special Attention to Establishing Need in the Body*

A common complaint about proposals is that writers fail to establish the need for what is being proposed. As any good salesperson knows, customers must feel that they need your product, service, or idea before they can be convinced to purchase or support it. In other words, do not simply try to dazzle readers with the good sense and quality of what you are proposing. Instead, lay the groundwork for acceptance by first showing the readers that a strong need exists.

Establishing need is most crucial in unsolicited proposals, of course, when readers may not be psychologically prepared to accept a change that will cost them money. Even in proposals that have been solicited, however, you should give some attention to restating the basic needs of the readers. If nothing else, this special attention shows your understanding of the problem.

■ Informal Proposal Guideline 8: Focus Attention in Your Conclusion

Called "conclusion" or "closing," this section gives the opportunity to control the readers' last impression. It also helps you avoid the awkwardness of ending proposals with the statement of costs, which is usually the last section in the discussion. In this closing section you can:

- Emphasize a main benefit or feature of your proposal
- Restate your interest in doing the work
- Indicate what should happen next

Regarding the last point, sometimes you may ask readers to call if they have questions. In other situations, however, it is appropriate to say that you will follow up the proposal with a phone call. This approach leaves you in control of the next step.

Incidentally, for informal sales proposals, there is a special technique that can push the proposal one step closer to approval. After the signature section, place an "acceptance block." As shown in the following example, this item makes it as easy as a signature for the reader to accept your proposal, rather than his or her having to write a return letter.

ACCEPTED BY LMN DEVELOPMENT, INC.

By: _____

Title: _____

Date: _____

■ Informal Proposal Guideline 9: Use Attachments for Less Important Details

Remember that the text of informal proposals is usually less than 10 pages. That being the case, you may have to put supporting data or illustrations in attachments that follow the conclusion. Cost and schedule information, in particular, is best placed at the end in well-labeled sections.

Make sure that the proposal text includes clear references to these visuals. If you have more than one attachment, give each one a letter and a title (for example,

"Attachment A: Project Costs"). If you have only one attachment, include the title but no letter (for example, "Attachment: Resumes").

■ *Informal Proposal Guideline 10: Edit Carefully*

In the rush of completing proposals, some writers fail to edit carefully. That is a big mistake. Make sure to build in enough time for a series of editing passes, preferably by different readers. There are two reasons why proposals of all kinds deserve this special attention.

1. They can be considered contracts in a court of law. If you make editing mistakes that alter meaning (such as an incorrect price figure), you could be bound to the error.

2. Proposals often present readers with their first impression of you. If the document is sloppy, they can make assumptions about your professional abilities as well.

GUIDELINES FOR FORMAL PROPOSALS

Sometimes the complexity of the proposal may be such that a formal response is best. As always, the final decision about format should depend upon the needs of your readers. Ask yourself questions like these in deciding whether to write an informal or a formal proposal:

- Is there too much detail for a letter or memo?
- Is a table of contents needed so that sections can be found quickly?
- Will the professional look of a formal document lend support to the cause?
- Are there so many attachments that a series of lengthy appendices would be useful?
- Are there many different readers with varying needs, such that there should be different sections for different people?

If you answer yes to one or more of these questions, give careful consideration to writing a formal proposal. This long format is most common in external sales proposals; however, important in-house proposals may sometimes require the same approach—especially in large organizations in which you may be writing to unknown persons in distant departments. Both in-house and sales examples follow the writing guidelines given next.

Formal proposals can be long and complex, so this part of the chapter treats each proposal section separately—from title page through conclusion. Two points will become evident as you use these guidelines. First, formal reports and formal proposals are a lot alike. A quick look at the last chapter will show you the similarities in format. Second, a formal proposal—like *all* technical writing described in this text—follows the basic ABC format described in chapter 3. Specifically, the parts of the formal proposal fit the pattern in this way:

ABC Format: Formal Proposal
Abstract

- Cover/Title Page
- Letter of Transmittal
- Table of Contents
- List of Illustrations
- Executive Summary
- Introduction

Body

- Technical information
- Management information
- Cost information
- [Appendices—appear after text but support Body section]

Conclusion

- Conclusion

As you read through and apply these guidelines, refer to Models 10–3 on pages 316–325 and Model 10–4 on pages 326–336 for annotated examples of the formal proposal. Note that each major section in the model proposals starts on a new page. Another alternative is to run most sections together, changing pages only at the end of the letter of transmittal and executive summary. Minor format variations abound, of course, but this chapter's guidelines will stand you in good stead throughout your career.

Cover/Title Page

Like formal reports, formal proposals usually are bound documents with a cover. The cover will include one or more of the items listed here for inclusion on the title page. Just as important, the cover should be designed to attract the reader's interest—with good page layout and perhaps even a graphic. Remember—proposals are sales documents. No one *has* to read them.

Inside the cover is the title page, which contains these four pieces of information:

- **Project title,** preceded by "Proposal for" or similar wording
- **Your reader's name** ("Prepared for . . .")
- **Your name or the name of your organization** spelled out in full ("Prepared by . . .")
- **Date of submission**

The title page gives clients their first impression of you. For that reason, consider using some tasteful graphics to make the proposal stand out from those of your competitors. For sales proposals, a particularly persuasive technique is to place the logo of the *client's company* on the cover or title page. In this way, you imply your interest in linking up with that firm *and* your interest in satisfying its needs, rather than simply selling your products or services.

Letter/Memo of Transmittal

Internal proposals have memos of transmittal; external sales proposals have letters of transmittal. These letters or memos must grab the reader's interest. The guidelines for format and organization presented here will help you write attention-getting prose. In particular, note that the letter or memo should be in single-spaced, ragged-right-edge format, even if the rest of the proposal is double-spaced copy with right-justified margins.

For details of letter and memo format, see chapter 7. The guidelines for the letter/memo of transmittal for formal reports also apply to formal proposals (see Transmittal Guideline 3 in chapter 9). For now, here are some highlights of format and content that apply especially to letters and memos of transmittal:

1. Use short beginning and ending paragraphs (about three to five lines each).
2. Use a conversational style, with little or no technical jargon. Avoid stuffy phrases such as "per your request."
3. Use the first paragraph for introductory information, mentioning what your proposal responds to (for example, a formal RFP, a conversation with the client, or your perception of a need).
4. Use the middle of the letter to emphasize one main benefit of your proposal, though the executive summary and proposal proper will mention benefits in detail. Stress what you can do to solve a problem, using the words "you" and "your" as much as possible (rather than "I" and "we").
5. Use the last paragraph to retain control by orchestrating the next step in the proposal process. When appropriate, indicate that you will call the client soon to follow up on the proposal.
6. Follow one of the letter formats described in chapter 7. Following are some exceptions, additions, or restrictions.

   - Use single-spaced, ragged-right-edge copy, which will make your letter stand out from the proposal proper.
   - Keep the letter on one page—a two-page letter loses that crisp, concise impact you want a letter to make.
   - Place the company proposal number (if there is one) at the top, above the date. Exact placement of both number and date depends on your letter style.
   - Include the client's company name or personal name on the first line of the inside address, followed by the mailing address that will be on the envelope. If you use a company name, place an "attention" line below the inside address. Include the full name (and title, if appropriate) of your contact person at the client firm. If you use a personal name, follow the last line of the inside address with a conventional greeting ("Dear Mr. Adams:").
   - (Optional) Include the project title below the attention line, using the exact wording that appears on the title page.
   - Close with "Sincerely" and your name at the bottom of the page. Also include your company affiliation.

Table of Contents

Create a very readable table of contents by spacing items well on the page. List all proposal sections, subsections, and their page references. At the end, list any appendices that may accompany the proposal.

Given the tight schedule on which most proposals are produced, errors can be introduced at the last minute because of additions or revisions. Therefore, take time to proofread the table of contents carefully. In particular, make sure to follow these guidelines:

- Wording of headings should match within the proposal text.
- Page references should be correct.
- All headings of the same order should be parallel in grammatical form.

List of Illustrations

Usually, the list of illustrations appears on a separate page after the table of contents. When there are few entries, however, the illustrations may be listed at the end of the table of contents page. In either case, the list should include the number, title, and page number of every illustration appearing in the body of the text. (If there is only one illustration, a number need not be included.) You may divide the list into tables and figures if many of both appear in your proposal.

Executive Summary

Executive summaries are the most frequently read parts of proposals. That fact should govern the time and energy you put into their preparation. Often read by decision-makers in an organization, the summary should present a concise one-page overview of the proposal's most important points. It should also accomplish these objectives:

- Avoid technical language
- Be as self-contained as possible
- Make brief mention of the problem, proposed solution, and cost
- Emphasize the main benefits of your proposal

Start the summary with one or two sentences that command readers' attention and engage their interest. Then focus on just a few main selling points (three to five is best). You might even want to highlight these benefits with indented lead-ins such as "Benefit 1, Benefit 2." When possible, use the statement of benefits to emphasize what is unique about your company or your approach so that your proposal will attract special attention. Finally, remember to write the summary after you have completed the rest of the proposal. Only at this point do you have the perspective to sit back and develop a reader-oriented overview.

Introduction

The introduction provides background information for both nontechnical and technical readers. Although the content will vary from proposal to proposal, some general guidelines apply. Basically, you should include information on the (1) purpose, (2) description of the problem to which you are responding, (3) scope of the proposed study, and (4) format of the proposal. (A *lengthy* problem or project statement should be placed in the first discussion section of the proposal, not in the introduction.)

- Use subheadings if the introduction goes over a page. In this case, begin the section with a lead-in sentence or two that mention the sections to follow.
- Start with a purpose statement that concisely states the reason you are writing the proposal.
- Include a description of the problem or need to which your proposal is responding. Use language directly from the request for proposal or other document the reader may have given you so that there is no misunderstanding. For longer problem or need descriptions, adopt the alternative approach of including a separate needs section or problem description after the introduction.
- Include a scope section in which you briefly describe the range of proposed activities covered in the proposal, along with any research or preproposal tasks that have already been completed.
- Include a proposal format section if you feel the reader would benefit from a listing of the major proposal sections that will follow.

Discussion Sections

Aim the discussion or body toward readers who need supporting information. Traditionally, the discussion of a formal sales proposal contains three basic types of information: (1) technical, (2) management, and (3) cost. Here are some general guidelines for presenting each type. Remember that the exact wording of headings and subheadings will vary, depending on proposal content.

1. Technical Sections

- Respond thoroughly to the client's concerns, as expressed in writing or meetings.
- Follow whatever organization plan that can be inferred from the request for proposal.
- Use frequent subheadings with specific wording.
- Back up all claims with facts.

2. Management Sections

- Describe who will do the work.
- Explain when the work will be done.
- Display schedule information graphically.
- Highlight personnel qualifications (but put resumes in appendices).

3. Cost Section

- Make costs extremely easy to find.
- Use formal or informal tables when possible.
- Emphasize value received for costs.
- Be clear about add-on costs or options.
- Always total your costs.

Conclusion

Formal proposals should always end with a section labeled "Conclusion" or "Closing." This final section of the text gives you the chance to restate a main benefit, summarize the work to be done, and assure clients that you plan to work with them closely to satisfy their needs. Just as important, this brief section helps you end on a positive note. You come back full circle to what you stressed at the beginning of the document—benefits to the client and the importance of a strong personal relationship. (Without the conclusion, the client's last impression would be made by the cost section in the discussion.)

Appendices

Because formal proposals are so long, readers sometimes have trouble locating information they need. Headings help, but they are not the whole answer. Another way you can help readers is by transferring technical details from the proposal text into appendices. The proposal still will contain detail—for technical readers who want it—but detail will not intrude into the text. This technique can save you or your employer considerable time by permitting you to develop standard appendices (also called *boilerplate*) to be used in later proposals.

Any supporting information can be placed in appendices, but here are some common items included there:

- Resumes
- Organization charts
- Company histories
- Detailed schedule charts
- Contracts
- Cost tables
- Detailed options for technical work
- Summaries of related projects already completed
- Questionnaire samples

Boilerplate is often taken right off the shelf and thus is not paged in sequence with your text. Instead, it is best to use individual paging within each appendix. For example, pages in an Appendix B would be numbered B-1, B-2, B-3.

GUIDELINES FOR FEASIBILITY STUDIES

Much like recommendation reports (see chapter 8), feasibility studies guide readers toward a certain line of action. Another similarity is that both report types can be either in-house or external. Yet most feasibility studies have these five distinctive features that justify their being considered separately here:

1. They are *always* solicited by the reader, usually for the purpose of deciding on the best course of action.

2. They *always* assume one of these two patterns of organization:

   - An analysis of the advantages and disadvantages of one course of action, product, or idea
   - A comparison of two or more courses of action, products, or ideas

3. They *always* are intended to help managers and other decision-makers vote for or against an idea or select among several alternatives.
4. They *usually* "nudge" (as opposed to "urge" or "push") the reader toward a decision. That is, they are supposed to be written in such a way that the facts speak for themselves.
5. They *often* are preceded by a proposal.

In some ways, feasibility studies could be viewed as a cross between technical reports and proposals. As a writer you are expected to deal with the topic objectively and honestly, yet you are also expected to express your point of view. The *American Heritage Dictionary* defines "feasible" in this way: "capable of being accomplished or brought about." Thus a feasibility study determines if some course of action is practical. For example, it may be *desirable* for a student to quit work and return to college full-time. Yet if that same student has hefty car and apartment payments, the only *feasible* alternative may be part-time course work.

The following guidelines will help you prepare the kinds of feasibility studies requested by your boss (if the study is in-house) or by your client (if the study is external). In either case, your study may be used as the basis for a major decision. Refer to Model 10–5 on pages 337–338 as you read and apply these guidelines to your own writing.

■ *Feasibility Study Guideline 1: Choose Format Carefully*

In deciding whether to use the format of an informal (letter or memo) or formal document, use the same criteria mentioned earlier in the chapter with regard to proposals. As always, the central questions concern your readers:

- What format will give them easiest access to the data, conclusions, and recommendations of your study?
- Are there enough pages to suggest need of a table of contents (that is, a formal report)?
- What is the format preference of your readers?
- What has been the format of previous feasibility studies written for the same organization?

■ *Feasibility Study Guideline 2: Use the ABC Format*

Like other forms of technical writing, good feasibility studies have this basic three-part structure: **a**bstract, **b**ody, and **c**onclusion. As in other documents, the exact side headings you choose may vary from report to report. Yet, the overall structure should be as follows:

ABC Format: Feasibility Study

- **Abstract:** Capsule summary of information for the most important readers (that is, the decision-makers)
- **Body:** Details that support whatever conclusions and recommendations the study contains, working logically from fact toward opinion
- **Conclusion:** Wrap-up in which you state conclusions and recommendations resulting from study

---

The following guidelines examine specific sections of feasibility studies, along with details of content and tone.

■ Feasibility Study Guideline 3: Call Your Abstract an Introductory Summary

This section provides information that the most important readers would want if they were in a rush to read your study. With that criterion in mind, consider including these items:

- Brief statement about who has authorized the study and for what purpose
- Brief mention of the criteria used during the evaluation
- Brief reference to your recommendation

The last item is important, for it saves readers the frustration of having to wade through the whole document in search of the answers to the questions "Is this a practical idea?" or "Which alternative is best?" It is best to mention the recommendation up front, giving readers a frame through which to see the entire report.

■ Feasibility Study Guideline 4: Organize the Body Well

More than anything else, readers of feasibility studies expect an unbiased presentation. That means the midsection of your report must clearly and logically work from facts toward recommendations. Here is one approach that works:

1. **Describe evaluation criteria used during your study,** if readers need more detail than was presented in the introductory summary.
2. **Describe exactly WHAT was evaluated and HOW,** especially if you are comparing several items.
3. **Choose criteria that are most meaningful to the readers,** such as:
   - Cost
   - Practicality of implementing idea
   - Changes that may be needed in personnel
   - Effect on growth of organization
   - Effects on day-to-day operations

Of course, exact criteria will depend upon the precise topic you are investigating.

4. **Discuss both advantages and disadvantages** when you are evaluating just one item. Move from advantages to disadvantages. The conclusion will allow you to come back around to supporting points.

5. **Follow organization guidelines for comparisons** when evaluating several alternatives (see chapter 5). You can discuss one item at a time OR you can discuss one criterion at a time.

■ *Feasibility Study Guideline 5: Use the Conclusion for Detailed Conclusions and Recommendations*

Here you get the opportunity to state (or restate) the conclusions evident from data you have presented in the discussion. First state conclusions, and then state your recommendations. Use listings for three or more points, to make this last section of the study as easy as possible to read.

■ *Feasibility Study Guideline 6: Use Graphics for Comparisons*

When comparing several items, you need to consider most readers' preference for tabulated information. Tables can appear either in the discussion section or in attachments. In both cases, follow graphics guidelines explained in chapter 11.

■ *Feasibility Study Guideline 7: Offer to Meet With the Readers*

Most readers have many questions after reading a feasibility study, even if that study has been quite thorough. You score points for eagerness and professionalism if you anticipate needs and express your willingness to meet with readers later. Such meetings give you another opportunity to demonstrate your understanding of the topic.

CHAPTER SUMMARY

Proposals and feasibility studies stand out as documents that aim to *convince* readers. In the case of proposals, you are writing to convince someone inside or outside your organization to adopt an idea, a product, or a service. In the case of feasibility studies, you are marshalling facts to support the practicality of one approach to a problem — sometimes in comparison with other approaches. Both documents can be either informal or formal, depending on length, complexity, or reader preference.

This chapter includes lists of writing guidelines for informal proposals, formal proposals, and feasibility studies. For informal proposals, follow these basic guidelines:

1. Plan well before you write.
2. Use letter or memo format.

3. Make text visually appealing.
4. Use the ABC format for organization.
5. Use the heading "introductory summary" for the generic abstract section.
6. Put important details in the body.
7. Give special attention to establishing need in the body.
8. Focus attention in your conclusion.
9. Use attachments for less important details.
10. Edit carefully.

In formal proposals, abide by the same general format presented in chapter 9 for formal reports. To be sure, formal proposals have a different tone and substance because of their more persuasive purpose. Yet they do have the same basic parts, with minor variations: cover/title page, letter/memo of transmittal, table of contents, list of illustrations, executive summary, introduction, discussion sections, conclusion, and appendices.

Feasibility studies demonstrate that an idea is or is not practical. Also, they may compare several alternatives. Follow these basic writing guidelines:

1. Choose format carefully.
2. Use the ABC format.
3. Call your abstract an introductory summary.
4. Organize the body well.
5. Use the conclusion for detailed conclusions and recommendations.
6. Use graphics for comparisons.
7. Offer to meet with the readers.

ASSIGNMENTS

The assignments in Part 1 and Part 2 can be completed either as individual projects or as group-writing projects. If your instructor assigns group projects, review the information in chapter 1 on group writing.

Part 1: Short Assignments

These short assignments require either that you write parts of informal or formal proposals *or* that you evaluate the effectiveness of an informal proposal included here.

1. **Introductory Summary.** For this assignment, select one of the seven project sheets in the color insert. Now assume that you were responsible for writing the proposal that resulted in the project about which the sheet is written. In other words, work backwards from the project sheet to the informal proposal that McDuff used to get the work. Write a short introductory summary for the original proposal. Pay special attention to information in the project sheet on technical tasks, for the proposal probably would have highlighted some of these tasks. Focus on the main reason you think the client would have for hiring McDuff. If necessary, invent additional information to complete this assignment successfully.

2. **Needs Section.** As this chapter suggests, informal proposals—especially those that are unsolicited—must make a special effort to establish the need for the product or service being proposed. Assume that you are writing an informal proposal to suggest a change in procedures or equipment at your college. Keep the proposal limited to a small change;

you may even see a need in the classroom where you attend class (audiovisual equipment? lighting? heating or air systems? aesthetics? soundproofing?). Write the needs section that would appear in the body of the informal proposal.

3. **Conclusion or Closing.** For this assignment, as with Assignment 1, select a project sheet from the color insert. Assume that you were the McDuff employee responsible for writing the informal proposal that resulted in the work described in the project sheet. Write an effective conclusion or closing for the proposal.

4. **Evaluation—Informal Proposal.** Review the informal proposal that follows, submitted by MainAlert Security Systems to the McDuff, Inc., office in Atlanta. Evaluate the effectiveness of every section of the proposal.

---

200 Roswell Road
Marietta, Georgia 30062
(404) 555-2000

September 15, 1993

Mr. Bob Montrose
Operations Manager
McDuff, Inc.
3295 Peachtree Road
Atlanta, Georgia 30324

Dear Bob,

Thank you for giving MainAlert Security Systems an opportunity to submit a proposal for installation of an alarm system at your new office. The tour of your nearly completed office in Atlanta last week showed me all I need to know to provide you with burglary and fire protection. After reading this proposal, I think you will agree with me that my plan for your security system is perfectly suited to your needs.

This proposal describes the burglary and fire protection system I've designed for you. This proposal also describes various features of the alarm system that should be of great value. To provide you with a comprehensive description of my plan, I have assembled this proposal in several main sections:

1. Burglary Protection System
2. Fire Protection System
3. Arm/Disarm Monitoring
4. Installation Schedule
5. Installation and Monitoring Costs

BURGLARY PROTECTION SYSTEM

The burglary protection system would consist of a 46-zone MainAlert alarm control set, perimeter protection devices, and interior protection devices. The alarm system would have a strobe light and a siren to alert anyone nearby of a burglary in progress. Our system also includes a two-line dialer to alert our central station personnel of alarm and trouble conditions.

Alarm Control Set

The MainAlert alarm control set offers many features that make it well suited for your purposes. Some of these features are as follows:

1. Customer-programmable keypad codes
2. Customer-programmable entry/exit delays
3. Zone bypass option

4. Automatic reset feature
5. Point-to-point annunciation

I would like to explain the point-to-point annunciation feature, since the terminology is not as self-explanatory as the other features are. Point-to-point annunciation is a feature that enables the keypad to display the zone number of the point of protection that caused the alarm. This feature also transmits alarm point information to our central station. Having alarm point information available for you and the police can help prevent an unexpected confrontation with a burglar.

Interior and Perimeter Protection

The alarm system I have designed for you uses both interior and perimeter protection. For the interior protection, I plan to use motion detectors in the hallways. The perimeter protection will use glass break detectors on the windows and door contacts on the doors.

There are some good reasons for using both interior and perimeter protection:
1. Interior and perimeter protection used together provide you with two lines of defense against intrusion.
2. A temporarily bypassed point of protection will not leave your office vulnerable to an undetected intrusion.
3. An employee who may be working late can still enjoy the security of the perimeter protection while leaving the interior protection off.

Although some people select only perimeter protection, it is becoming more common to add interior protection for the reasons I have given. Interior motion detection, placed at carefully selected locations, is a wise investment.

Local Alarm Signaling

The local alarm-signaling equipment consists of a 40-watt siren and a powerful strobe light. The siren and strobe will get the attention of any passerby and unnerve the most brazen burglar.

Remote Alarm Signaling

Remote alarm signaling is performed by a two-line dialer that alerts our central station to alarm and trouble conditions. The dialer uses two telephone lines so that a second line is available if one of the lines is out. Any two existing phone lines in your office can be used for the alarm system. Phone lines dedicated for alarm use are not required.

FIRE PROTECTION SYSTEM

My plan for the fire protection system includes the following equipment:
1. Ten-zone fire alarm panel
2. Eight smoke detectors
3. Water flow switch
4. Water cutoff switch
5. Four Klaxon horns

The ten-zone fire alarm panel will monitor one detection device per zone. Because each smoke detector, the water flow switch, and the water cutoff switch have a separate zone, the source of a fire alarm can be determined immediately.

To provide adequate local fire alarm signaling, this system is designed with four horns. Remote signaling for the fire alarm system is provided by the MainAlert control panel. The fire alarm would report alarm and trouble conditions to the MainAlert control panel. The MainAlert alarm control panel would, in turn, report fire alarm and fire trouble signals to our central station. The MainAlert alarm panel would not have to be set to transmit fire alarm and fire trouble signals to our central station.

ARM/DISARM MONITORING

Since 20 of your employees would have alarm codes, it is important to keep track of who enters and leaves the office outside of office hours. When an employee would arm or disarm the alarm system, the alarm would send a closing or opening signal to our central station. The central station would keep a record of the employee's identity and the time the signal was received. With the arm/disarm monitoring service, our central station would send you opening/closing reports on a semi-monthly basis.

INSTALLATION SCHEDULE

Given the size of your new office, our personnel could install your alarm in three days. We could start the day after we receive approval from you. The building is now complete enough for us to start anytime. If you would prefer for the construction to be completed before we start, that would not present any problems for us. To give you an idea of how the alarm system would be laid out, I have included an attachment to this proposal showing the locations of the alarm devices.

INSTALLATION AND MONITORING COSTS

Installation and monitoring costs for your burglary and fire alarm systems as I have described them in this proposal will be as follows:

- $8200 for installation of all equipment
- $75 a month for monitoring of burglary, fire, and opening/closing signals under a two-year monitoring agreement

The $8200 figure covers the installation of all the equipment I have mentioned in this proposal. The $75-a-month monitoring fee also includes opening/closing reports.

CONCLUSION

The MainAlert control panel, as the heart of your alarm system, is an excellent electronic security value. The MainAlert control panel is unsurpassed in its ability to report alarm status information to our central station. The perimeter and interior protection offers complete building coverage that will give you peace of mind.

The fire alarm system monitors both sprinkler flow and smoke conditions. The fire alarm system I have designed for you can provide sufficient warning to allow the fire department to save your building from catastrophic damage.

The arm/disarm reporting can help you keep track of employees who come and go outside of office hours. It's not always apparent how valuable this service can be until you need the information it can provide.

I'll call you early next week, Bob, in case you have any questions about this proposal. We will be able to start the installation as soon as you return a copy of this letter with your signature in the acceptance block.

Sincerely,

Anne Rodriguez Evans
Commercial Sales

Enc.

ACCEPTED by McDuff, Inc.

By:_____

Title:_____

Date:_____

ALARM SYSTEM LAYOUT FOR
McDUFF INC. – ATLANTA, GA.

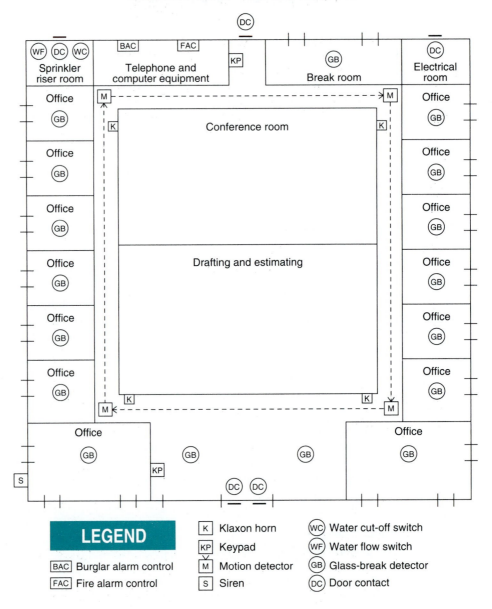

LEGEND		
	K Klaxon horn	WC Water cut-off switch
BAC Burglar alarm control	KP Keypad	WF Water flow switch
FAC Fire alarm control	M Motion detector	GB Glass-break detector
	S Siren	DC Door contact

Part 2: Longer Assignments

For each of these assignments, complete a copy of the Planning Form included at the end of the book.

5. **Informal Proposal—McDuff.** Choose Option A or Option B. Remember that informal proposals should be fairly limited in scope, given their length and format.

Option A: In-House

- Use your past or present work experience to write a memo proposal suggesting a change at McDuff, Inc. Possible topic areas include changes in operating procedures, revisions to company policies, additions to the work force, alterations of the physical plant, or purchase of products or services.
- Place yourself in the role of an employee of McDuff. The proposal may be solicited or unsolicited, whatever best fits your situation.
- Make sure that your proposal topic is limited enough in scope to be covered fully in an informal proposal with memo format.
- Choose at least two levels of readers who could conceivably be decision-makers about a proposal such as the one you are writing—for example, branch or corporate managers. Review chapter 2 if necessary.

Option B: Sales

- Select a product or service (1) with which you are reasonably familiar (on the basis of your work experience, research, or other interests) and (2) that could conceivably be purchased by a company like McDuff.
- Put yourself in the role of someone representing the company that makes the product or provides the service.
- Write an informal sales proposal in which you propose purchase of the product or service by a representative of McDuff.

6. **Formal Proposal—McDuff.** Choose Option A, B, or C. Make sure that your topic is more complex than the one you would choose for the preceding informal-proposal assignments.

Option A: Community-Related

- Write a formal proposal in which you propose a change in (1) the services offered by a city or town (for example, mass transit or waste management) or (2) the structure or design of a building, garden, parking lot, shopping area, school, or other civic property.
- Select a topic that is reasonably complex and yet one about which you can locate information.
- Place yourself in the role of an outside consultant with a division of McDuff, Inc., who is proposing the change.
- Choose either an unsolicited or a solicited context.
- Write to an audience that could actually be the readers. Do enough research to identify at least two levels of audience.

Option B: School-Related

- Write a proposal in which you propose a change in some feature of a school you attend or have attended.
- Choose from topics such as operating procedures, personnel, curricula, activities, and physical plant.

- Select an audience that would actually make decisions on such a proposal.
- Give yourself the role of an outside consultant working for McDuff, Inc.

Option C: Work-Related

- Write a proposal in which you, as a representative of McDuff, propose purchase of a product or service by another firm.
- Choose a topic about which you have work experience, research knowledge, or keen interest—and one that could conceivably be offered in one of McDuff's project areas. Make sure you have good sources of information.
- Choose either a solicited or an unsolicited context.

7. **Feasibility Study—McDuff.**

- Choose any one of the proposal assignments that you completed as part of the preceding assignments. (Or for this assignment, you can use a proposal completed by one of your classmates.)
- Take yourself out of the role of proposal writer. Instead, consider yourself to be someone assigned (or hired) to complete the task of evaluating the practicality of the proposal, after it has been received.
- If appropriate, choose several alternatives to evaluate.

Professional
Documentation, Inc.

3450 Jones Mill Road, Norcross, Georgia 30092
(404) 555-8438

January 15, 1994

Mr. David Barker
Technical Communications Manager
Real Big Professional Software
P.O. Box 123456
Atlanta, Georgia 30339

Dear David:

I enjoyed meeting with you and learning about your new *General Ledger* software product. Because you require a March release, I can understand why you want to choose an approach to documentation and get the project started.

This proposal describes a strategy for completing the documentation in the 10 weeks between now and your March deadline. Included are these main sections:

1. Selection of the Best Format
2. Adoption of a Publication Plan
3. Control of Costs
4. Conclusion

SELECTION OF THE BEST FORMAT

I think your customers will be best served by a combination installation and user's guide. It uses a functional approach to show how *General Ledger* works. My assessment results from these completed steps:

- Interviews with support staffers responsible for providing technical support to customers using the company's other accounting products
- Interviews with programmers developing *General Ledger*, who have an intimate knowledge of how it works
- Conversations with you that clarified your organization's general expectations for the documentation

The assessment is also based on my experience developing documentation for other products. I strive to use clear, concise prose and ample white space to provide a visually appealing text. The text will be enhanced and supplemented with graphics depicting *General Ledger*'s feature screens. The screens themselves will be captured directly

continues

MODEL 10–1
Letter proposal

311

Shows under-standing of cli-ent's main concern— *scheduling.*

Asserts ability to meet scheduling need.

Gives helpful overview of sec-tions to follow.

Uses list to itemize impor-tant points—that is, the *basis* for his assessment.

from the program and inserted into the text by your staff using your in-house publishing system.

This approach will yield a thorough and easy-to-use document that will allow your customers to take full advantage of General Ledger's many innovative features.

As you know, writing documentation is a cooperative effort. Each member of the General Ledger product team will play key roles during the development process. To keep us all on track, I have put together a publication plan that shows how the project will progress from beginning to end.

ADOPTION OF A PUBLICATION PLAN

The publication plan shows how we can have the documentation ready for General Ledger's March unveiling. The four major steps are described here.

Define the Project

Much of this work has already been accomplished as a result of doing the research for this proposal. As a preliminary step, we will meet and review the project's scope and priority within the organization. We will detail the resources that will be available to complete the project. Most important, we will look at expectations: management's, yours, and the customers'.

Develop a Schedule

This step is the key to the publication process and ensures a common understanding of what has to be done and in what period of time. It has three basic steps:

- We define the tasks that are part of the project.
- We define the resources we have available to deal with the identified task.
- We assign tasks to the most appropriate individuals.

Manage the Project

What is good project management? In this plan, good management is essentially good communications. In the first three steps, we define the information that project members must have to understand how the document will be produced and their roles in that process. Ongoing management of the project will be a matter of keeping the channels of communication open.

Perform a Postmortem

The last step is an evaluation of the effectiveness of the publication plan. It provides the opportunity for us to learn how to do future documentation better. It is important to look back at what went right and what went wrong during a project and to share this information with the others. You will get a complete postmortem report from me after the project is completed.

MODEL 10–1, *continued*

Continues emphasis on benefits to reader.

Leads in smoothly to next section.

Starts with *overview* of sections to follow.

Organizes paragraph around three *main points*.

Uses bulleted list for primary *steps* in project.

Introduces section with *question* to attract attention to passage.

Throughout the project, this management system will guide us in completing General Ledger's documentation on time and within budget.

CONTROL OF COSTS

Good documentation helps to sell software. By working smart, we can develop documentation that will enhance General Ledger's appeal, and we can do it at a reasonable cost.

My experience in this area and the management system described here will reduce waste and duplication of effort, two factors that affect cost. This savings means I can bring the project in within the 200-hour cap you mentioned.

This estimate assumes that three of the program's four main features are in a complete, or "fixed," state and that the fourth main feature is about 50 percent complete. This estimate also assumes that all programming will be finished by March 5, which will allow time to put the guide through final review and production.

CONCLUSION

The functional approach, which describes a product in terms of its operations, is the documentation format that will best serve General Ledger customers. Your goal of having the documentation ready by March will be aided by adopting a four-step publication plan. The plan will define the strategy for writing the documentation and will help keep costs down.

I'll call you in a few days, David, to answer any questions you might have about this proposal. I can begin work on the documentation as soon as you sign the acceptance block and return a copy of this letter to me.

Sincerely,

Steven Nickels

Steven Nickels
Documentation Specialist

Enc.

ACCEPTED by Real Big Professional Software

By:_____

Title: _____

Date:_____

Shows interest in *following* through.

Places benefit in heading.

Shows he can meet project criteria—but also clarifies the *assumptions* he is making.

Returns to *main concern* of reader— scheduling.

Retains *control* of next step.

Includes acceptance block to simplify approval process.

MODEL 10–1, *continued*

DATE: October 4, 1994
TO: Gary Lane
FROM: Jeff Bilstrom
SUBJECT: Creation of Logo for Montrose Service Center

Gives concise
view of
problem—*and*
his proposed
solution.

Part of my job as director of public relations is to get the Montrose name firmly entrenched in the minds of metro Atlanta residents. Having recently reviewed the contacts we have with the public, I believe we are sending a confusing message about the many services we offer retired citizens in this area.

To remedy the problem, I propose we adopt a logo to serve as an umbrella for all services and agencies supported by the Montrose Service Center. This proposal gives details about the problem and the proposal solution, including costs.

The Problem

Includes effec-
tive lead-in.

The lack of a logo presents a number of problems related to marketing the center's services and informing the public. Here are a few:

Uses bulleted
list to highlight
main difficulties
posed by cur-
rent situation.

- The letterhead mentions the organization's name in small type, with none of the impact that an accompanying logo would have.
- The current brochure needs the flair that could be provided by a logo on the cover page, rather than just the page of text and headings that we now have.
- Our 14 vehicles are difficult to identify because there is only the lettered organization name on the sides without any readily identifiable graphic.
- The sign in front of our campus, a main piece of free advertising, could better spread the word about Montrose if it contained a catchy logo.
- Other signs around campus could display the logo, as a way of reinforcing our identity and labeling buildings.

Ends section
with good *transi-
tion* to next
section.

It's clear that without a logo, the Montrose Service Center misses an excellent opportunity to educate the public about its services.

The Solution

Starts with *main
point*—need for
logo.

I believe a professionally designed logo could give the Montrose Service Center a more distinct identity. Helping to tie together all branches of our operation, it would give the public an easy-to-recognize symbol. As a result, there would be a stronger awareness of the center on the part of potential users and financial contributors.

MODEL 10–2
Memo proposal

The new logo could be used immediately to:

- Design and print letterhead, envelopes, business cards, and a new brochure.
- Develop a decal for all company vehicles that would identify them as belonging to Montrose.
- Develop new signs for the entire campus, to include a new sign for the entrance to the campus, one sign at the entrance to the Blane Workshop, and one sign at the entrance to the Administration Building.

Cost

Developing a new logo can be quite expensive. However, I have been able to get the name of a well-respected graphic artist in Atlanta who is willing to donate his services in the creation of a new logo. All that we must do is give him some general guidelines to follow and then choose among eight to ten rough sketches. Once a decision is made, the artist will provide a camera-ready copy of the new logo.

• Design charge	$ 0.00
• Charge for new letterhead, envelopes, business cards, and brochures (min. order)	545.65
• Decal for vehicles 14 @ $50.00 + 4%	728.00
• Signs for campus	415.28
Total Cost	$1,688.93

Conclusion

As the retirement population of Atlanta increases in the next few years, there will be a much greater need for the services of the Montrose Service Center. Because of that need, it's in our best interests to keep this growing market informed about the organization.

I'll stop by later this week to discuss any questions you might have about this proposal.

Margin notes:
- Focuses on *benefits* of proposed change.
- Emphasizes *benefit* of possible price break.
- Uses *listing* to clarify costs.
- Closes with major benefit to reader and urge to action.
- Keeps control of next step.

MODEL 10–2, *continued*

PROPOSAL FOR SUPPLYING
TEAK CAM CLEAT SPACERS

Prepared by
Totally Teak, Inc.

Prepared for
John L. Riggini
Bosun's Locker Marine Supply

August 22, 1994

MODEL 10–3
Formal proposal (external)

Totally Teak, Inc.
6543 Amster Avenue, N.W.
Atlanta, Georgia 30308
(404) 555-9425

August 22, 1994

In this example, the letter of transmittal appears immediately following the title page. It can also appear before the title page (see Model 10–4).

John L. Riggini, President
Bosun's Locker Marine Supply
38 Oakdale Parkway
Norcross, OH 43293

Dear Mr. Riggini:

I enjoyed talking with you last week about inventory needs at the 10 Bosun's stores. In response to your interest in our products, I'm submitting this proposal to supply your store with our Teak Cam Cleat Spacers.

Establishes link with previous client contact.

This proposal outlines the benefits of adding Teak Cam Cleat Spacers to your line of sailing accessories. The potential for high sales volume stems from the fact that the product satisfies two main criteria for any boat owner:

Stresses two main benefits.

1. It enhances the appearance of the boat.
2. It makes the boat easier to handle.

Your store managers will share my enthusiasm for this product when they see the response of their customers.

Says he will call *(rather than asking client to call).*

I'll give you a call next week to answer any questions you have about this proposal.

Sincerely,

William G. Rugg

William G. Rugg,
President
Totally Teak, Inc.

WR/rr

MODEL 10–3, *continued*

continues

TABLE OF CONTENTS

Organizes entire proposal around *benefits.*

MODEL 10–3, *continued*

318

LIST OF ILLUSTRATIONS

continues

MODEL 10−3, *continued*

319

EXECUTIVE SUMMARY

This proposal outlines features of a custom-made accessory designed for today's sailors—whether they be racers, cruisers, or single-handed skippers. The product, Teak Cam Cleat Spacers, has been developed for use primarily on the Catalina 22, a boat owned by many customers of the 10 Bosun's stores. However, it can also be used on other sailboats in the same class.

The predictable success of Teak Cam Cleat Spacers is based on two important questions asked by today's sailboat owners:

- Will the accessory enhance the boat's appearance?
- Will it make the boat easier to handle and, therefore, more enjoyable to sail?

This proposal answers both questions with a resounding affirmative by describing the benefits of teak spacers to thousands of people in your territory who own boats for which the product is designed. This potential market, along with the product's high profit margin, will make Teak Cam Cleat Spacers a good addition to your line of sailing accessories.

2

MODEL 10–3, *continued*

The purpose of this proposal is to show that Teak Cam Cleat Spacers will be a practical addition to the product line at the Bosun's Locker Marine Supply stores. This introduction highlights the need for the product, as well as the scope and format of the proposal.

Background

Sailing has gained much popularity in recent years. The high number of inland impoundment lakes, as well as the vitality of boating on the Great Lakes, has spread the popularity of the sport. With this increased interest, more and more sailors have become customers for a variety of boating accessories.

What kinds of accessories will these sailors be looking for? Accessories that (1) enhance the appearance of their sailboats (2) make their sailboats easier to handle and, consequently, more enjoyable to sail. With these customer criteria in mind, it is easy to understand the running joke among boat owners (a profitable joke among marine supply dealers): "A boat is just a hole in the water that you pour your money into."

The development of this particular production originated from our designers' firsthand sailing experiences on the Catalina 22 and knowledge obtained during manufacture (and testing) of the first prototype. In addition, we conducted a survey of owners of boats in this general class. The results showed that winch and cam cleat designs are major concerns.

Proposal Scope and Format

The proposal focuses on the main advantages that Teak Cam Cleat Spacers will provide your customers. These six sections follow:

1. Practicality
2. Suitability for a Variety of Sailors
3. High-Quality Construction and Appearances
4. Dealer Benefits
5. Sizable Potential Market
6. Affordable Price

3

MODEL 10–3, *continued*

continues

Makes clear the proposal's purpose.

Gives lead-in about sections to follow.

Establishes *need* for product.

Shows his understanding of *need* (personal experience of designing owner survey).

Gives list of sections to follow, to reinforce organization of proposal.

FEATURES AND BENEFITS

Teak Cam Cleat Spacers offer Bosun's Locker Marine Supply the best of both worlds. On the one hand, the product solves a nagging problem for sailors. On the other, it offers your store managers a good opportunity for profitability. Described here are six main benefits for you to consider.

Practicality

This product is both functional and practical. When installed in the typical arrangement shown in the figure below, the Teak Cam Cleat Spacer raises the height of the cam cleat, thereby reducing the angle between the deck and the sheet as it feeds downward from the winch. As a result of this increased height, a crewmember is able to cleat a sheet with one hand instead of two.

SIDE VIEWS OF CLEATING ARRANGEMENT

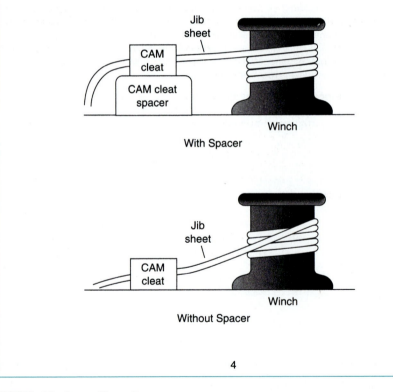

4

MODEL 10–3, *continued*

Such an arrangement allows a skipper to maintain steerage of the boat, keeping one hand on the helm while cleating the sheet with the other. Securing a sheet in this manner can be done more quickly and securely. Also, this installation reduces the likelihood of a sheet "popping out" of the cam cleat during a sudden gust of wind.

Suitability for a Variety of Sailors

For the racer, cruiser, and single-handed sailor, sailing enjoyment is increased as sheets and lines become easier to handle and more secure. In a tight racing situation, these benefits can be a deciding factor. The sudden loss of sail tension at the wrong moment as a result of a sheet popping out of the cam cleat could make the difference in a close race.

A cruising sailor is primarily concerned with relaxation and pleasure. A skipper in this situation wants to reduce his or her workload as much as possible. In the instance of a sheet popping loose, the sudden chaos of a sail flapping wildly interrupts an otherwise tranquil atmosphere. Teak Cam Cleat Spacers reduce the chance of this happening.

A cruising sailor often has guests aboard. In this situation, as well as in a race, the skipper wants to maintain a high level of seamanship, especially where the control of the boat and the trim of its sails are concerned.

The single-handed sailor derives the greatest benefit from installing Teak Cam Cleat Spacers. Without crew nearby to assist with handling lines or sheets, anything that makes work easier for this skipper is welcome.

High-Quality Construction and Appearance

The teakwood frame from which this product is manufactured is well suited for use around water, since teak will not rot. It also looks nice when oiled or varnished.

The deck of most sailboats is made primarily of fiberglass. The appearance of such a boat can be significantly enhanced by the addition of some teak brightwork.

Each spacer is individually handcrafted by Totally Teak, Inc., to guarantee a consistent level of high quality.

Dealer Benefits

Teak Cam Cleat Spacers make a valuable addition to the dealer's product line. They complement existing sailing accessories as well as provide the customer with the convenience of a readily available prefabricated product.

A customer who comes in to buy a cam cleat is a ready prospect for the companion spacer. Such a customer will likely want to buy mounting hardware as well.

With this unique teak product readily available, a dealer can save the customer the time and trouble of fabricating makeshift spacers.

5

MODEL 10–3, *continued*

continues

Includes number
of owners to
emphasize *po-
tential sales.*

Keeps price in-
formation *short*
and *clear.*

Sizable Potential Market

These Teak Cam Cleat Spacers are designed with a large and growing potential market in mind. They are custom-made for the Catalina 22, one of the most popular sailboats in use today. Over 13,000 of these model sailboats have been manufactured to date. These spacers are also well suited for other similar-class sailboats.

Affordable Price

The Teak Cam Cleat Spacers made by Totally Teak, Inc., wholesale for $3.95/pair. Suggested retail is $6.95/pair. This low price is easy on the skipper's wallet and should help this product move well. And, of course, the obviously high profit margin should provide an incentive to your store managers.

6

MODEL 10–3, *continued*

CONCLUSION

Why should a marine supply dealer consider carrying Teak Cam Cleat Spacers? This product satisfies two common criteria of sailboat owners today: it enhances the appearance of any sailboat, and it makes the boat easier to handle. The potential success of this product is based on its ability to meet these criteria and the following features and benefits:

1. It is practical, allowing quick, one-handed cleating.
2. It is ideally suited for a variety of sailors, whether they are racing, cruising, or sailing single-handedly.
3. It is a high-quality, handcrafted product that enhances the appearance of any sailboat.
4. It is a product that benefits the dealer by making a valuable addition to her or his product. It complements existing sail accessories and satisfies a customer need.
5. It is geared toward a sizable potential market. Today there are thousands of sailboats in the class for which this accessory is designed.
6. It is affordably priced and provides a good profit margin.

Links list of benefits with order of same in discussion— drives home advantages of product to user and dealer.

7

MODEL 10–3, *continued*

NAVAL AIR STATION

There is no slack . . . in light Attack!

Marietta, North Carolina 27456
(919) 555-1050

February 24, 1994

Lt. Dennis Smoot
Maintenance Officer
ATKRON Two Zero Five
Naval Air Station
Marietta, NC 27456

Dear Lt. Smoot:

My five years' work at the Naval Air Station has shown me that we have a very professional maintenance staff. I believe we can further increase the professionalism by starting an incentive awards program.

The Silver Wrench Award program proposed here will offer many benefits for our operation, with a very low cost. Possible benefits include better aircraft availability, a more professional atmosphere, and increased morale in the maintenance department.

Please review the enclosed proposal at your convenience. I will call you next week to discuss the prospects for starting this program.

Sincerely,

Jim Barnes

AEC (AW) Jim Barnes
Maintenance Coordinator

MODEL 10–4
Formal proposal (internal)

THE SILVER WRENCH AWARD
INCENTIVE PROGRAM

Prepared for
LT Dennis Smoot
Maintenance Officer
ATKRON Two Zero Five

by
AEC (AW) Jim Barnes

February 24, 1994

MODEL 10–4, *continued*

continues

TABLE OF CONTENTS

LIST OF ILLUSTRATIONS

Reveals *organization* of entire proposal through table of contents.

Lists table numbers *and* titles.

MODEL 10–4, *continued*

EXECUTIVE SUMMARY

Maintenance personnel are often required to work long hours with little or no reward for excellent performance. The constant tempo of flight operations demands that our aircraft and their electronic systems be maintained in top condition. Without the efforts of our dedicated maintenance workers, mission objectives could not be achieved. To keep their momentum going, I have developed this proposal for a program that will reward a superior work center each quarter.

You may ask, "How will the squadron benefit?" The answer is that for a minimum investment, we will accomplish these objectives:

- Increase our productivity by making more aircraft available
- Promote a professional atmosphere
- Improve morale

In short, the Silver Wrench Award will encourage maintenance employees to be their best at all times.

As you know, meeting our production goals is much easier if maintenance troops take pride in their work and willingly perform arduous assignments. We have many objectives to achieve this year, so we must act now to take full advantage of this proposed program. With your support, we will be recognized as the finest attack squadron in both the navy and the naval reserve.

2

*Uses very *first* sentence to indicate need.*

*Connects understood *need* with his *proposal*.*

*Uses leadoff question to engage reader's *interest*.*

*Highlights *benefits* by use of *bullets*.*

*Appeals to reader's own self-interest— *reputation* of squadron.*

MODEL 10–4, *continued*

continues

INTRODUCTION

Today's high-tech navy requires optimum performance from all personnel to maintain a high state of readiness and achieve its objectives. This fact is especially true in naval aviation where complex aircraft systems must be kept in superior condition. Our squadron is no exception to this requirement.

Gives brief *background* *statement*.

Our maintenance personnel have received excellent training and have demonstrated on many occasions that they can get the job done. However, the rigors of maintaining an ever-increasing flight program can sometimes tax the professional capabilities of top-notch maintenance personnel. The Silver Wrench Award program would induce personnel to be their best at all times.

Purpose

Concisely states *why* the proposal has been written.

This proposed program aims to increase production at the work center, instill pride among maintenance personnel, and encourage high levels of professionalism throughout the maintenance department.

Scope

Acknowledges debt to similar program.

Outlines *scope* of program.

This idea has been adapted from the COMNAVAIRESFOR Golden Wrench Award program. The Golden Wrench Award program permits formal, yearly recognition of an individual squadron type (such as fighter, attack, or patrol) for maintenance excellence. As this proposal will show, the Silver Wrench Award program would formally recognize an individual work center each quarter. To choose a winner, information would be gathered from these sources:

- Maintenance data reporting
- Work-center audits
- Monitor programs
- Maintenance CPO

Proposal Format

This proposal is organized into three main sections, as follows:

Summarizes content of three main sections to follow in discussion.

The management section assigns the responsibilities for program operations to the maintenance officer, quality-assurance supervisor, and data analyst. Also, it sets up specific guidelines for collecting data and maintaining the plaque.

Detailed information about point assignments is presented in the data-compilation section. Quality-assurance analyst and maintenance CPO data are fully explained and then rated on a point scale.

The financial section gives all cost and manpower requirements for program support. For your convenience, two tables are provided for easy interpretation of data. In addition, several sources of financial support for the award program are listed.

3

MODEL 10–4, *continued*

Discusses pro-
gram details in
context of *who*
will be respon-
sible.

Mentions two
subsections to
follow.

Gives brief de-
scription of *how*
and *when* pro-
gram will be im-
plemented by
MO.

PROGRAM MANAGEMENT

The maintenance officer (MO) will oversee the Silver Wrench Award program, and the quality-assurance/analysis (QA/A) supervisor will have main responsibility for collecting all data.

Maintenance Officer

The primary responsibilities of the maintenance officer are to ensure that award guidelines (see Appendix) are strictly followed and that the award is presented on time. The Silver Wrench should be presented during the months of January, April, July, and October.

The MO will carefully review all point totals and select a winning work center. However, he or she should be careful not to select the work center with the most points if that shop has been involved in a recent malpractice incident: lost tool, improper maintenance procedures, low work standards, and so forth.

Quality-Assurance/Analysis Supervisor

The QA/A supervisor is ultimately responsible for collecting all point totals. These totals will be recorded on a data-collection form and presented to the maintenance officer for work-center selection.

Data collection should begin at least 15 days before the close of the award period. This date will coincide with the delivery of the monthly data reports. After the selection is made, the QA/A supervisor will have a brass plate engraved with the appropriate information:

- Work-center name
- Supervisor name
- Year and quarter selected
- Maintenance CPO

The QA/A supervisor has the responsibility for maintaining the plaque.

4

MODEL 10–4, *continued*

continues

COMPILATION OF AWARD DATA

The main data for award selection will be obtained from the Quality-Assurance/ Analysis Division and the individual maintenance chief petty officers. Although the MO and the maintenance CPO ultimately decide the award winners, they must have access to accurate and carefully tabulated data to ensure fairness in the selection process.

Quality-Assurance Information

Quality-assurance workers will provide most of the required data for this award. This information will be obtained from work-center audits and monitor programs. Collected data will be passed to the quality-assurance supervisor and placed on the data-collection form.

Work -Center Audits. Quality-assurance personnel will review the audits from the previous quarter and select the top four work centers. Every work center is audited once each quarter to ensure compliance with the maintenance standards established by OPNAVINST 4790.2D. These audits provide a thorough review of all work centers and indicate any shortcomings that exist. Each of the top work centers will receive four points.

Monitor Programs. Monitor-program forms from the previous quarter will be reviewed for both good and poor performance. Programs that will be checked include the following:

- Tool control
- Hydraulic contamination
- Nitrogen servicing
- Oxygen servicing
- Corrosion control
- Foreign object damage
- Nondestructive testing
- Safety
- Calibration
- Oil analysis
- Fuel contamination
- Training
- Ejection seat safety

Each good report will receive plus one point, and each poor report will receive minus one point.

Analyst Information

The squadron analyst will provide data on the documentation habits of each work center. The areas to be examined by the analyst are as follows:

- Documentation accuracy
- Corrosion documentation
- End-of-month closeouts

Point totals shall be presented to the quality-assurance supervisor for tabulation.

5

MODEL 10–4, *continued*

332

MDR Documentation. The analyst will select the four work centers with the lowest documentation error rate during the past quarter. Each work center will receive one point. If complete reports are unavailable, the three latest reports will be used.

Corrosion Documentation. The analyst will review the MDR-4 report and determine which work centers will receive credit for documenting corrosion prevention and treatment. Work centers will get one point per month for each month of prevention documented and one point per month for each month of treatment.

Closeouts. Each work center that does not submit or is late in submitting its end-of-month closeouts will be penalized two points per month per quarter.

Maintenance CPO Information

Every maintenance chief petty officer stays informed of the events that occur throughout the squadron. Many times, a maintenance CPO will see top performers doing their jobs above and beyond normal expectations. This performance is often undocumented and requires some method of accountability.

To fill this need, the quality-assurance supervisor will solicit data from each maintenance CPO. These data will be rated at two points each. This information will be added to the collection form and submitted to the maintenance chief and maintenance officer.

Gives thorough analysis of *proposed procedure.*

6

MODEL 10–4, *continued*

continues

FINANCIAL REQUIREMENTS

The Silver Wrench Award program is an inexpensive method of increasing productivity. Table 1 and Table 2 indicate the low program costs and small man-hour requirements.

After the initial outlay of nine man-hours and $34.15, program upkeep will involve about six man-hours and $5.00 a quarter. This cost equates to five minutes and three cents a day for a two-year period to promote a program that will reap many benefits to the squadron.

TABLE 1: EXPENSES

ITEM	COST
Plaque	$18.00
Brass Plates	4.50
Engraving (initial)	5.45
Engraving (quarterly)	5.00
Wrench	1.20
Total (initial)	$34.15
Total (quarterly)	$ 5.00

TABLE 2: MAN-HOUR REQUIREMENTS

ITEM	MAN-HOURS
Purchase materials	1.0
Engrave plates	1.0
Assemble plaque	2.0
Compile data	4.0
Select winner	1.0
Total (initial)	9.0
Total (quarterly)	6.0

Financial backing can be provided from several sources:

- Squadron coffee mess
- Officer fund
- CPO fund
- Personal donations

The plaque suggested here is designed to cover a two-year period. However, larger and more expensive plaques can be purchased to cover longer periods.

7

MODEL 10–4, *continued*

CONCLUSION

The Silver Wrench Award should become a valuable part of our maintenance program. This award can produce many benefits for the squadron at a very low cost. These benefits consist of the following:

Returns to major benefits.

- Reduced aircraft downtime
- More personal professionalism
- Higher morale
- Increased work-center production
- Reduced audit discrepancies
- Increased attention to details
- Safer maintenance practices
- Improved control of aircraft corrosion

Leaves reader with impression that everyone will benefit from program.

Extreme importance is placed on a squadron's ability to exceed its flight-hour program and to complete every mission. With the start of the Silver Wrench Award program, our squadron is sure to advance to its highest potential.

8

MODEL 10–4, *continued*

continues

APPENDIX: SILVER WRENCH AWARD GUIDELINES

The maintenance officer (MO) and the maintenance chief will select one work center per quarter for the Silver Wrench Award. Their selection will be based on the following information:

Quality-Assurance/Analysis

Audits. Select the four best work centers based on audit performance. Each will be rated at four points each.

Monitor Programs. Provide data based on the monitor-program forms from the past quarter. Each will be rated at PLUS one point for every satisfactory monitor and MINUS one point for every unsatisfactory monitor.

MDR Documentation. Select the four work centers that have submitted the most accurate documentation of MDR data during the last 3 months of available MDR-2 reports. Each work center will receive one point per work center per quarter.

Corrosion Documentation. Provide information on preventing and treating corrosion at work centers. This information will be scored at the following rates:

Corrosion Prevention. (code A-04 series). Data will be rated at one point per month per quarter of documentation.

Corrosion Treatment. (code Z-170 series). Data will be rated at one point per month per quarter of documentation.

MDR Closeouts. Provide information on delinquent, or nonsubmission of, MDR closeouts, to be rated at MINUS two points per month per quarter.

Maintenance Department CPOs

Have each CPO select one work center with the best performance during the past quarter (to be rated at two points).

MO and Maintenance Chief

Collect information from QA/A and maintenance chiefs and select the winner of the Silver Wrench Award. A high score should not be a substitute for sound judgment during the selection process.

9

MODEL 10−4, *continued*

Mc Duff, Inc.

MEMORANDUM

DATE: July 22, 1991
TO: Greg Bass
FROM: Mike Tran *MT*
SUBJECT: Replacement of In-House File Server

INTRODUCTORY SUMMARY

Gives context for feasibility study.

Summarizes conclusion of report.

The purpose of this feasibility study is to determine if the NTR PC905 would make a practical replacement for our in-house file server. As we agreed in our weekly staff meeting, our current file-serving computer is damaged beyond repair and must be replaced by the end of the week. This study shows that the NTR PC905 is a suitable replacement that we can purchase within our budget and install by Friday afternoon.

FEASIBILITY CRITERIA

Pinpoints three criteria to be discussed.

There are three major criteria that I addressed. First, the computer we buy must be able to perform the tasks of a file-serving computer on our in-house network. Second, it must be priced within our $4,000 budget for the project. Third, it must be delivered and installed by Friday afternoon.

Performance

As a file server, the computer we buy must be able to satisfy these criteria:

- Store all programs used by network computers
- Store the source code and customer-specific files for Xtracheck
- Provide fast transfer of files between computers while serving as host to the network
- Serve as the printing station for the network laser printer

Shows how NTR PC905 will fulfill performance criteria.

The NTR PC905 comes with a 120MB hard drive. This capacity will provide an adequate amount of storage for all programs that will reside on the file server. Our requirements are for 30MB of storage for programs used by network computers and 35MB of storage for source code and customer-specific programs. The 120MB drive will leave us with 55MB of storage for future growth and work space.

Only covers advantages because there are *no* disadvantages to buying PC905.

The PC905 can transfer files and execute programs across our network. It can perform these tasks at speeds up to five times faster than our current file server. Productivity should increase because the time spent waiting for transfer will decrease.

continues

MODEL 10-5
Feasibility study (one alternative)

The computer we choose as the file server must also serve as the printing station for our network laser printer. The PC905 is compatible with our Hewy Packer laser printer. It also has 2.0MB more memory than our current server. As a result, it can store larger documents in memory and print them with greater speed.

Budget

The budget for the new file server is $4,000. The cost of the PC905 is as follows:

PC905 with 120MB Hard Drive	$2,910
Keyboard	112
Monitor	159
Total	$3,181

No new network boards need to be purchased because we can use those that are in the current server. We also have all additional hardware and cables that will be required for installation. Thus the PC905 can be purchased for $800 under budget.

Time Frame

Our sales representative at NTR guarantees that we can have delivery of the system by Friday morning. Given this assurance, we can have the system in operation by Friday afternoon.

Additional Benefits

We are currently using NTR PCs at our customer sites. I am very familiar with the setup and installation of these machines. By purchasing a brand of computer currently in use, we will not have to worry about additional time spent learning new installation and operation procedures. In addition, we know that all our software is fully compatible with NTR products.

The warranty on the PC905 is for one year. After the warranty period, the equipment is covered by the service plan that we have for all our other computers and printers.

CONCLUSION

I recommend that we purchase the NTR PC905 as the replacement computer for our file server. It meets or exceeds all criteria for performance, price, and installation.

Makes costs easy to find with simple table.

Highlights major goal—quick installation.

Ends with "extras"—that is, benefits not among major criteria but still useful.

Restates significant point already noted in introductory summary.

MODEL 10–5, *continued*

PART III

11 | *Graphics*

McDuff employees like these often use computers to produce logos, charts, designs, and other graphics.

Computers have radically changed the world of graphics. Now, almost anyone with a keyboard, a mouse, and the right software can quickly produce illustrations that used to take hours to construct. As a result, today there are sophisticated graphics in every medium—newspapers, magazines, television, and, of course, technical communication.

Because readers *expect* graphics to accompany text, you as a technical professional must respond to this need. Well-designed and well-placed graphics will keep you competitive. Your graphics do not have to be fancy, however. Nor is it true that adding graphics will necessarily improve a document. Readers are impressed by visuals only when they are well done and appropriate.

Fortunately, you do not have to be a computer expert to understand the fundamentals of graphics. In fact, the availability of high-tech graphics has made it even more important that technical professionals first understand the basics of graphics before applying sophisticated techniques. To emphasize these basics, this chapter (1) defines some common graphics terms, (2) explains the main reasons to use graphics and gives some general guidelines, (3) lists specific guidelines for eight common graphics, and (4) shows you how to avoid graphics misuse. Although the chapter does include some production tips, the main emphasis is on *why* and *when* to use graphics.

TERMS IN GRAPHICS

Terminology for graphics is not uniform in the professions. That fact can lead to some confusion. For the purposes of this chapter, however, some common definitions are adopted and listed here:

- **Graphics:** This generic term refers to any nontextual portion of documents or oral presentations. It can be used in two ways: (1) to designate the field (for example, "Graphics is an area in which he showed great interest") or (2) to name individual graphical items ("She placed three graphics in her report").
- **Illustrations, visual aids:** Used synonymously with "graphics," these terms also can refer to all nontextual parts of a document. The term *visual aids,* however, often is limited to the context of oral presentations.
- **Tables and figures:** These terms name the two subsets of graphics.
  Tables refers to illustrations that place numbers or words in columns or rows or both.
  Figures refers to all graphics other than tables. Examples include charts (pie, bar, line, flow, and organization), engineering drawings, maps, and photographs.
- **Charts, graphs:** A subset of "figures," these synonymous terms refer to a type of graphic that displays data in visual form—as with bars, pie shapes, or lines on graphs. Chart is the term used most often in this text.
- **Technical drawing:** Another subset of "figures," a technical drawing is a representation of a physical object. Such illustrations can be drawn from many perspectives and can include "exploded" views.

Of course, you may see other graphics terms. For example, some technical companies use the word *plates* for figures. Be sure to know the terms your readers understand and the types of graphics they use.

BACKGROUND

Although the technology for producing graphics continuously gets more complex, the reasons for using them remain the same. Before exploring specific types of illustrations, this section covers some fundamentals. Why do readers like graphics to accompany text? What basic guidelines should you follow with all illustrations?

Reasons for Using Graphics

Before deciding whether to use a pie chart, table, or any other graphic, you need to know what graphics do for your writing. Here are four main reasons for using them.

■ *Reason 1: Graphics Simplify Ideas*

Readers know less about the subject than you—that is why they need your proposal, report, memo, or letter. Graphics help them cut through details and grasp basic ideas. For example, a simple illustration of a laboratory instrument, such as a Bunsen burner, makes the description of a lab procedure much easier to understand. In a more complex example, Figure 11−1 uses a group of four different charts to convey the one main point—that McDuff's new Equipment Development group lags behind the company's other profit centers. A quick look at the charts tells the story of the group's difficulties much better than would several hundred words of text.

Problems in McDuff's E. D. Group

1990 McDuff Sales (for 7 project areas)

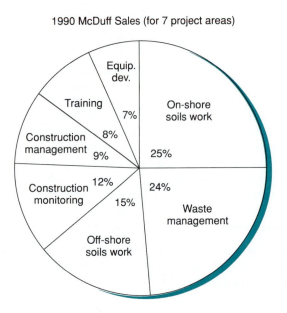

1988-1990 Profits for
Equipment Development (ED)

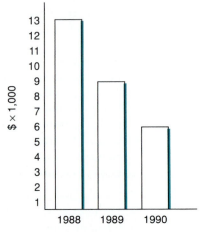

$ Value of ED Contracts (next 6 months)

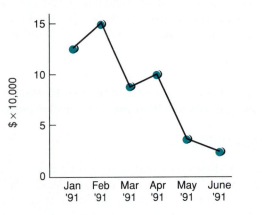

Billable vs. Non-billable Time
(average employee/based on 40 hr week)

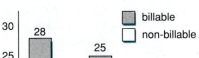
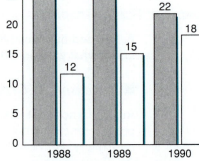

FIGURE 11–1
Graphics used to simplify ideas

■ *Reason 2: Graphics Reinforce Ideas*

When a point really needs emphasis, create a graphic. For example, you might draw a map to show where computer terminals will be located within a building, or use a pie chart to show how a budget will be spent, or include a drawing that indicates how to operate a VCR. In all three cases, the graphic would reinforce points made in the accompanying text.

■ *Reason 3: Graphics Create Interest*

Graphics are "grabbers." That means they can be used to entice readers into the text, just as they engage readers' interest in magazines and journals. If your customers have three reports on their desks and must quickly decide which one to read first, they probably will pick up the one with an engaging picture or chart on the cover page. It may be something as simple as (1) a map outline of the state, county, or city where you will be doing a project, (2) a picture of the product or service you are providing, or (3) a symbol of the purpose of your writing project. Whether on the cover or in the text, graphics attract attention.

Figure 11−2 shows how an outside consultant used a well-known Leonardo da Vinci drawing to attract attention to his proposal to McDuff. The drawing helps to (1) add a classical touch to the cover, (2) focus on the human side of employee testing, and (3) associate the innovation of Infinite Vision, Inc., with the creativity of da Vinci.

■ *Reason 4: Graphics Are Universal*

Some people wrongly associate the growing importance of graphics with today's reliance on television and other popular media—as if graphics pander to less-intellectual instincts. While visual media such as television obviously rely on pictures, the fact is that graphics have been mankind's universal language since cave drawings. A picture, drawing, or chart makes an immediate emotional impact that can help or hurt your case. Advertisers know the power of images, but few writers of technical documents have learned to merge the force of graphics with their text.

General Guidelines

A few basic guidelines apply to all graphics. Keep these fundamentals in mind as you move from one type of illustration to another.

■ *Graphics Guideline 1: Refer to All Graphics in the Text*

With a few exceptions—such as cover illustrations used to grab attention—graphics should be accompanied by clear references within your text. Specifically, you should follow these rules:

- Include the graphic number in Arabic, not Roman, when you are using more than one graphic.
- Include the title, and sometimes the page number, if either is needed for clarity or emphasis.
- Incorporate the reference smoothly into text wording.

Improving Productivity at McDuff, Inc.
An Innovative Approach to Employee Testing

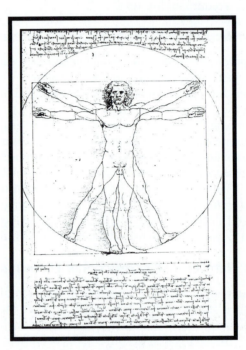

Prepared for Robert S. McDuff
President, McDuff, Inc.

by
James H. Stephens
Infinite Vision, Inc.

February 22, 1994

FIGURE 11–2
Graphics used to create interest

Here are two ways to phrase and position a graphics reference. In Example 1, there is the additional emphasis of the graphics title, whereas in Example 2, the title is left out. Also, note that you can draw more attention to the graphic by placing the reference at the start of the sentence in a separate clause. Or you can relegate the reference to a parenthetical expression at the end or middle of the passage. Choose the option that best suits your purposes.

- **Example 1:** In the past five years, 56 businesses in the county have started in-house recycling programs. The result has been a dramatic shift in the amount of property the county has bought for new waste sites, as shown in Figure 5 ("Landfill Purchases, 1985–1990").
- **Example 2:** As shown in Figure 5, the county has purchased much less land for landfills during the last five years. This dramatic reduction results from the fact that 56 businesses have started in-house recycling programs.

■ *Graphics Guideline 2: Think About Where to Put Graphics*

In most cases, locate a graphic close to the text in which it is mentioned. This immediate reinforcement of text by an illustration gives graphics their greatest strength. Variations of this option, as well as several other possibilities, are presented here:

- **Same page as text reference:** A simple visual, such as an informal table, should go on the same page as the text reference if you think it too small for a separate page.
- **Page opposite text reference:** A complex graphic, such as a long table, that accompanies a specific page of text can go on the page opposite the text—that is, on the opposite page of a two-page spread. Usually this option is exercised *only* in documents that are printed on both sides of the paper throughout.
- **Page following first text reference:** Most text graphics appear on the page after the first reference. If the graphic is referred to throughout the text, it can be repeated at later points. (Remember—readers prefer to have graphics positioned exactly where they need them, rather than their having to refer to another part of the document.)
- **Attachments or appendices:** Graphics can go at the end of the document in two cases: first, if the text contains so many references to the graphic that placement in a central location, such as an appendix, would make it more accessible; and second, if the graphic contains less important supporting material that would only interrupt the text.

■ *Graphics Guideline 3: Position Graphics Vertically When Possible*

Readers prefer graphics they can view without having to turn the document sideways. However, if the table or figure cannot fit vertically on a standard 8½" × 11" page, either use a foldout or place the graphic horizontally on the page. In the latter case, position the illustration so that the top is on the left margin. (In other words, the page must be turned clockwise to be viewed.)

■ *Graphics Guideline 4: Avoid Clutter*

Let simplicity be your guide. Readers go to graphics for relief from, or reinforcement of, the text. They do not want to be bombarded by visual clutter. Omit information that is not relevant to your purpose, while still making the illustration clear and self-contained. Also, use enough white space so that the readers' eyes are drawn to the graphic. The final section of this chapter discusses graphics clutter in more detail.

■ *Graphics Guideline 5: Provide Titles, Notes, Keys, and Source Data*

Graphics should be as self-contained and self-explanatory as possible. Moreover, they must note any borrowed information. Follow these basic rules for format and acknowledgement of sources:

- **Always use a title.** Follow the graphic number with a short, precise title— either on the line below the number *or* on the same line after a colon (for example, ''Figure 3: Salary Scales'').

  In **tables,** the number and title go at the top. (As noted in Table Guideline 1 on page 368, one exception is informal tables. They have no table number or title.)

  In **figures,** the number and title can go either above or below the illustration. Center titles or place them flush with the left margin.
- **Include notes for explanation.** When introductory information for the graphic is needed, place a note directly underneath the title *or* at the bottom of the graphic.
- **Use keys or legends for simplicity.** If a graphic needs many labels, consider using a legend or key, which lists the labels and corresponding symbols on the graphic. For example, a pie chart might have the letters *A, B, C, D,* and *E* printed on the pie pieces, while a legend at the top, bottom, or side of the figure would list what the letters represent.
- **Place complete source information at the bottom.** You have a moral, and sometimes legal, obligation to cite the person, organization, or publication from which you borrowed information for the figure. Precede the description with the word ''Source'' and a colon. Or, if you borrowed just part of a graphic, introduce the citation with ''Adapted from.'' Besides citing the source, it is usually necessary to request permission to use copyrighted or proprietary information, depending on your use and the amount you are borrowing. (A prominent exception is most information provided by the federal government. Most government publications are not copyrighted.) Consult a reference librarian for details about seeking permission.

SPECIFIC GUIDELINES FOR EIGHT GRAPHICS

Illustrations come in many forms; almost any nontextual part of your document can be placed under the umbrella term *graphic*. Among the many types, these eight are often used in technical writing: (1) pie charts, (2) bar charts, (3) line charts, (4) schedule charts, (5) flowcharts, (6) organization charts, (7) technical drawings, and (8) tables. This section of the chapter highlights their different purposes and gives guidelines for using each type.

Pie Charts

Familiar to most readers, pie charts show relationships between the parts and the whole—when just approximate information is needed. Their simple circles with clear labels can provide comforting simplicity within even the most complicated report. Yet the simple form keeps them from being useful when you need to reveal detailed information. Here are specific guidelines for constructing pie charts.

■ Pie Chart Guideline 1: Use No More Than 10 Divisions

To make pie charts work well, limit the number of pie pieces to no more than 10. In fact, the fewer the better. This approach lets the reader grasp major relationships, without having to wade through the clutter of tiny divisions that are difficult to read. In Figure 11–3, for example, McDuff's client can readily see that the project staff will come from the three McDuff offices closest to the project site.

FIGURE 11–3
Pie chart with as few pieces as possible. (Chart shows McDuff work force breakdown for the offshore Atlantic project. McDuff can draw most project workers from its East Coast offices.)

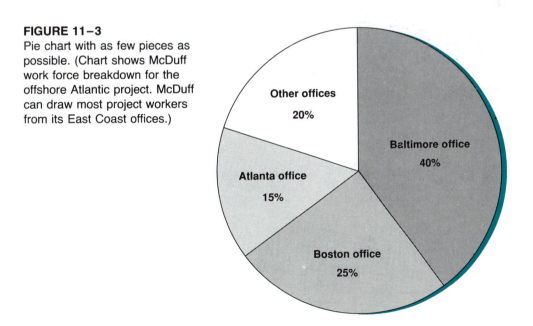

Other offices
20%

Baltimore office
40%

Atlanta office
15%

Boston office
25%

■ Pie Chart Guideline 2: Move Clockwise From 12:00, From Largest to Smallest Wedge

Readers prefer pie charts oriented like a clock—with the first wedge starting at 12:00. Also, moving from largest to smallest wedge provides a convenient organizing principle.

Make exceptions to this design only for good reason. In Figure 11−3, for example, the last wedge represents a greater percentage than the previous wedge. In this way, it does not break up the sequence the writer wants to establish by grouping the three McDuff offices with the three largest percentages of project workers.

■ Pie Chart Guideline 3: Use Pie Charts Especially for Percentages and Money

Pie charts catch the reader's eye best when they represent items divisible by 100, as with percentages and dollars. Figure 11−3 shows percentages, and Figure 11−4 shows money. Using the pie chart for money breakdowns is made even more appropriate by the coinlike shape of the chart.

FIGURE 11−4
Pie chart showing money break-down for average deductions from a McDuff paycheck

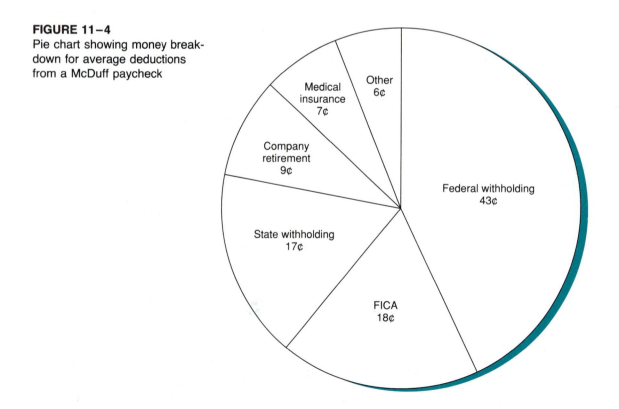

■ *Pie Chart Guideline 4: Be Creative, But Stay Simple*

Figure 11−5 shows that you can emphasize one piece of the pie by (A) shading a wedge, (B) removing a wedge from the main pie, or (C) placing related pie charts in a three-dimensional drawing. Today there are graphics software packages that can create these and other variations for you, so experiment a bit. Of course, always make sure to keep your charts from becoming too detailed. Pie charts should stay simple.

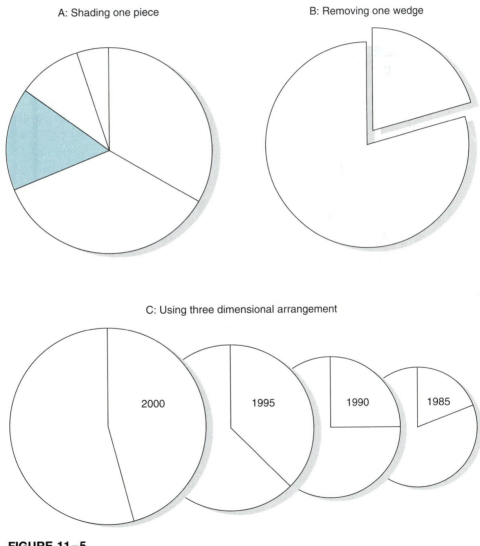

FIGURE 11−5
Techniques for emphasis in pie charts
Source: William S. Pfeiffer, *Proposal Writing* (Columbus, Ohio: Merrill, 1989), 145.

▨   *Pie Chart Guideline 5: Draw and Label Carefully*

The most common pie chart errors are (1) wedge sizes that do not correspond correctly to percentages or money amounts and (2) pie sizes that are too small to accommodate the information placed in them. Here are some suggestions for avoiding these mistakes:

- **Pie size:** Make sure the chart occupies enough of the page. On a standard 8½" × 11" sheet with only one pie chart, your circle should be from 3" to 6" in diameter—large enough not to be dwarfed by labels and small enough to leave sufficient white space in the margins.
- **Labels:** Place the wedge labels either inside the pie or outside, depending on the number of wedges, the number of wedge labels, or the length of the labels. Choose the option that produces the cleanest-looking chart.
- **Conversion of percentages:** If you are drawing the pie chart by hand, not using a computer program, use a protractor or similar device. One percent of the pie equals 3.6 degrees (3.6 × 100% = 360 degrees in a circle). With that formula as your guide, you can convert percentages or cents to degrees. Remember, however, that a pie chart does not reveal fine distinctions very well; it is best used for showing larger differences.

Bar Charts

Like pie charts, bar charts are easily recognized, for they are seen every day in newspapers and magazines. Unlike pie charts, however, bar charts can accommodate a good deal of technical detail. Comparisons are provided by means of two or more bars running either horizontally or vertically on the page. Follow these five guidelines to create effective bar charts.

▨   *Bar Chart Guideline 1:*
Use a Limited Number of Bars

Though bar charts can show more information than pie charts, both types of illustrations have their limits. Bar charts begin to break down when there are so many bars that information is not easily grasped. The maximum bar number can vary according to chart size, of course. Figure 11−6 shows several multibar charts from the 1992 McDuff annual report. The impact of the charts is enhanced by the limited number of bars.

▨   *Bar Chart Guideline 2: Show Comparisons Clearly*

Bar lengths should be varied enough to show comparisons quickly and clearly. Avoid using bars that are too close in length, for then readers must study the chart before understanding it. Such a chart lacks immediate visual impact.

Also, avoid the opposite tendency of using bar charts to show data that are much different in magnitude. To relate such differences, some writers resort to the dubious technique of inserting "break lines" (two parallel lines) on an axis to reflect breaks in scale (see Figure 11−7). Although this approach at least reminds

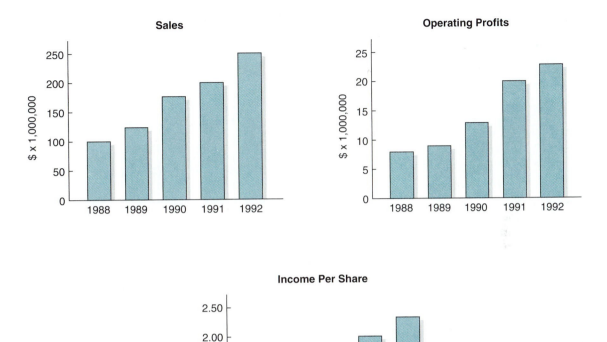

FIGURE 11–6
Bar charts from McDuff annual report (1992)

readers of the breaks, it is still deceptive. For example, note that Figure 11–7 provides no *visual* demonstration of the relationship between 50 and 2800. The reader must think about these differences before making sense out of the chart. In other words, the use of hash marks runs counter to a main goal of graphics— creating an immediate and accurate visual impact.

▪ Bar Chart Guideline 3: Keep Bar Widths Equal and Adjust Space Between Bars Carefully

While bar length varies, bar width must remain constant. As for distance between the bars, following are three options (along with examples in Figure 11–8):

▪ **Option A: Use no space** when there are close comparisons or many bars, so that differences are easier to grasp.

FIGURE 11–7

Hash marks on bar charts—a technique that can lead to mis-understanding

Adapted from William S. Pfeiffer, *Proposal Writing* (Columbus, Ohio: Merrill, 1989), 147.

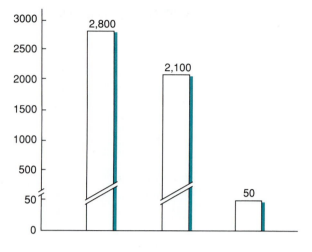

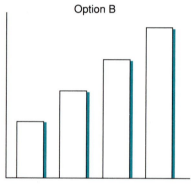

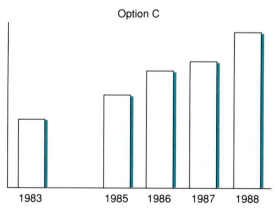

FIGURE 11–8

Bar chart variations

Source: William S. Pfeiffer, *Proposal Writing* (Columbus, Ohio: Merrill, 1989), 148.

- **Option B: Use equal space, but less than bar width** when bar height differences are great enough to be seen in spite of the distance between bars.
- **Option C: Use variable space** when gaps between some bars are needed to reflect gaps in the data.

■ *Bar Chart Guideline 4: Carefully Arrange the Order of Bars*

The arrangement of bars is what reveals meaning to readers. Here are two common approaches:

- **Sequential:** used when the progress of the bars shows a trend—for example, McDuff's increasing number of environmental projects in the last five years
- **Ascending or descending order:** used when you want to make a point by the rising or falling of the bars—for example, the rising profits of McDuff's waste management division over the last seven years

■ *Bar Chart Guideline 5: Be Creative*

Figure 11–9 shows two bar chart variations that help display multiple trends. The *segmented bars* in Option A produce four types of information: the total sales (A + B + C) and the individual sales for A, B, and C. The *grouped bars* in Option B show

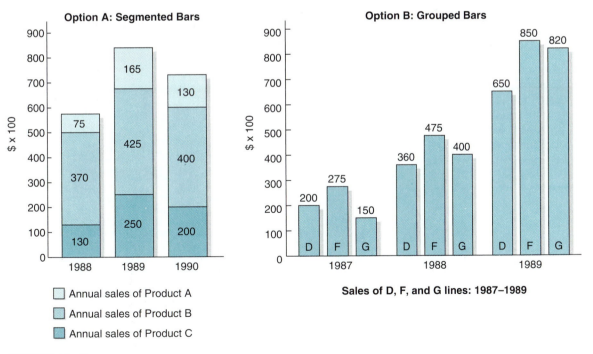

FIGURE 11–9

Bar chart variations for multiple trends

Adapted from William S. Pfeiffer, *Proposal Writing* (Columbus, Ohio: Merrill, 1989), 150.

the individual sales trends for D, F, and G, along with a comparison of all three by year. Note that the amounts are written on the bars to highlight comparisons.

Although these and other bar chart variations may be useful, remember to retain the basic simplicity of the chart.

Line Charts

Line charts are a common graphic. Almost every newspaper contains a few charts covering topics such as stock trends, car prices, or weather. More than other graphics, line charts telegraph complex trends immediately. They work by using vertical and horizontal axes to reflect quantities of two different variables. The vertical (or y) axis usually plots the dependent variable; the horizontal (or x) axis usually plots the independent variable. (The dependent variable is affected by changes in the independent variable.) Lines then connect points that have been plotted on the chart. When drawing line charts, follow these five main guidelines:

■ Line Chart Guideline 1: Use Line Charts for Trends

Readers are affected by the direction and angle of the chart's line(s), so take advantage of this persuasive potential. In Figure 11−10, for example, the writer wants to show the feasibility of adopting a new medical plan for McDuff. Including a line chart in the study gives immediate emphasis to the most important issue—the effect the new plan would have on stabilizing the firm's medical costs.

FIGURE 11−10
Line chart used to show effect of proposed medical plan on McDuff health costs

Adapted from William S. Pfeiffer, *Proposal Writing* (Columbus, Ohio: Merrill, 1989), 151.

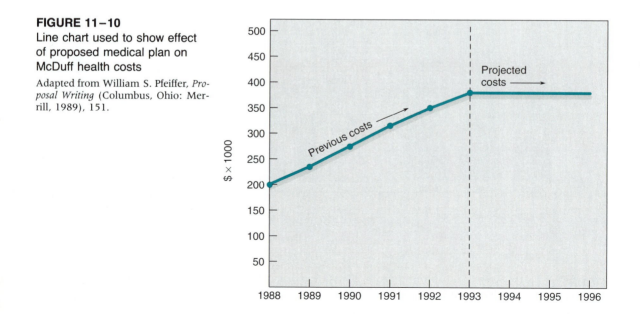

■ *Line Chart Guideline 2:*
Locate Line Charts With Care

Given their strong impact, line charts can be especially useful as attention-grabbers. Consider placing them (1) on cover pages (to engage reader interest in the document), (2) at the beginning of sections that describe trends, and (3) in conclusions (to reinforce a major point of your document).

■ *Line Chart Guideline 3:*
Strive for Accuracy and Clarity

Like bar charts, line charts can be misused or just poorly constructed. Be sure that the line or lines on the graph truly reflect the data from which you have drawn. Also, select a scale that does not mislead readers with visual gimmicks. Here are some specific suggestions to keep your line charts accurate and clear:

- Start all scales from zero to eliminate the possible confusion of breaks in amounts (see Bar Chart Guideline 2).
- Select a vertical-to-horizontal ratio for axis lengths that is pleasing to the eye (three vertical to four horizontal is common).
- Make chart lines as thick as or thicker than the axis lines.
- Use shading under the line when it will make the chart more readable.

■ *Line Chart Guideline 4: Do Not*
Place Numbers on the Chart Itself

Line charts derive their main effect from the simplicity of lines that show trends. Avoid cluttering the chart with a lot of numbers that only detract from the visual impact.

■ *Line Chart Guideline 5: Use*
Multiple Lines With Care

Like bar charts, line charts can show multiple trends. Simply add another line or two. If you place too many lines on one chart, however, you run the risk of confusing the reader with too much data. Use no more than four or five lines on a single chart (see Figure 11–11).

Schedule Charts

Many documents, especially proposals and feasibility studies, include a special kind of chart that shows readers when certain activities will be accomplished. This kind of chart usually highlights tasks and times already mentioned in the text. Often called a milestone or Gantt chart (after Henry Laurence Gantt, 1861–1919), it usually comprises these parts (see Figure 11–12):

- **Vertical axis,** which lists the various parts of the project, in sequential order
- **Horizontal axis,** which registers the appropriate time units
- **Horizontal bar lines** (Gantt) or separate markers (milestone), which show the starting and ending times for each task

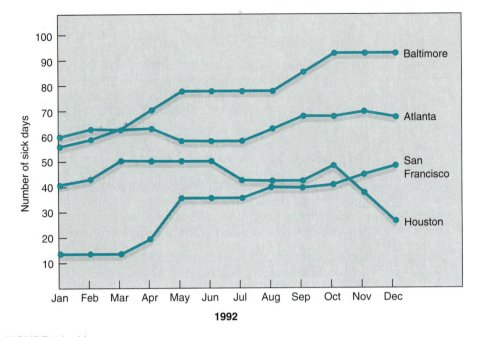

FIGURE 11–11
Line chart using multiple lines to show number of sick days taken at four McDuff offices: 1992

Adapted from William S. Pfeiffer, *Proposal Writing* (Columbus, Ohio: Merrill, 1989), 152.

Follow these basic guidelines for constructing effective schedule charts in your proposals, feasibility studies, or other documents.

■ *Schedule Chart Guideline 1:*
Include Only Main Activities

Keep readers focused on a maximum of 10 or 15 main activities. If they need more detail, construct a series of schedule charts linked to the main "overview" chart.

■ *Schedule Chart Guideline 2: List Activities in*
Sequence, Starting at the Top of the Chart

As shown in Figure 11–12, the convention is to list activities from the top to the bottom of the vertical axis. Thus the reader's eye moves from the top left to the bottom right of the page, the most natural flow for most readers.

■ *Schedule Chart Guideline 3:*
Run Labels in the Same Direction

If readers have to turn the chart sideways to read labels, they may lose interest.

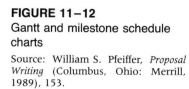

FIGURE 11–12
Gantt and milestone schedule charts

Source: William S. Pfeiffer, *Proposal Writing* (Columbus, Ohio: Merrill, 1989), 153.

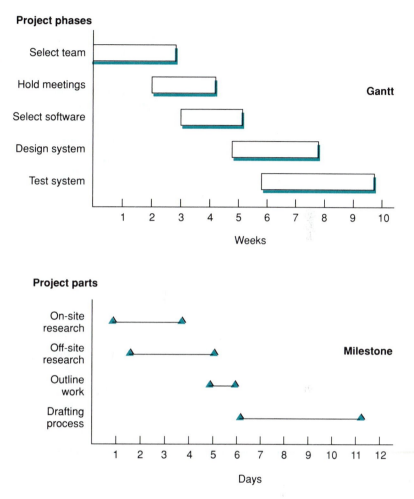

Schedule Chart Guideline 4: Create New Formats When Needed

Figure 11–12 shows only two common types of schedule charts; you should devise your own hybrid form when it suits your purposes. Your goal is to find the simplest format for telling your reader when a product will be delivered, a service completed, and so forth. Figure 11–13 includes one such variation.

Schedule Chart Guideline 5: Be Realistic About the Schedule

Schedule charts can come back to haunt you if you are not realistic at the outset. As you set dates for activities, be forthright with yourself and your reader about the likely time something can be accomplished. Your managers and clients understand delays

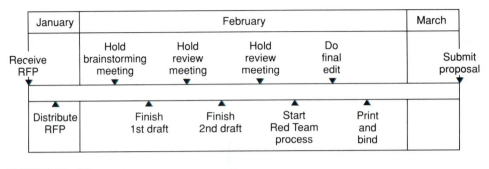

FIGURE 11–13
Schedule chart variation

caused by weather, equipment breakdowns, and other unforeseen events. However, they will be less charitable about schedule errors that result from sloppy planning.

Flowcharts

Flowcharts tell a story about a process, usually by stringing together a series of boxes and other shapes that represent separate activities (see Figure 11–14). Because they have a reputation for being hard to read, you need to take extra care in designing them. These five guidelines will help.

■ Flowchart Guideline 1: Present Only Overviews

Readers usually want flowcharts to give them only a capsule version of the process, not all the details. Reserve your list of particulars for the text or the appendices, where readers expect it.

■ Flowchart Guideline 2: Limit the Number of Shapes

Flowcharts rely on rectangles and other shapes to relate a process—in effect, to tell a story. Different shapes represent different types of activities. This variety helps in describing a complex process, but it can also produce confusion. For the sake of clarity and simplicity, limit the number of different shapes in your flowcharts. Figure 11–15 includes a flowchart that is complex but still readable. Note that the writer has modified geometric shapes to match what they represent (for example, note the tractor-feed holes on the sides of the "print out" block).

■ Flowchart Guideline 3: Provide a
Legend When Necessary

Simple flowcharts often need no legend. The few shapes on the chart may already be labeled by their specific steps. When charts get more complex, however, include a legend that identifies the meaning of each shape used.

FIGURE 11–14
Flowchart for basic McDuff project

Adapted from William S. Pfeiffer, *Proposal Writing* (Columbus, Ohio: Merrill, 1989), 155.

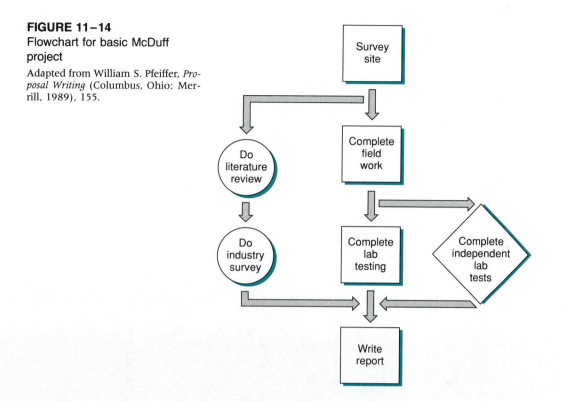

■ *Flowchart Guideline 4: Run the Sequence From Top to Bottom or From Left to Right*

Long flowcharts like the one in Figure 11–15 may cover the page with several columns or rows. Yet they should always show some degree of uniformity by assuming either a basically vertical or horizontal direction.

■ *Flowchart Guideline 5: Label All Shapes Clearly*

Besides a legend that defines meanings of different shapes, the chart usually includes a label for each individual shape or step. Follow one of these approaches:

- Place the label inside the shape.
- Place the label immediately outside the shape.
- Put a number in each shape and place a legend for all numbers in another location (preferably on the same page).

Organization Charts

Organization charts reveal the structure of a company or other organization—the people, positions, or work units. The challenge in producing this graphic is to make sure that the arrangement of information accurately reflects the organization.

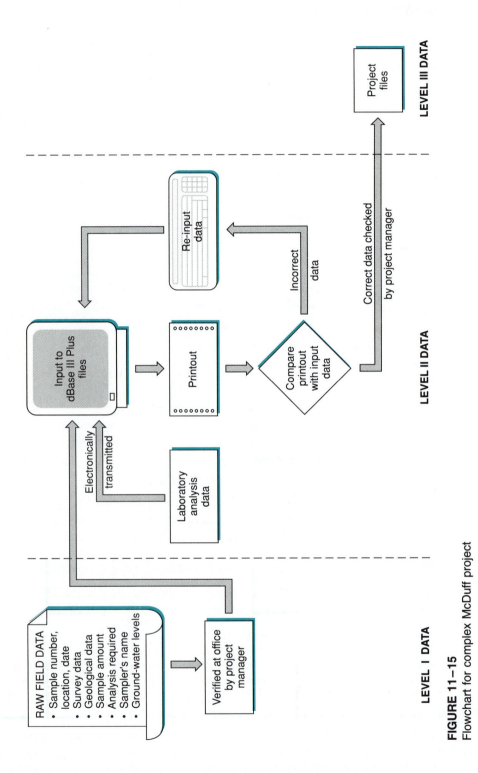

FIGURE 11–15
Flowchart for complex McDuff project

■ *Organization Chart Guideline 1: Use the Linear "Boxes" Approach to Emphasize High-Level Positions*

This traditional format uses rectangles connected by lines to represent some or all of the positions in an organization (see Figure 11–16). Because high-level positions usually appear at the top of the chart, where the attention of most readers is focused, this design tends to emphasize upper management.

■ *Organization Chart Guideline 2: Connect Boxes With Solid or Dotted Lines*

Solid lines show direct reporting relationships; dotted lines show indirect or staff relationships (see Figure 11–16).

■ *Organization Chart Guideline 3: Use a Circular Design to Emphasize Mid- and Low-Level Positions*

This arrangement of concentric circles gives more visibility to workers outside upper management. These are often the technical workers most deeply involved in the details of a project. For example, Figure 11–17 draws attention to the project engineers perched on the chart's outer ring.

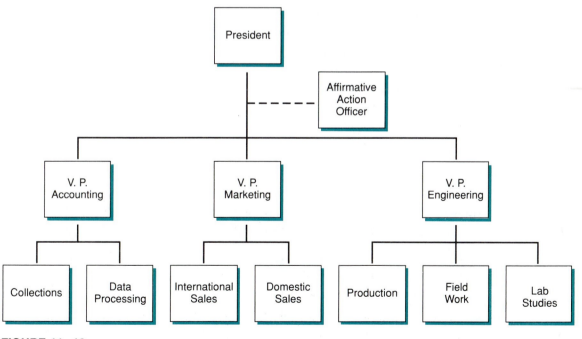

FIGURE 11–16
Basic organization chart
Adapted from William S. Pfeiffer, *Proposal Writing* (Columbus, Ohio: Merrill, 1989), 157.

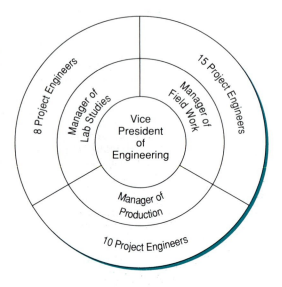

FIGURE 11–17
Concentric organization chart
Source: William S. Pfeiffer, *Proposal Writing* (Columbus, Ohio: Merrill, 1989), 158.

■ *Organization Chart Guideline 4:*
Use Varied Shapes Carefully

Like flowcharts, organization charts can use different shapes to indicate different levels or types of jobs. However, beware of introducing more complexity than you need. Use more than one shape only if you are convinced this approach is needed to convey meaning to the reader.

■ *Organization Chart Guideline 5: Be Creative*

When standard forms will not work, create new ones. For example, Figure 11–18 uses an organization chart as the vehicle for showing the lines of responsibility in a specific project.

Technical Drawings

Technical drawings are important tools of companies that produce or use technical products. These drawings can accompany documents such as instructions, reports, sales orders, and proposals. They are preferred over photographs when specific views are more important than photographic detail. Whereas all drawings used to be produced mainly by hand, now they are more and more frequently created by CAD (computer-assisted design) systems. Follow these guidelines for producing technical drawings that complement your text.

■ *Drawing Guideline 1: Choose the*
Right Amount of Detail

Keep drawings as simple as possible. Use only that level of detail that serves the purpose of your document and satisfies your reader's needs. For example, Figure

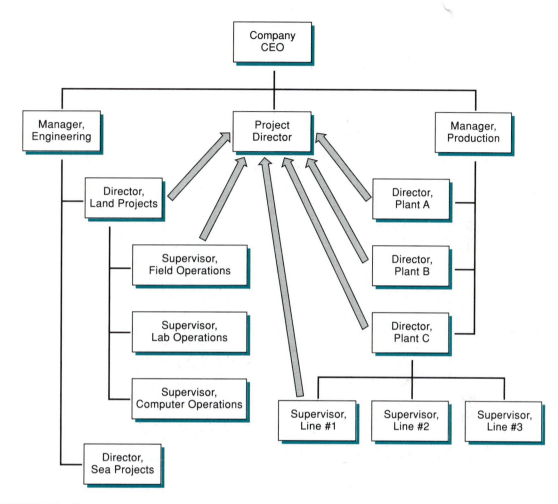

FIGURE 11–18
Organization chart focusing on project (indicates individuals most involved with upcoming project)
Source: William S. Pfeiffer, *Proposal Writing* (Columbus, Ohio: Merrill, 1989), 159.

11–19 will be used in a McDuff public service brochure on maintaining home heating systems. Its intention is to focus on just one main part of the thermostat— the lever and attached roller. Completed on a CAD system, this drawing presents an exploded view so that the location of the arm can be easily discerned by the reader of the brochure.

■ *Drawing Guideline 2: Label Parts Well*

A common complaint of drawings is that parts included in the illustration are not carefully or clearly labeled. Be sure to place labels on every part you want your reader to see. (Conversely, you can choose *not* to label those parts that are irrelevant to your purpose.)

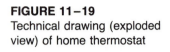

FIGURE 11–19
Technical drawing (exploded
view) of home thermostat

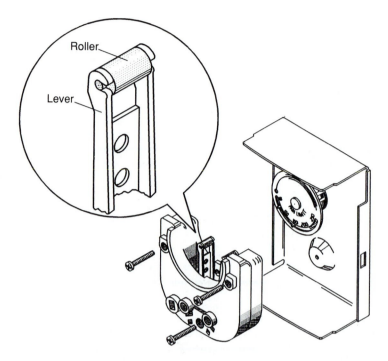

When you label parts, use a typeface large enough for easy reading. Also,
arrange labels so that (1) they are as easy as possible for your reader to locate and
(2) they do not detract from the importance of the drawing itself. The simple
labeling in Figure 11–19 fulfills these objectives.

■ Drawing Guideline 3: Choose the Most Appropriate View

As already noted, illustrations—unlike photographs—permit you to choose the
level of detail that is needed. In addition, drawings offer you a number of options
for perspective or view:

- **Exterior view** (shows surface features with either a two- or three-dimensional
  appearance—see Figure 11–20)
- **Cross-section view** (shows a "slice" of the object so that interiors can be
  viewed)
- **Exploded view** (shows relationship of parts to each other by "exploding" the
  mechanism—see Figure 11–19)

■ Drawing Guideline 4: Use Legends When There Are Many Parts

In complex drawings, avoid cluttering the illustration with many labels. Figure
11–20, for example, places all labels in one easy-to-find spot, rather than leaving
them on the drawing.

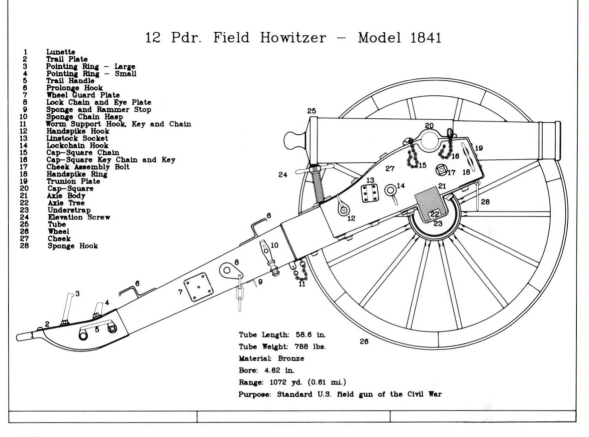

12 Pdr. Field Howitzer — Model 1841

1	Lunette
2	Trail Plate
3	Pointing Ring — Large
4	Pointing Ring — Small
5	Trail Handle
6	Prolonge Hook
7	Wheel Guard Plate
8	Lock Chain and Eye Plate
9	Sponge and Rammer Stop
10	Sponge Chain Hasp
11	Worm Support Hook, Key and Chain
12	Handspike Hook
13	Linstock Socket
14	Lockchain Hook
15	Cap-Square Chain
16	Cap-Square Key Chain and Key
17	Cheek Assembly Bolt
18	Handspike Ring
19	Trunion Plate
20	Cap-Square
21	Axle Body
22	Axle Tree
23	Understrap
24	Elevation Screw
25	Tube
26	Wheel
27	Cheek
28	Sponge Hook

Tube Length: 58.6 in.

Tube Weight: 788 lbs.

Material: Bronze

Bore: 4.62 in.

Range: 1072 yd. (0.61 mi.)

Purpose: Standard U.S. field gun of the Civil War

FIGURE 11–20
Technical drawing using CAD system

Assume that this illustration is part of a McDuff report to the National Park System. A group of specialists just completed a restoration project at a Civil War battlefield. Among many other tasks, the company (1) stopped erosion that had been destroying several hilly sites, (2) moved five howitzer cannon to permanent sites on a mountain ridge, where they were located during the war, and (3) built walking paths that would allow some public access to the battle locations, without damaging the terrain. Given the importance of the cannon to the project, a McDuff CAD draftsperson completed a technical illustration. The complete drawing, with labels, appears in the text of the report. A reduced-size version appears on the cover of the report.

Tables

Tables present your readers with raw data, usually in the form of numbers but sometimes in the form of words. Tables are classified as either formal or informal:

- **Informal tables:** limited data arranged in the form of either rows or columns
- **Formal tables:** data arranged in a grid, always with both horizontal rows and vertical columns

These five guidelines will help you make decisions about designing and positioning tables within the text of your documents.

■ Table Guideline 1: Use Informal Tables as Extensions of Text

Informal tables are usually merged with the text on a page, rather than isolated on a separate page or attachment. As such, an informal table usually has (1) no table number or title and (2) no listing in the list of illustrations in a formal report or proposal.

Example:
Our project in Alberta, Canada, will involve engineers, technicians, and salespeople from three offices, in these numbers:

San Francisco Office	45
St. Louis Office	34
London Office	6
Total	85

■ Table Guideline 2: Use Formal Tables for Complex Data Separated From Text

Formal tables may appear on the page of text that includes the table reference, on the page following the first text reference, or in an attachment or appendix. In any case, you should:

- Extract important data from the table and highlight them in the text
- Make every formal table as clear and visually appealing as possible

■ Table Guideline 3: Use Plenty of White Space

Used around and within tables, white space guides the eye through a table much better than do black lines. Avoid putting complete boxes around tables. Instead, leave one inch more of white space than you would normally leave around text.

■ Table Guideline 4: Follow Usual Conventions for Dividing and Explaining Data

Figure 11–21 shows a typical formal table. It satisfies the overriding goal of being clear and self-contained. To achieve that objective in your tables, follow these guidelines:

1. **Titles and numberings:** Give a title to each formal table, and place title and number above the table. Number each table if the document contains two or more tables.
2. **Headings:** Create short, clear headings for all columns and rows.
3. **Abbreviations:** Include in the headings any necessary abbreviations or symbols, such as lb or %. Spell out abbreviations and define terms in a key or footnote if any reader may need such assistance.

TABLE 6: McDuff's Employee Retirement Fund			
	Book Value	**Market Value**	**% of Total Market Value**
Temporary Securities	$ 434,084	434,084	5.9%
Bonds	3,679,081	3,842,056	52.4
Common Stocks	2,508,146	3,039,350	41.4
Mortgages	18,063	18,063	.3
Real Estate	1,939	1,939	nil
Totals	$6,641,313	$7,335,492	100.0%

Note: This table contrasts the book value versus the market value of the McDuff Employee Retirement Fund, as of December 31, 1990.

FIGURE 11–21
Example of formal table
Source: McDuff's accounting firm of Bumble and Bumble, Inc.

4. **Numbers:** Round off numbers when possible, for ease of reading. Also, align multidigit numbers on the right edge, or at the decimal when shown.
5. **Notes:** Place any necessary explanatory headnotes either between the title and the table (if the notes are short) *or* at the bottom of the table.
6. **Footnotes:** Place any necessary footnotes below the table.
7. **Sources:** Place any necessary source references below the footnotes.
8. **Caps:** Use uppercase and lowercase letters, rather than full caps.

■ *Table Guideline 5: Pay Special Attention to Cost Data*

Most readers prefer to have complicated financial information placed in tabular form. Given the importance of such data, edit cost tables with great care. Devote extra attention to these two issues:

▪ Placement of decimals in costs
▪ Correct totals of figures

Documents like proposals can be considered contracts in some courts of law, so there is no room for error in relating costs.

MISUSE OF GRAPHICS

Computers have revolutionized the world of graphics by placing sophisticated tools in the hands of many writers. Yet this largely positive event has its dark side. Today you will see many more graphics that—in spite of their slickness—distort data and misinform the reader. The previous sections of this chapter have established principles and guidelines to help writers avoid such distortion and misinformation. This last section shows what can happen to graphics when sound design principles are *not* applied.

Description of the Problem

The popular media give a good glimpse into the problem of faulty graphics. One observer has used newspaper reports about the October 19, 1987, stock market plunge as one indication of the problem. Writing in *Aldus Magazine,* Daryl Moen noted that 60 percent of U.S. newspapers included charts and other graphics about the market drop the day after it occurred.[1] Moen's study revealed that one out of eight had data errors, and one out of three distorted the facts with visual effects. That startling statistic suggests that faulty illustrations are a genuine problem.

Edward R. Tufte analyzes graphics errors in more detail in his excellent work, *The Visual Display of Quantitative Information.* In setting forth his main principles, Tufte notes that "graphical excellence is the well designed presentation of interesting data—a matter of *substance* of *statistics,* and of *design.*" He further contends that graphics must "give to the viewer the greatest number of ideas in the shortest time with the least ink in the smallest space."[2]

One of Tufte's main criticisms is that charts are often disproportional to the actual differences in the data represented. The next subsection shows some specific ways that this error has worked its way into contemporary graphics.

Examples of Distorted Graphics

There are probably as many ways to distort graphics as there are graphical types. This section gets at the problem of misrepresentation by showing several examples and describing the errors involved. None of the examples commits major errors, yet each one fails to represent the data accurately.

■ Example 1: Faulty Comparisons on Modified Bar Chart

Figure 11–22 accompanied a newspaper article about changes in mailing costs and service. The problem here is that the chart's decoration—the mailboxes—inhibits rather than promotes clear communication. Although the writer intends to use mailbox symbolism in lieu of precise bars, the height of the mailboxes does not correspond to the *actual* increase in second-class postage rates.

A revised graph should include either (1) mailboxes that correctly approximate the actual differences in second-class rates or (2) a traditional bar chart without the mailboxes.

---

[1]Daryl Moen, "Misinformation Graphics," *Aldus Magazine* (January/February 1990), 64.

[2]Edward R. Tufte, *The Visual Display of Quantitative Information* (Cheshire, Conn.: Graphics Press, 1983), 51.

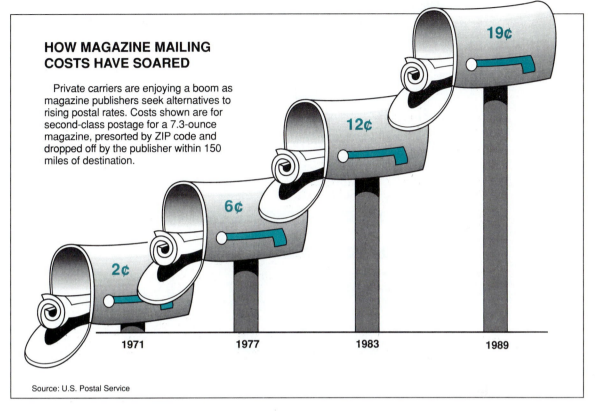

HOW MAGAZINE MAILING COSTS HAVE SOARED

Private carriers are enjoying a boom as magazine publishers seek alternatives to rising postal rates. Costs shown are for second-class postage for a 7.3-ounce magazine, presorted by ZIP code and dropped off by the publisher within 150 miles of destination.

19¢

12¢

6¢

2¢

1971   1977   1983   1989

Source: U.S. Postal Service

FIGURE 11–22
Faulty comparisons on modified bar chart
Source: *Atlanta Constitution*, 30 Nov. 1989, p. H–1. Used by permission.

■ *Example 2: "Chartjunk" That Confuses the Reader*

Figure 11–23 concludes a report from a county government to its citizens. Whereas the dollar backdrop is meant to reinforce the topic—that is, the use to which tax funds are put—in fact, it impedes communication. Readers cannot quickly see comparisons. Instead, they must read the entire list below the illustration, mentally rearranging the items into some order.

At the very least, the expenditures should have been placed in sequence, from least to greatest percentage or vice versa. Even with this order, however, one could argue that the dollar bill is a piece of "chartjunk" that fails to display the data effectively.

■ *Example 3: Confusing Pie Charts*

The pie chart in Figure 11–24 (1) omits percentages that should be attached to each of the budgetary expenditures, (2) fails to move in a largest-to-smallest, clock-

WHAT YOUR GENERAL FUND TAX DOLLAR PROVIDES
The General fund is the county's primary operating fund, used to account for the revenues and expenditures necessary to carry out the basic governmental activities of the county. Revenues are derived primarily from taxes, license and permit fees, and service charges. The expenditures incurred are for current day-to-day expenses and operating equipment.

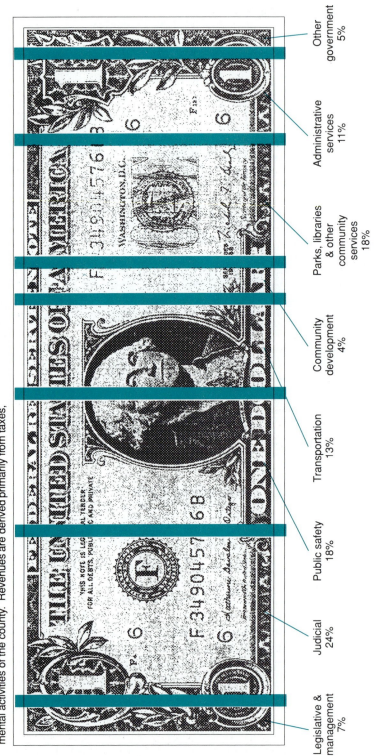

Other government 5%

Administrative services 11%

Parks, libraries & other community services 18%

Community development 4%

Transportation 13%

Public safety 18%

Judicial 24%

Legislative & management 7%

FIGURE 11-23
"Chartjunk" that confuses the reader

Source: Cobb County 1988–1989 Annual Report (Cobb County, Ga.). Used by permission.

	FUND NAME	DESCRIPTION	FY '89 BUDGET
A	General Fund	Basic government activities	$112,895,822
B	Transit Fund	Implementation of bus system	10,812,522
C	Fire District Fund	Operation of Fire Department	21,253,523
D	Bond Funds	General obligation bond issue proceeds	11,073,371
E	Road Sales Tax Fund	1% special purpose sales tax for road improvements	116,869,904
F	Water & Pollution Control Fund	Daily water system operation	60,572,506
G	Debt Service Fund	Principal & interest payments for general obligation bonds	8,240,313
H	Water RE&I Fund	Maintenance of existing facilities	27,365,744
I	Solid Waste RE&I Fund	Maintenance of existing facilities	1,111,237
J	Solid Waste Disposal Facilities	Landfill operations	6,003,367
K	Water Construction Fund	Construction of new facilities	110,884,507
L	Other Uses*		18,384,364
		SUB-TOTAL	$505,467,180
		LESS INTERFUND ACTIVITY	– 23,393,042
		TOTAL EXPENDITURES	**$482,074,138**

*Other Uses includes: Community Service Block Grants, Law Library, Claims Fund, Capital Projects, Senior Services, Community Development Block Grant, Grant Fund

In addition to the General Fund, the county budgets a number of other specialized funds. These include the Fire District Fund, Transit Fund, Road Sales Tax Fund, and enterprise funds such as Water and Pollution Control, and Solid Waste Disposal Facilities.

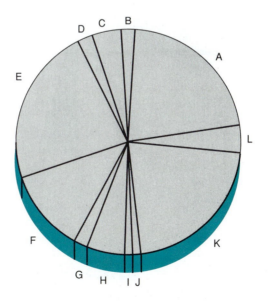

FIGURE 11–24
Confusing pie chart
Source: Cobb County 1988–89 Annual Report (Cobb County, Ga.), 14. Used by permission.

wise sequence, (3) includes too many divisions, many of which are about the same size and thus difficult to distinguish, and (4) introduces a third dimension that adds no value to the graphic.

Figure 11–25 attempts the visual strategy of alternating shades, but it only succeeds in overloading the chart with too many small percentage divisions in uncertain order. Moreover, the reader cannot easily see how the pie slices would be grouped under the three headings listed below the chart. A grouped bar chart would have better served the purpose, with "Southeast," "New England," "Mid-Atlantic," and "Other" providing the groupings.

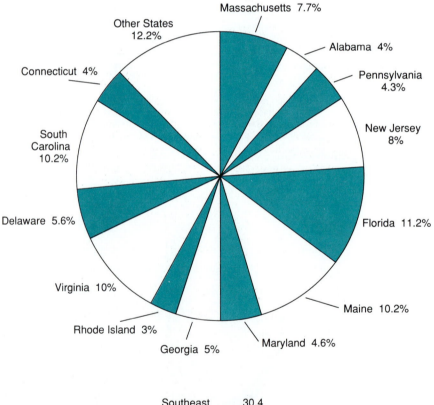

Location of All Ding-Dong Convenience Stores

Southeast	30.4
New England	24.9
Mid-Atlantic	32.5
Other	12.2
	100%

FIGURE 11–25
Confusing pie chart

FIGURE 11–26
Confusing pie chart

ASSETS OF JONES RETIREMENT FUND

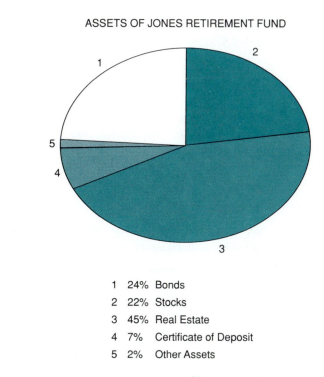

1 24% Bonds
2 22% Stocks
3 45% Real Estate
4 7%   Certificate of Deposit
5 2%   Other Assets

Figure 11–26 negates the value of the pie chart by assuming an oblong shape, rather than a circle. This distortion can make it difficult for the reader to distinguish among sections that are similar in size, such as Sections 1 and 2 in Figure 11–26. The pie chart should be a perfect circle, should have percentages on the circle, and should move in large-to-small sequence from the 12:00 position.

CHAPTER SUMMARY

More than ever before, readers of technical documents expect good graphics to accompany text. Graphics (also called illustrations or visual aids) can be in the form of (1) tables (rows and/or columns of data) or (2) figures (a catchall term for all nontable illustrations). Both types are used to simplify ideas, reinforce points made in the text, generate interest, and create a universal appeal.

Eight common graphics used in technical writing are pie charts, bar charts, line charts, schedule charts, flowcharts, organization charts, technical drawings, and tables. As detailed in this chapter, you should follow specific guidelines in constructing each type. These basic guidelines apply to all graphics:

1. Refer to all graphics in the text.
2. Think about where to put graphics.
3. Position graphics vertically when possible.
4. Avoid clutter.
5. Provide titles, notes, keys, and source data.

ASSIGNMENTS

Your instructor may want you to practice graphics in the context of some of the writing assignments in this textbook, especially in chapters 8, 9, and 10. Here are a few additional exercises.

1. **Pie, Bar, and Line Charts.** Figure 11–27 shows total energy production and consumption from 1960 through 1987, while also breaking down both into the four categories of coal, petroleum, natural gas, and "other." Use those data to complete the following charts:

No. 926. Energy Production and Consumption, by Major Source: 1960 to 1987

[Btu=British thermal unit. For Btu conversion factors, see text, section 19. See also *Historical Statistics, Colonial Times to 1970,* series M 76–92]

YEAR	Total pro-duction (quad. Btu)	PERCENT OF PRODUCTION				Total con-sumption (quad. Btu)	PERCENT OF CONSUMPTION				Consump-tion/produc-tion ratio
		Coal	Petro-leum [1]	Natu-ral gas [2]	Other [3]		Coal	Petro-leum [1]	Natu-ral gas [2]	Other [3]	
1960	41.5	26.1	36.0	34.0	3.9	43.8	22.5	45.5	28.3	3.8	1.06
1961	42.0	24.9	36.2	34.9	4.0	44.5	21.6	45.5	29.1	3.8	1.06
1962	43.6	25.0	35.6	35.1	4.2	46.5	21.3	45.2	29.5	4.0	1.07
1963	45.9	25.8	34.8	35.4	4.0	48.3	21.5	44.9	29.8	3.7	1.05
1964	47.7	26.2	33.9	35.8	4.0	50.5	21.7	44.2	30.3	3.8	1.06
1965	49.3	26.5	33.5	35.8	4.3	52.7	22.0	44.1	29.9	4.0	1.07
1966	52.2	25.8	33.7	36.4	4.1	55.7	21.8	43.8	30.5	3.8	1.07
1967	55.0	25.1	33.9	36.5	4.4	57.6	20.7	43.9	31.2	4.2	1.05
1968	56.8	24.0	34.0	37.6	4.4	61.0	20.2	44.2	31.5	4.1	1.07
1969	59.1	23.5	33.1	38.7	4.8	64.2	19.3	44.1	32.2	4.4	1.09
1970	62.1	23.5	32.9	38.9	4.7	66.4	18.5	44.4	32.8	4.3	1.07
1971	61.3	21.5	32.7	40.5	5.3	67.9	17.1	45.0	33.1	4.8	1.11
1972	62.4	22.6	32.1	39.7	5.6	71.3	16.9	46.2	31.9	5.0	1.14
1973	62.1	22.5	31.4	39.9	6.2	74.3	17.5	46.9	30.3	5.3	1.20
1974	60.8	23.1	30.5	38.9	7.4	72.5	17.5	46.1	30.0	6.5	1.19
1975	59.9	25.0	29.6	36.8	8.6	70.5	17.9	46.4	28.3	7.4	1.18
1976	59.9	26.1	28.8	36.4	8.6	74.4	18.3	47.3	27.4	7.1	1.24
1977	60.2	26.2	29.0	36.4	8.5	76.3	18.2	48.7	26.1	7.0	1.27
1978	61.1	24.4	30.2	35.6	9.9	78.1	17.6	48.6	25.6	8.1	1.28
1979	63.8	27.5	28.4	35.0	9.1	[4] 78.9	19.1	47.1	26.2	7.7	1.24
1980	64.8	28.7	28.2	34.2	8.9	76.0	20.3	45.0	26.8	7.8	1.17
1981	64.4	28.5	28.2	34.2	9.1	74.0	21.5	43.2	26.9	8.4	1.15
1982	63.9	29.2	28.7	32.0	10.2	70.8	21.6	42.7	26.1	9.6	1.11
1983	61.2	28.2	30.1	30.6	11.2	70.5	22.6	42.6	24.6	10.2	1.15
1984	[5] 65.8	30.0	28.6	30.7	10.7	74.1	23.0	41.9	25.0	10.0	1.13
1985	64.8	29.8	29.3	29.6	11.3	74.0	23.6	41.8	24.1	10.4	1.14
1986	64.3	30.4	28.6	29.0	12.0	74.3	23.2	43.4	22.5	10.9	1.16
1987	64.6	31.2	27.3	29.5	12.0	76.0	23.7	42.9	22.6	10.8	1.18

[1] Production includes crude oil and lease condensate. Consumption includes domestically produced crude oil, natural gas liquids, and lease condensate, plus imported crude oil and products.   [2] Production includes natural gas liquids; consumption excludes natural gas liquids.   [3] Comprised of hydropower, nuclear power, geothermal energy and other.   [4] Represents peak year for U.S. energy consumption.   [5] Represents peak year for U.S. energy production.

Source: U.S. Energy Information Administration, *Annual Energy Review,* and unpublished data.

FIGURE 11–27
Reference for Assignment 1

Source: U.S. Department of Commerce, Bureau of the Census, *Statistical Abstract of the United States, 1989* (Washington, 1989), 554.

- A pie chart that shows the four groupings of energy consumption in 1987
- A bar chart that shows the trend in total consumption during these six years: 1960, 1965, 1970, 1975, 1980, and 1985
- A segmented bar chart that shows the total energy production, and the four percent-of-production subtotals, for 1960 and 1980
- A single-line chart showing energy production from 1965 through 1975
- A multiple-line chart that contrasts the coal, petroleum, and natural gas percent of production for any 10-year span on the table

2. **Schedule Charts.** Using any options discussed in this chapter, draw a schedule chart that reflects your work on one of the following:

- A project at work
- A laboratory course at school
- A lengthy project in a course such as this one

3. **Flowcharts.** Select a process with which you are familiar because of work, school, home, or other interests. Then draw a flowchart that outlines the main activities involved in this process.
4. **Organization Chart.** Select an organization with which you are familiar, or one about which you can find information. Then construct a linear flowchart that would help an outsider understand the management structure of all or part of the organization.
5. **Technical Drawing.** Drawing freehand or using computer-assisted design, produce a simple technical drawing of an object with which you are familiar through work, school, or home use.
6. **Tables.** Using the map in Figure 11–28, draw an informal table correlating the five main groupings with the number of states in each.
7. **Misuse of Graphics.** Find three deficient graphics in newspapers, magazines, reports, or other technical documents. Submit copies of the graphics along with a written critique that (1) describes in detail the deficiencies of the graphics and (2) offers suggestions for improving them.
8. **Misuse of Graphics.** Analyze the graphics in Figures 11–29, 11–30, and 11–31. Describe any deficiencies and offer suggestions for improvement.

Hazardous waste sites—June 1988

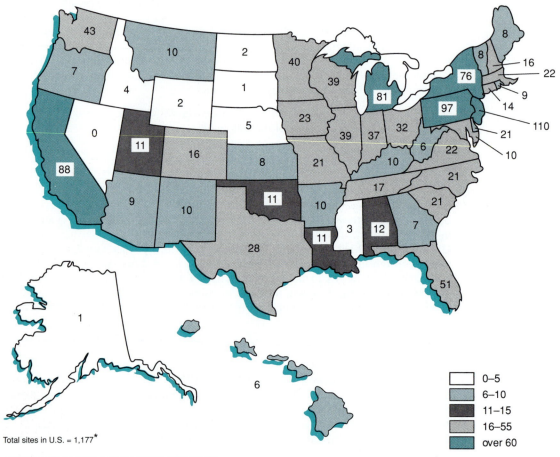

Total sites in U.S. = 1,177*

Represents final and proposed sites on National Priority List
*Includes nine in Puerto Rico and one in Guam
  Source: U.S. Environmental Protection Agency, National Priorities List Fact Book.

Legend:
- 0–5
- 6–10
- 11–15
- 16–55
- over 60

FIGURE 11–28
Reference for Assignment 6

Source: U.S. Department of Commerce, Bureau of the Census, *Statistical Abstract of the United States, 1989* (Washington, 1989), 220.

FIGURE 11–29

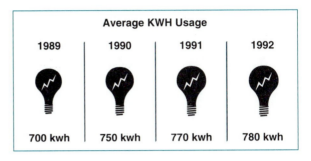

Average KWH Usage

1989	1990	1991	1992
700 kwh	750 kwh	770 kwh	780 kwh

FIGURE 11–30

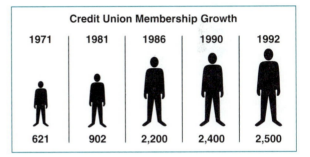

Credit Union Membership Growth

1971	1981	1986	1990	1992
621	902	2,200	2,400	2,500

FIGURE 11–31

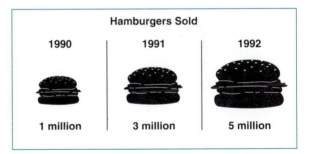

Hamburgers Sold

1990	1991	1992
1 million	3 million	5 million

12 | *Oral Communication*

A manager at McDuff's Munich office presents an overview of the firm to some prospective clients.

*Y*our career will present you with many opportunities for oral presentations, both formal and informal. At the time they arise, of course, you may not consider them to be "opportunities." Instead, they may seem to loom on the horizon as stressful obstacles. That response is normal. The purpose of this chapter is to provide the tools that will help oral presentations contribute to your self-esteem and career success. You will find guidelines for preparation and delivery, techniques for dealing with anxiety, and an example of a technical presentation. Also, the chapter addresses the related topic of running effective meetings.

The entire chapter is based on one simple principle: *Anyone can become a good speaker.* Put aside the myth that competent speakers are born with the talent, that "either they have it or they don't." Certainly some people have more natural talent at thinking on their feet or have a more resonant voice. But success at speaking can come to all speakers, whatever their talent, if they follow the "3 Ps":

Step 1: **P**repare carefully
Step 2: **P**ractice often
Step 3: **P**erform with enthusiasm

These steps form the foundation for all specific guidelines that follow. Before presenting these guidelines, this chapter examines specific ways that formal and informal presentations become part of your professional life.

PRESENTATIONS AND YOUR CAREER

Some oral presentations you will choose to give; others will be "command performances" thrust upon you. Using McDuff, Inc., as a backdrop, the following examples present some realistic situations in which the ability to speak well can lead to success for you and for your organization:

- **Getting hired:** As a job applicant with a business degree, you are asked to present several McDuff managers with a 10-minute summary of your education, previous experience, and career goals.
- **Getting customers:** As coordinator of a McDuff proposal team, you have just been informed that McDuff made the "shortlist" of companies bidding on the contract to manage a large construction project. You and your three team members must deliver a 20-minute oral presentation that highlights the written proposal. To be given at the client's office in Grand Rapids, Michigan, in five days, the presentation will begin and end with comments by you, in your role as coordinator. Your three colleagues will each contribute a five-minute talk.
- **Keeping customers:** As a field engineer at McDuff's St. Louis office, you recently submitted a report on your evaluation of a 50-year-old dam in the Ozarks. Now your clients, the commissioners of the county that owns the dam, have asked you to attend their monthly meeting to present an overview of your findings and respond to questions.
- **Contributing to your profession:** As a laboratory supervisor for McDuff, you belong to a professional society that meets yearly to discuss issues in your field. This year you have been asked to deliver a 15-minute presentation on new procedures for testing toxic-waste samples in the laboratory.
- **Contributing to your community:** As an environmental scientist at McDuff's San Francisco office, you have been asked to speak at the quarterly meeting of OceanSave, an activist environmental organization. The group suggests that you speak for half an hour on environmental threats to aquatic life. You accept the invitation because you know that McDuff management encourages such community service.

As you can see from this list, oral presentations are defined quite broadly. Usually they can be classified according to criteria like these:

1. **Format:** from informal question/answer sessions to formal speeches
2. **Length:** from several-minute overviews to long sessions of an hour or more
3. **Number of presenters:** from solo performances to group presentations
4. **Content:** from a few highlights to detailed coverage

Throughout your career, you will speak to different-sized groups, on diverse topics, and in varied formats. The next two sections provide some common guidelines on preparation, delivery, and graphics.

GUIDELINES FOR PREPARATION AND DELIVERY

The goal of most oral presentations is quite simple: You must present a few basic points, in a fairly brief time, to an interested but usually impatient audience. Simplicity, brevity, and interest are the keys to success. If you deliver what *you* expect when *you* hear a speech, then you will give good presentations yourself.

Although the guidelines here apply to any presentation, they relate best to those that precede or follow a written report, proposal, memo, or letter. Few career presentations are isolated from written work. With this connection in mind, you will note many similarities between the guidelines for good speaking and those for good writing covered in earlier chapters—especially the importance of analyzing the needs of the audience.

■ *Presentation Guideline 1: Know Your Listeners*

Before you write one word of a presentation, recognize these features of most listeners:

- They cannot "rewind the tape" of your presentation, as they can skip back and forth through the text of a report.
- They are impatient after the first few minutes, particularly if they do not know exactly where a speech is going.
- They will daydream a bit and need their attention brought back to the matter at hand (expect a 30-second attention span).
- They have heard so many disappointing presentations that they might not have high expectations for yours.

To respond to these realities, you must discover as much as possible about your listeners. For example, you can (1) consider what you already know about your audience, (2) talk with colleagues who have spoken to the same group, (3) find out which listeners make the decisions, and (4) remember that everyone prefers simplicity.

Most important, make sure not to talk over anyone's head. If there are several levels of technical expertise represented by the group, find the lowest common denominator and decrease the technical level of your presentation accordingly. Remember—decision-makers are often the ones without current technical experience. Also, most listeners, even the most technical ones, want only highlights. Later they can review written documents for details or solicit more technical information during the question-and-answer session after you speak.

■ *Presentation Guideline 2: Use the Preacher's Maxim*

The well-known preacher's maxim goes like this: "First you tell 'em what you're gonna tell 'em, then you tell 'em, and then you tell 'em what you told 'em." Why should most speakers follow this plan? Because it gives the speech a simple three-part structure that most listeners can grasp easily. Here is how your speech should be organized (note that it corresponds to the ABC format used throughout this text for writing):

1. Abstract (beginning of presentation): Right at the outset, you should (1) get the listeners' interest (with an anecdote, a statistic, or other technique), (2) state the exact purpose of the speech, and (3) list the main points you will cover. However, make sure not to try the patience of your audience with an extended introduction. Use no more than a minute.

    Example: "Last year, Jones Engineering had 56 percent more field accidents than the year before. This morning, I'll examine a proposed safety plan that aims to solve this problem. My presentation will focus on three main benefits of the new plan: lower insurance premiums, less lost time from accidents, and better morale among the employees."

2. Body (middle of presentation): Here you discuss the points mentioned briefly in the introduction, in the same order that they were mentioned. Provide the kinds of obvious transitions that help your listener stay on track.

    Example: "The final benefit of the new safety plan will be improved morale among the field workers at all our job sites. . . ."

3. Conclusion (end of presentation): In the conclusion, you should review the main ideas covered in the body of the speech and specify actions you want to occur as a result of your presentation.

    Example: "Jones Engineering can benefit from this new safety plan in three main ways. . . . If Jones implements the new plan next month, I believe you will see a dramatic reduction in on-site accidents during the last half of the year."

This simple three-part plan for all presentations gives listeners the handle they need to understand your speech. First, there is a clear "road map" in the introduction so that they know what lies ahead in the rest of the speech. Second, there is an organized pattern of ideas in the body, with clear transitions between points. And third, there is a strong finish that brings the audience back full circle to the main thrust of the presentation.

■ *Presentation Guideline 3: Stick to a Few Main Points*

Readers and listeners are always searching for ways to group information into a few related points that are easy to remember. Respond to this need by providing just a couple of main points; these may be all that your listeners recall a day or even an hour after the speech. When you must necessarily present many points, group them into a few categories in the same way you would outline a document.

    Short-term memory holds limited items. It follows that listeners are most attentive to speeches that are organized around a few major points. In fact, a good argument can be made for organizing information in groups of *threes* whenever possible. For reasons that are not totally understood, listeners seem to remember groups of three items more than they do any other size groupings—perhaps because the number is simple, perhaps because it parallels the overall three-part structure of most speeches and documents (beginning, middle, end), or perhaps because so many good speakers have used triads (Winston Churchill's "Blood, sweat, and tears," Caesar's "I came, I saw, I conquered," etc.). Whatever the reason, groupings of three will make your speech more memorable to the audience.

■ *Presentation Guideline 4: Put Your Outline on Cards, Paper, or Overheads*

The best presentations are extemporaneous. That is, the speaker shows great familiarity with the material but uses notes for occasional reference. Avoid the two extremes of (1) reading a speech verbatim (which many listeners consider the ultimate insult) or (2) memorizing a speech (which can make your presentation seem somewhat wooden and artificial).

Ironically, you appear more natural if you refer to notes during a presentation. Such extemporaneous speaking allows you to make last-minute changes in phrasing and emphasis that may improve delivery, rather than locking you into specific phrasing that is memorized or written out word for word.

Depending on your personal preference, you may choose to write speech notes on (1) index cards, (2) a sheet or two of paper, or (3) overhead transparencies. Here are the main advantages and disadvantages of each:

Notes on Cards (3" × 5" or 4" × 6")
Advantages

- Are easy to carry in a shirt pocket, coat, or purse
- Provide way to organize points, through ordering of cards
- Can lead to smooth delivery in that each card contains only one or two points that are easy to view
- Can be held in one hand, allowing you to move away from lectern while speaking

Disadvantages

- Keep you from viewing outline of entire speech
- Require that you flip through cards repeatedly in speech
- Can limit use of gestures with hands
- Can cause confusion if they are not in correct order

Notes on Sheets of Paper
Advantages

- Help you quickly view outline of entire speech
- Leave your hands free to use gestures
- Are less obvious than note cards, for no flipping is needed

Disadvantages

- Tend to tie you to lectern, where the sheets lie
- May cause slipups in delivery if you lose your place on page

Notes on Overhead Transparencies
Advantages

- Introduce variety to audience
- Give visual reinforcement to audience
- Can be turned on and off

Disadvantages

- Cannot be altered as easily at last minute
- Must be presented neatly and in parallel style
- Involve the usual risks that reliance on machinery always introduces into your presentation

Presentation Guideline 5: Practice, Practice, Practice

Many speakers prepare a well-organized speech but then fail to add the essential ingredient: practice. Constant practice distinguishes superior presentations from mediocre ones. It also helps to eliminate the nervousness that most speakers feel at one time or another.

In practicing your presentation, make use of four main techniques. They are listed here, from least effective to most effective:

- **Practice before a mirror:** This old-fashioned approach allows you to hear and see yourself in action. The drawback, of course, is that it is difficult to evaluate your own performance while you are speaking. Nevertheless, such run-throughs definitely make you more comfortable with the material.
- **Use of audiotape:** Most presenters have access to a tape player, so this approach is quite practical. The portability of the machines allows you to practice almost anywhere. Although taping a presentation will not improve gestures, it will help you discover and eliminate verbal distractions such as filler words (*uhhhh, um, ya know*).
- **Use of live audience:** Groups of your colleagues, friends, or family—simulating a real audience—can provide the kinds of responses that approximate those of a real audience. In setting up this type of practice session, however, make certain that observers understand the criteria for a good presentation and are prepared to give an honest, forthright critique.
- **Use of videotape:** This practice technique allows you to see and hear yourself as others do. Your careful review of the tape, particularly when done with another qualified observer, can help you identify and eliminate problems with posture, eye contact, vocal patterns, and gestures. At first it can be a chilling experience, but soon you will get over the awkwardness of seeing yourself on film.

Presentation Guideline 6: Speak Vigorously and Deliberately

"Vigorously" means with enthusiasm; "deliberately" means with care, attention, and appropriate emphasis on words and phrases. The importance of this guideline becomes clear when you think back to how you felt during the last speech you heard. At the very least, you expected the speaker to show interest in the subject and to demonstrate enthusiasm. Good information is not enough. You need to deliver this information in a way that arouses the interest of the listeners.

You may wonder, "How much enthusiasm is enough?" The best way to answer this question is to hear or (preferably) watch yourself on tape. Your delivery

should incorporate just enough enthusiasm so that it sounds and looks a bit un-
natural to you. Few if any listeners ever complain about a speech being too en-
thusiastic or a speaker being too energetic. But many, many people complain about
dull speakers who fail to show that they themselves are excited about the topic.
Remember—every presentation is, in a sense, "show time." Your enthusiasm will
help reduce the natural lethargy that listeners develop in their usually passive role
as an audience.

■ Presentation Guideline 7: Avoid Filler Words

Avoiding filler words presents a tremendous challenge to most speakers. When
they think about what comes next or encounter a break in the speech, they may
tend to fill the gap with filler words and phrases such as these:

> uhhhhh. . .
>
> ya know. . .
>
> okay. . .
>
> well. . .uh. . .
>
> like . . .
>
> I mean. . .
>
> umm. . .

These gap-fillers are a bit like spelling errors in written work: Once your listeners
find a few, they start looking for more and thus are distracted from your presentation. To eliminate such distractions, follow these three steps:

Step 1: **Use pauses to your advantage.** Constantly remind yourself that
there is nothing wrong with short gaps or pauses in the midst of
your speech. In fact, such gaps help indicate that you are shifting
from one point to another. In signaling a transition, a pause serves
to draw attention to the point you make right after the pause. Once
you realize the strategic importance of these pauses, you will not
feel the need to plug them up with filler words.

Step 2: **Practice with tape.** Tape is brutally honest: When you play it
back, you will become instantly aware of fillers that occur more
than once or twice. As you watch and listen, keep a tally sheet of
the fillers you use and their frequency. Your goal will be to reduce
this frequency with every practice session.

Step 3: **Ask for help from others.** After working with tape machines in
Step 2, give your speech to an individual who has been instructed
to stop you after each filler. This technique gives immediate reinforcement.

■ Presentation Guideline 8: Use Rhetorical Questions

Enthusiasm, of course, is your best delivery technique for capturing the attention of
the audience. Another technique is the use of rhetorical questions at pivotal points
in your presentation.

Rhetorical questions are those you ask to get listeners thinking about a topic, not those that you would expect them to answer out loud. For example, here is a rhetorical question used by a computer salesperson in proposing a purchase by one of McDuff's small offices: "I've discussed the three main advantages that a centralized word-processing center would provide your office staff. But is this an approach that you can afford at this point in the company's growth?" Then the speaker would follow the question with remarks supporting the position that the system is affordable.

"What if" scenarios provide another way to introduce rhetorical questions. They gain the listeners' attention by having them envision a situation that might occur. For example, a safety engineer could use this kind of rhetorical question in proposing McDuff's asbestos-removal services to a regional bank: "What if you repossessed a building that contained dangerous levels of asbestos? Do you think that your bank would then be liable for removing all the asbestos?" Again, the question pattern heightens listener interest.

You will discover quite quickly in practice sessions that rhetorical questions do not come naturally. Instead, you must make a conscious effort to insert them at important points in your speech. If you are successful in this effort, rhetorical questions offer two main advantages. First, they prod listeners to think about your point and set up an expectation that important information will follow. Second, they break the monotony of standard declarative sentence patterns.

Use rhetorical questions at points when it is most important to gain or redirect the attention of the audience. Three particularly effective uses follow:

1. **As a grabber at the beginning of a speech:** "Have you ever wondered how you might improve the productivity of your word-processing staff?"
2. **As a transition between major points:** "We've seen that centralized word processing can improve the speed of report production, but will it require any additions to your staff?"
3. **As an attention-getter right before your conclusion:** "Now that we've examined the features of centralized word processing, what's the next step you should make at McDuff?"

■ *Presentation Guideline 9: Maintain Eye Contact*

Your main goal—always—is to keep listeners interested in what you are saying. This goal requires that you maintain control, using whatever techniques you can employ to direct the attention of the audience. Frequent eye contact is one good strategy.

The simple truth is that listeners pay closer attention to what you are saying when you look at them. Think of how you react when a speaker makes constant eye contact with you. If you are like most people, you feel as if the speaker cares about you personally—even if there are 100 people in the audience. Also, you tend to feel more obligated to listen when you know that the speaker's eyes will be meeting yours throughout the presentation. Here are some ways you can make eye contact a natural part of your own strategy for effective oral presentations:

- **With audiences of about 30 or less:** Make regular eye contact with everyone in the room. Be particularly careful not to ignore members of the audience who are seated to your far right and far left (see Figure 12–1). Many speakers tend to focus on the listeners within Section B. Instead, make wide sweeps so that listeners in Sections A and C get equal attention.

■ **With large audiences:** There may be too many people or too large a room for you to make individual eye contact with all listeners. In this case, focus on just a few people in all three sections of the audience noted in Figure 12–1. This approach gives the appearance that you are making eye contact with the entire audience.

■ **With any size audience:** Occasionally look away from the audience—either to your notes or toward a part of the room where there are no faces looking back. In this way, you avoid the appearance of staring too intensely at your audience. Also, these breaks give you the chance to collect your thoughts or check your notes.

■ *Presentation Guideline 10: Use Appropriate Gestures and Posture*

Speaking is only one part of giving a speech; another is adopting appropriate posture and using gestures that will complement what you are saying. If you watch good speakers carefully, you will see that they are much more than "talking heads" before a lectern. Instead, they:

1. Use their hands and fingers to emphasize major points
2. Stand straight, without leaning on or gripping the lectern
3. Step out from behind the lectern on occasion, to decrease the distance between them and the audience
4. Point toward visuals on screens or charts, without losing eye contact with the audience

The audience will be judging you by what you say *and* what they see, a fact that again makes videotaping a crucial part of your preparation. With work on this facet of your presentation, you can avoid problems like keeping your hands constantly in your pockets, rustling change (remove pocket change and keys beforehand), tapping a pencil, scratching nervously, slouching over a lectern, and shifting from foot to foot.

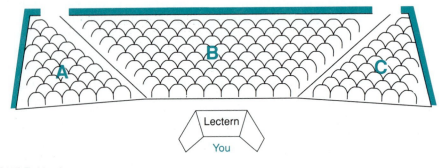

FIGURE 12–1
Audience sections

GUIDELINES FOR PRESENTATION GRAPHICS

More than ever before, listeners expect good graphics during oral presentations. Much like gestures, graphics transform the words of your presentation into true communication with the audience. To emphasize the importance of speech graphics, here is an incident from industry once reported to the author:

> Several years ago the competitors for designing a large city's football stadium had been narrowed to three firms. Firm A, a large and respected company, had done some preliminary design work on the project and was expected to get the contract. Firm B, another large and respected firm, was competing fiercely for the job. And Firm C, a small and fairly new company, was considered by all concerned to be a genuine long shot. Yet it had submitted an interesting-enough proposal to be chosen as a finalist.
>
> All three firms were invited to make 10- or 15-minute presentations on their proposals. The presentations by Firms A and B were professional, conventional, and predictable. Firm C, however, took a different and riskier approach. Its presentation was barely 5 minutes long and was given simultaneously with a videotape. As expected, the speech itself stressed the benefits of Firm C's design for the stadium. The accompanying videotape, however, was quite unconventional. It interspersed drawings of Firm C's design with highlights of that city's football team scoring touchdowns, catching passes, and making game-winning tackles.
>
> Firm C's effort to associate the winning football team with its proposed design worked. Shortly after the presentations, the selection committee chose the dark horse, Firm C, to design the new stadium.

The lesson here is *not* that fancy visuals always win a contract. Instead, the point is that innovative graphics, in concert with a solid proposal presentation, can set you apart from the competition. Firm C had a sound stadium design reinforced by an unusual visual display. Granted, it walked the fine line between effective and manipulative graphics. Yet even the fanciest visuals cannot disguise a bad idea to a discerning audience. The Firm C presenters won because they found an effective way to present their proposal. They incorporated graphics into their presentation to reinforce main selling points.

■ *Graphics Guideline 1: Discover Listener Preferences*

Some professionals prefer simple speech graphics, such as the conventional flip chart. Others prefer more sophisticated equipment, such as videotape and slide-tape machines. Your listeners are usually willing to indicate their preferences when you call on them. Contact the audience ahead of time and make some inquiries.

■ *Graphics Guideline 2: Think About Graphics Early*

Graphics done as an afterthought usually look "tacked on." Plan graphics while you prepare the text so that the final presentation will seem fluid. This guideline holds true especially if you rely upon specialists to prepare your visuals. These professionals need some lead time to do their best work. Also, they can often provide helpful insights about how visuals will enhance the presentation—*if* you consult them early enough and *if* you make them a part of your presentation team.

Remember—the goal is to use graphics of which you can be proud. Never, never put yourself in the position of having to apologize for the quality of your graphic material. If an illustration is not up to the quality your audience would expect, do *not* use it.

■ *Graphics Guideline 3: Keep the Message Simple*

Listeners can be suspicious of overly slick visual effects. Most people prefer the simplicity of simple overhead transparencies and flip charts, for example. If you decide to use more sophisticated techniques, make sure they fit your context. In the previous stadium example, Firm C took a risk by using videotaped football highlights to sell its stadium design. Complex graphics such as videotape can sometimes overshadow the proposal itself. In Firm C's case, however, the tape's purpose was kept quite simple. It simply associated the success of the team with the potential success of the proposed stadium design.

■ *Graphics Guideline 4: Make Any Wording Brief and Visible*

The best graphics rely on visual image, not words. Avoid cluttering them with language. Instead, provide necessary explanations during the presentation. When you do need to put words in a visual—perhaps in a list of major points—pare them down to the bare minimum. Single words or phrases can then be elaborated on in your speech text.

Equally important, be certain that all wording is visible from the *back of the room*. Nothing is more irritating than a poster, an overhead, or a slide that cannot be read. Prevent this problem by asking beforehand about the room size and arrangement; then adjust letter size and thickness accordingly. Incidentally, standard type is too small to use effectively on overhead transparencies. When using overheads, have the originals typeset in large print or prepared on a word-processing system with oversized type.

■ *Graphics Guideline 5: Use Colors Carefully*

Colors can add flair to visuals. Follow these simple guidelines to make colors work for you:

- Have a good reason for using color (such as the need to highlight three different bars on a graph with three distinct colors).
- Use only dark, easily seen colors, and be sure that a color contrasts with its background (for example, yellow on white would not work well).
- Use no more than three or four colors in each graphic (to avoid a confused effect).
- For variety, consider using white on a black or dark green background.

■ *Graphics Guideline 6: Leave Graphics Up Long Enough*

Because graphics reinforce text, they should be shown only while you address the particular point at hand. For example, reveal a graph just as you are saying, "As you can see from the graph, the projected revenue reaches a peak in 1995." Then pause and leave the graph up a bit longer for the audience to absorb your point.

How long is *too* long? A graphic outlives its usefulness when it remains in sight after you have moved on to another topic. Listeners will continue to study it and ignore what you are now saying. If you use a graphic once and plan to return to it, take it down after its first use and show it again later.

■ Graphics Guideline 7: Avoid Handouts

Because timing is so important in your use of speech graphics, handouts are usually a bad idea. Readers move through a handout at their own pace, rather than at the pace the speaker might prefer. Thus handouts cause you to lose the attention of your audience. Use them only if (1) no other visual will do, (2) your listener has requested them, or (3) you distribute them as reference material *after* you have finished talking.

■ Graphics Guideline 8: Maintain Eye Contact While Using Graphics

Do not let your own graphics entice you away from the audience. Maintain control of listeners' responses by looking back and forth from the visual to faces in the audience. To point to the graphic aid, use the hand closest to the visual. Using the opposite hand causes you to cross over your torso, forcing you to turn your neck and head away from the audience.

■ Graphics Guideline 9: Include All Graphics in Your Practice Sessions

Dry runs before the actual presentation should include every graphic you plan to use, in its final form. This is good reason to prepare graphics as you prepare text, rather than as an afterthought. Running through a final practice without graphics would be much like doing a dress rehearsal for a play without costumes and props—you would be leaving out the parts that require the greatest degree of timing and orchestration. Practicing with graphics helps you improve transitions.

■ Graphics Guideline 10: Use Your Own Equipment

Murphy's Law always seems to apply when you use another person's audiovisual equipment: Whatever can go wrong, will. For example, a new bulb with a 100-hour life decides to burn out, there is no extra bulb in the equipment drawer, the outlet near the projector does not work, an extension cord is too short, the screen does not stay down—the list goes on and on. Even if the equipment works, it often operates differently from what you are used to. About the only way to put the odds in your favor is to carry your own equipment and set it up in advance.

If you do have to rely on someone else's equipment, make special preparations. Here are a few ways to ward off disaster:

■ Find out exactly who will be responsible for providing the audiovisual equipment and contact that person in advance.

- Have some easy-to-carry backup supplies in your car—an extension cord, an overhead projector bulb, felt-tip markers, and chalk.
- Bring handout versions of your visuals as a last resort.

Remember—you never want to put yourself in the position of having to apologize for inadequate or nonexistent graphics. Plan well.

OVERCOMING NERVOUSNESS

The problem of nervousness deserves special mention because it is so common. Virtually everyone who gives speeches feels some degree of nervousness before "the event." An instinctive "fight or flight" response kicks in for the many people who have an absolute dread of presentations. In fact, surveys have determined that most of us rate public speaking at the top of our list of fears, even above sickness and death! Given this common response, this chapter considers the problem and offers suggestions for overcoming it.

Why Do We Fear Presentations?

Most of us feel comfortable with informal conversations, when we can voice our views to friends and indulge in impromptu exchanges. We are used to this type of casual presenting of our ideas. Formal presentations, however, put us into a more structured, more awkward, and thus more threatening environment. Despite the fact that we may know the audience is friendly and interested in our success, the structured context triggers nervousness that is sometimes difficult to control.

This nervous response is normal and even, to some degree, useful. It gets you "up" for the speech. That adrenaline pumping through your body can generate a degree of enthusiasm that propels the presentation forward and creates a lively performance. Just as veteran actors admit to some nervousness helping to improve their performance, excellent speakers usually can benefit from the same effect.

The problem occurs when nervousness felt before or during a speech becomes so overwhelming that it affects the quality of the presentation. Since sympathy is the last feeling a speaker wants the audience to have, it is worth spending time considering some techniques to combat nervousness.

A Strategy for Staying Calm

As the cliché goes, do not try to eliminate "butterflies" before a presentation—just get them to fly in formation. In other words, it is best to acknowledge that a certain degree of nervousness will always remain. Then go about the business of getting it to work for you. Here are a few suggestions:

■ No Nerves Guideline 1: Know Your Speech

The most obvious suggestion is also the most important one. If you prepare your speech well, your command of the material will help to conquer any queasiness you feel—particularly at the beginning of the speech, when nervousness is usually at its peak. Be so sure of the material that your listeners will overlook any initial discomfort you may feel.

■ *No Nerves Guideline 2: Prepare Yourself Physically*

Your physical well-being before the speech can have a direct bearing on anxiety. More than ever before, we understand the essential connection between mental and physical well-being. This connection suggests that you should take these precautions before your presentation:

- **Avoid caffeine or alcohol for at least several hours before you speak.** You surely do not need either the additional jitters brought on by caffeine or the false sense of ease brought on by consumption of alcohol.
- **Eat a light, well-balanced meal within a few hours of speaking.** However, do not overdo it—particularly if a meal comes right before your speech. If you are convinced that any eating will increase your anxiety, wait to eat until after speaking.
- **Practice deep-breathing exercises before you speak.** Inhale and exhale slowly, making your body slow down to a pace you can control. If you can control your breathing, you can probably keep the butterflies flying in formation.
- **Exercise normally the same day of the presentation.** A good walk will help invigorate you and reduce nervousness. However, do not wear yourself out by exercising more than you would normally.

■ *No Nerves Guideline 3: Picture Yourself Giving a Great Presentation*

Many speakers become nervous because their imaginations are working overtime. They envision the kinds of failures that almost never occur–passing out on stage, not being able to speak, forgetting lines, being verbally attacked by members of the audience. Instead, speakers should be constantly bombarding their psyches with images of success, not failure. Mentally take yourself through the following steps of the presentation:

- Arriving at the room
- Feeling comfortable at your chair
- Getting encouraging looks from your audience
- Giving an attention-getting introduction
- Presenting your supporting points with clarity and smoothness
- Ending with an effective wrap-up
- Fielding questions with confidence

Sometimes called "imaging," this technique helps to program success into your thinking and to control feelings of doom that pass through the minds of even the best speakers.

■ *No Nerves Guideline 4: Arrange the Room as You Want*

To control your anxiety, it helps to assert some control over the physical environment as well. You need everything going for you in a presentation if you are to feel at ease. Make sure that chairs are arranged to your satisfaction, that the lectern is positioned to your taste, that the lighting is adequate, and so on. These features of the setting can almost always be adjusted if you make the effort to ask. Again, it is a matter of your asserting control so that your overall confidence is increased.

■ *No Nerves Guideline 5: Have a Glass of Water Nearby*

Extreme thirst and a dry throat are physical symptoms of nervousness that can affect your delivery. There is nothing to worry about as long as you have some water available. Think about this need ahead of time so that you do not have to interrupt your presentation to pour a glass of water.

■ *No Nerves Guideline 6: Engage in Casual Banter Before the Speech*

If you have the opportunity, chat with members of the audience before the speech. This ice-breaking technique will reduce your nervousness and help start your relationship with the audience.

■ *No Nerves Guideline 7: Remember That You Are the Expert*

As a final "psyching up" exercise before you speak, remind yourself that you have been invited or hired to speak on a topic about which you have useful knowledge. Your listeners want to hear what you have to say and are almost always eager for you to succeed in providing useful information to them. So tell yourself, "I'm the expert here!"

■ *No Nerves Guideline 8: Do Not Admit Nervousness to the Audience*

No matter how anxious you may feel, never admit it. First of all, you do not want listeners to feel sorry for you—that is not an emotion that will lead to a positive critique of your speech. Second, nervousness is almost never apparent to the audience. Your heart may be pounding, your knees may be shaking, and your throat may be dry, but few if any members of the audience can see these symptoms. Why draw attention to the problem by admitting to it? Third, you can best defeat initial anxiety by simply pushing right on through.

■ *No Nerves Guideline 9: Slow Down*

Some speakers who feel nervous tend to speed through their presentations. If you have prepared well and practiced the speech on tape, you are not likely to let this happen. Having heard yourself on tape, you will be better able to sense that the pace is too quick. As you speak, constantly remind yourself to maintain an appropriate pace. If you have had this problem before, you might even write "Slow down!" in the margin of your notes.

■ No Nerves Guideline 10: Join a Speaking Organization

The previous nine guidelines will help reduce your anxiety about a particular speech. To help solve the problem over the long term, however, consider joining an organization like Toastmasters International, which promotes the speaking skills of all its members. Like some other speech organizations, Toastmasters has chapters that meet at many companies and campuses. These meetings provide an excellent, nonthreatening environment in which all members can refine their speaking skills.

EXAMPLE OF MCDUFF ORAL PRESENTATION

This section presents the text and visuals of a short presentation given by Kim Mason, an environmental expert for McDuff's Atlanta office. She has been invited to speak at the monthly lunch meeting of an organization of building owners in the Atlanta region. The agreed-upon topic is the problem of asbestos contamination.

The members of Kim's audience have an obvious interest in the problem: They own buildings that are at risk. Yet they know little more about asbestos than that it is a health issue they must consider when they renovate. Kim's job is to inform them and heighten their awareness. She needs to cover only the highlights, however, because the presentation will be followed by a detailed question-and-answer session. Although some of these owners have been, and will be, clients of McDuff, she has an ethical obligation to avoid promoting McDuff during her presentation.

> Good evening. My name is Kim Mason, and I work for the asbestos-abatement division of McDuff, Inc., in Atlanta. I've been asked to give a short presentation on the problem of asbestos and then to respond to your questions about the importance of removing it from buildings. I'll focus on three main reasons why you, as building owners, should be concerned about the asbestos problem: (TRANSPARENCY 1 [SEE FIGURE 12–2, p. 398])
>
> 1. To prevent future health problems of your tenants
> 2. To satisfy regulatory requirements of the government
> 3. To give yourself peace of mind for the future
>
> Again, my comments will provide just an overview, serving as a basis for the question session that follows in a few minutes. (TRANSPARENCY 2 [SEE FIGURE 12–3, p. 399])
>
> Question: What is the most important reason you need to be concerned about asbestos? Answer: The long-term health of the tenants, workers, and other people in buildings that contain asbestos. Research has clearly linked asbestos with a variety of diseases, including lung cancer, colon cancer, and asbestosis (a debilitating lung disease). Although this connection was first documented in the 1920s, it has only been

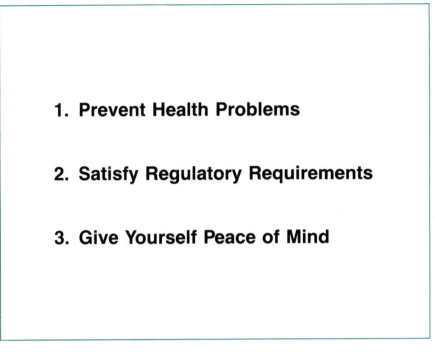

1. Prevent Health Problems

2. Satisfy Regulatory Requirements

3. Give Yourself Peace of Mind

FIGURE 12–2
Transparency 1 for sample presentation

taken seriously in the last few decades. Unfortunately, by that time asbestos had already been commonly used in many building materials that are part of many structures today.

Here's a list of some of the most common building products containing asbestos. (TRANSPARENCY 3 [SEE FIGURE 12–4, p. 400]) As you can see, asbestos was used in materials as varied as floor tiles, pipe wrap, roof felt, and insulation around heating systems. An abundant and naturally occurring mineral, asbestos was fashioned into construction materials through processes such as packing, weaving, and spraying. Its property of heat resistance, as well as its availability, was the main reason for such widespread use.

While still embedded in material, asbestos causes no real problems. However, when it deteriorates or is damaged, fibers may become airborne. In this state, they can enter the lungs and cause the health problems mentioned a minute ago. This risk prompted the Environmental Protection Agency in the mid-1970s to ban the use of certain asbestos products in most new construction. But today the decay and renovation of many asbestos-containing building materials may put many of our citizens at risk for years to come. (TRANSPARENCY 4 [SEE FIGURE 12–5, p. 401])

After your concern for occupants' health, what's the next best reason to learn more about asbestos? It's the *law*. Both the Occupational Safety and Health Administration (OSHA) and the Georgia Department of Natural Resources (DNR) require that you follow certain procedures when structures you own could endanger tenants and asbestos-removal workers with contamination. For example, when a structure

Prevent Health Problems

FIGURE 12–3
Transparency 2 for sample presentation

undergoes renovation that will involve any asbestos-containing material (ACM), the ACM must be removed by following approved engineering procedures. Also, the contaminated refuse must be disposed of in approved landfills. Considering the well-documented potential for health problems related to airborne asbestos, this legislative focus on asbestos contamination makes good sense.

By the way, both OSHA and DNR regulations require removal of asbestos by licensed contractors. These contractors, however, will assume liability only for what they have been told to remove. They may or may not have credentials and training in health and safety. Therefore, building owners should hire a firm with a professional who will (1) survey the building and present a professional report on the degree of asbestos contamination and (2) monitor the work of the contractor in removing the asbestos. By taking this approach, you as an owner stand a good chance of eliminating all problems with your asbestos.

Yes, it is *your* asbestos. As owner of a building, you also legally own the asbestos associated with that building–*forever*. For example, if a tenant claims to have been exposed to asbestos because of your abatement activity and then brings a lawsuit, you must have documentation showing that you contracted to have the work performed in a "state-of-the-art" manner. If, as recommended, you have hired a qualified monitoring firm and a reputable contracting firm, liability will be focused on the contractor and the monitoring firm–*not* on you. (TRANSPARENCY 5 [SEE FIGURE 12–6, p. 402])

Which brings me to the last reason for concerning yourself with any potential asbestos problem: *peace of mind.* If you examine and then effectively deal with any

> ### Some Building Materials Containing Asbestos
>
> **Plastic Products**
>
> **Floor Tile**
> **Coatings & Sealants**
>
> **Paper Products**
>
> **Roof Felt**
> **Gaskets**
> **Paper Pipe Wrap**
>
> **Insulating Products**
>
> **Sprayed Coating**
> **Preformed Pipe Wrap**
> **Insulation Board**
> **Boiler Insulation**

FIGURE 12–4
Transparency 3 for sample presentation

asbestos contamination that exists in your buildings, you will sleep better at night. For one thing, you will have done your level best to preserve the health of your tenants. For another, as previously noted, you will have shifted any potential liability from you to the professionals you hired to solve the problem–assuming you hired professionals. Your monitoring firm will have continuously documented the contractor's operations and will have provided you with reports to keep in your files, in the event of later questions by lawyers or regulatory agencies.

In just these few minutes, I have given only highlights about asbestos. It poses a considerable challenge for all of us who own buildings or work in the abatement business. Yet the current diagnostic and cleanup methods are sophisticated enough to suggest that this problem, over time, *will* be solved. Now I would be glad to answer questions.

RUNNING EFFECTIVE MEETINGS

Like formal presentations, meetings are a form of spoken communication that goes hand in hand with written work. Important reports and proposals–even many routine ones–often are followed or preceded by a meeting. For example, you may meet with your colleagues to prepare a group-written report, with your clients to discuss a proposal, or with your department staff to outline recommendations to

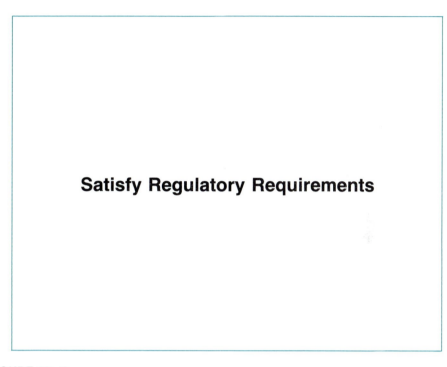

FIGURE 12–5
Transparency 4 for sample presentation

appear in a yearly report to management. This section will make you a first-class meeting leader by (1) highlighting some common problems with meetings, along with their associated costs to organizations, and (2) describing 10 guidelines for overcoming these problems.

Common Problems With Meetings

Here are six major complaints about meetings held in all types of organizations:

1. They start and end too late.
2. Their purpose is unclear.
3. Not everyone in the meeting really needs to be there.
4. Conversations get off track.
5. Some people dominate while others do not contribute at all.
6. Meetings end with no sense of accomplishment.

As a result of these frustrations, career professionals waste much of their time in poorly run meetings.

Because they waste participants' time, bad meetings also waste a lot of money. To find out what meetings cost an organization, do this rough calculation. Use information about an organization for which you work, or for which a friend or family member works.

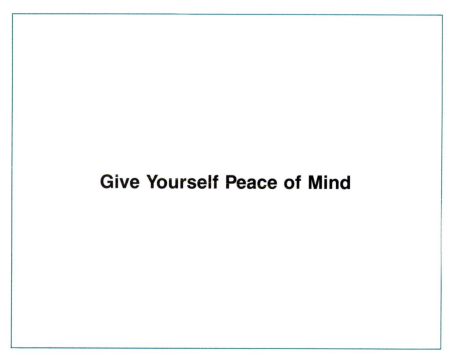

FIGURE 12–6
Transparency 5 for sample presentation

1. Take the average weekly number of meetings in an office.
2. Multiply that number by the average length of each meeting, in hours.
3. Multiply the result of step 2 by the average number of participants in each meeting.
4. Multiply the result of step 3 by the average hourly salary or billable amount of the participants.

The result, which may surprise you, is the average weekly cost of meetings in the office that you investigated. With these heavy costs in mind, the next section presents some simple guidelines for running good meetings.

Guidelines for Good Meetings

When you choose (or are chosen) to run a meeting, your professional reputation is at stake—let alone the costs just mentioned. Therefore, it is in your own best interests to make sure meetings run well. When you are a meeting participant, you also have an obligation to speak up and help accomplish the goals of the meeting.

The guidelines that follow will help create successful meetings. They fall into three main stages:

Stage 1: Before the meeting (Guidelines 1–4)

Stage 2: During the meeting (Guidelines 5–9)

Stage 3: After the meeting (Guideline 10)

Note that these 10 guidelines apply to *working* meetings—that is, those in which participants use their talents to accomplish specific objectives. Such meetings usually involve a lot of conversation. The guidelines do not apply as well to *informational* meetings wherein a large number of people are assembled only to listen to announcements.

■ Meeting Guideline 1: Involve Only Necessary People

"Necessary" means those people who, because of their position or knowledge, can contribute significantly to the meeting. Your goal should be a small working group—four to six people is ideal. If others need to know what occurs, send them a copy of the minutes after the meeting.

■ Meeting Guideline 2: Distribute an Agenda Beforehand

To be good participants, most people need to do some thinking before the meeting about the objectives of the session. The agenda also gives you, as leader, a way to keep the meeting on schedule. When you are worried about being able to cover the agenda items, consider attaching time limits to each item. That technique helps the meeting leader keep the discussion moving.

■ Meeting Guideline 3: Distribute Readings Beforehand

Jealously guard time at a meeting, making sure to use it for productive discussion. If any member has reading materials that committee members should review as a basis for these discussions, such readings should be handed out ahead of time. Do not use meeting time for reading. Even worse, do not refer to handouts that all members have not had the opportunity to go over.

■ Meeting Guideline 4: Have Only One Meeting Leader

To prevent confusion, one person should always be in charge. The meeting leader should be able to perform these tasks:

- **Listen carefully** so that all views get a fair hearing
- **Generalize accurately** so that earlier points can be brought back into the discussion when appropriate
- **Give credit to participants** so that they receive reinforcement for their efforts
- **Move toward consensus** so that the meeting does not wallow in endless discussion

■ Meeting Guideline 5: Start and End on Time

Nothing deadens a meeting more than a late start, particularly when it is caused by people arriving late. Tardy participants are given no incentive to arrive on time when a meeting leader waits for them. Even worse, prompt members become demoralized by such delays. Latecomers will mend their ways when you make a practice of starting right on time.

It is also important to set an ending time for meetings so that members have a clear view of the time available. Concerning meeting length, most people do their best work in the first hour. After that, productive discussion reaches a point of diminishing returns. If working meetings must last longer than an hour, make sure to build in short breaks and stay on the agenda.

■ *Meeting Guideline 6: Keep Meetings on Track*

By far the biggest challenge for a meeting leader is to encourage open discussion while still moving toward resolution of agenda items. As a leader, you need to be assertive, yet tactful, in your efforts to discourage these three main time-wasters:

- Long-winded digressions by the committee
- Domination by one or two outspoken participants
- Interruptions from outside the meeting

■ *Meeting Guideline 7: Strive for Consensus*

Consensus means agreement by all those present. Your goal should be to orchestrate a meeting such that all members, after a bit of compromise, feel comfortable with a decision. Such a compromise, when it flows from healthy discussion, is far preferable to a decision generated by voting on alternatives. After all, you are trying to reach a conclusion that everyone helped produce, rather than one toward which only part of the committee feels ownership.

■ *Meeting Guideline 8: Use Visuals*

Graphics always help make points more vivid at a meeting. They are especially useful for recording ideas that are being generated rapidly during a discussion. Toward this end, you may want someone from outside the discussion to write important points on a flip chart, a chalkboard, or an overhead transparency.

■ *Meeting Guideline 9: End With a Summary*

Before the meeting adjourns, take a few minutes to summarize what items have been discussed and agreed to. This wrap-up gives everyone the opportunity to clarify any point brought up during the meeting. Also, each member will know exactly what actions come next.

■ *Meeting Guideline 10: Distribute Minutes Soon*

Write and send out minutes within 48 hours of the meeting. For even routine meetings, it is important that there be a record of the meeting's accomplishments. Also, if any discussion items are particularly controversial, consider having committee members approve minutes with their signature and return them to you, before final distribution.

CHAPTER SUMMARY

The main theme of this chapter is that anyone can become a good speaker by preparing well, practicing often, and giving an energetic performance. This effort pays off richly by helping you deal effectively with employers, customers, and professional colleagues.

Ten guidelines for preparation and delivery will lead to first-class presentations:

1. Know your listeners.
2. Use the preacher's maxim.
3. Stick to a few main points.
4. Put your outline on cards, paper, or overheads.
5. Practice, practice, practice.
6. Speak vigorously and deliberately.
7. Avoid filler words.
8. Use rhetorical questions.
9. Maintain eye contact.
10. Use appropriate gestures and posture.

You should also strive to incorporate illustrations into your speeches by following these 10 guidelines:

1. Discover listener preferences.
2. Think about graphics early.
3. Keep the message simple.
4. Make any wording brief and visible.
5. Use colors carefully.
6. Leave graphics up long enough.
7. Avoid handouts.
8. Maintain eye contact while using graphics.
9. Include all graphics in your practice sessions.
10. Use your own equipment.

A major problem for many speakers is fear of giving presentations. You can control this fear by following these guidelines:

1. Know your speech!
2. Prepare yourself physically.
3. Picture yourself giving a great presentation.
4. Arrange the room as you want.
5. Have a glass of water nearby.
6. Engage in casual banter before the speech.
7. Remember that you are the expert.
8. Do not admit nervousness to the audience.
9. Slow down.
10. Join a speaking organization.

Besides conventional presentations, you often will be required to speak in meetings—as either a leader or a participant. Given the effect that poor meetings have on morale and profits, it is important to develop a strategy for running them well. Here are 10 guidelines:

1. Involve only necessary people.
2. Distribute an agenda beforehand.
3. Distribute readings beforehand.
4. Have only one meeting leader.
5. Start and end on time.
6. Keep meetings on track.
7. Strive for consensus.
8. Use visuals.
9. End with a summary.
10. Distribute minutes soon.

ASSIGNMENTS

In completing the first assignment, you may need to review material on McDuff contained in chapter 2.

1. **Presentation Based on Short Report.** Select any of the short written assignments in chapters that you have already completed. Prepare a five- to six-minute presentation based on the report you have chosen. Assume that your main objective is to present the audience with the major highlights of the written report, which they have all read. Use at least one visual aid.
2. **Presentation Based on Formal Report.** Prepare a 10- to 12-minute presentation based on any of the long-report assignments at the end of chapter 9. Assume that your audience has read or skimmed the report. Your main objective is to present highlights, along with some important supporting details. Use at least three visual aids.
3. **Presentation Based on Proposal.** Prepare a five- to six-minute presentation based on the proposal assignment at the end of chapter 10. Assume that your audience wants highlights of your written proposal, which they have read.
4. **Group Presentation.** Prepare a group presentation in the size groups indicated by your instructor. It may be related to a collaborative writing assignment in an earlier chapter or it may be done as a separate project. Review the chapter 1 guidelines for collaborative work. Though related to writing, some of these suggestions apply to any group work.

   Your instructor will set time limits for the entire presentation and perhaps for individual presentations. Make sure that your group's members move smoothly from one speech to the next; the individual presentations should work together for a unified effect.

13 | *Technical Research*

Solid research is vital to good technical writing. Here, a McDuff scientist uses a computerized data base to find information for a report.

*R*esearch writing does not end with the last college term paper. On the contrary, your career will often require you to gather technical information from libraries, interviews, questionnaires, and other sources. Such on-the-job research may help produce documents as diverse as reports, proposals, conference presentations, published papers, newspaper articles, or essays in company magazines. Your professional reputation may depend on the degree to which you can apply expert research skills to these tasks.

This chapter takes you on a brief tour of the research process as used in college courses and practiced on the job. Specifically, the chapter has these six objectives:

1. To offer strategies for getting started on any technical research
2. To describe techniques for gathering secondary research (that is, information provided by others in books, articles, newspapers, etc.)
3. To describe techniques for gathering primary research (that is, ''original'' research gained through sources such as questionnaires and interviews)
4. To explain the procedure for using borrowed information correctly
5. To highlight several documentation systems, with emphasis on one that places source information within parentheses in the text
6. To give guidelines for writing abstracts of research material

A common thread throughout the chapter is a McDuff case study. Tanya Grant, who works in the Marketing Department at McDuff's corporate office, is searching for information to complete a project for Rob McDuff, the president of the firm. We'll observe Tanya as she gathers secondary research and then primary research, in her effort to find what she needs.

Like Tanya, later you will have to apply the research process to a specific technical-writing task in college or on the job. That is when you will learn the most about the research process. It is one thing to read about doing research; it is quite another to dive into your project and work directly with bibliographies, data bases, books, periodicals, indexes, and other resources. Get firsthand research experience as soon as you can.

Finally, remember that the best research writing *smoothly* merges the writer's ideas with supporting data. Such writing should (1) impress readers with its clarity and simplicity and (2) avoid sounding like a strung-together series of quotations. Achieving these two goals is one of the greatest challenges in research writing.

GETTING STARTED

In chapter 1, you learned about the three phases of any writing project: planning, drafting, and revising. Performing research occurs in the planning stage, usually right before you complete an outline. Before starting your research, ask yourself questions like these that will help give direction to your work:

1. What *main questions* must be answered during the research phase?
2. What *types of information* might be most useful?
3. What *types of sources* would be most useful?
4. What *format* must be used in documenting any material borrowed from sources?

Let's discover ways to answer such questions by taking a look at the research project of McDuff's Tanya Grant.

Tanya has just been given an important mission. The company president, Rob McDuff, wants her to examine the possibility of the company publishing a newsletter exclusively on the subject of acid rain. First of all, McDuff believes the acid-rain problem is so important that a separate newsletter—written from a private company's perspective—will provide a service in reporting current research. Second, he hopes his firm becomes a major player in helping solve the acid-rain problem. In short, the newsletter could serve both as a public service and also as a marketing device.

Rob McDuff has told Tanya that the newsletter will seek two types of readers:(1) well-educated consumers who want more information and (2) current and prospective clients whose business interests compel them to learn more about acid rain, as well as other environmental issues.

Tanya's task is to write a report on the practicality of starting the environmental newsletter McDuff has proposed. Some of her report will investigate start-up costs, employee needs, and other in-house matters. Her other research effort, however, will be to (1) see what current journals cover the topic, and (2) find out what readers, especially clients, would want covered in such a magazine. The following outline shows how Tanya would probably answer the four questions previously noted, as she begins her research:

1. *Main question:* Is there a need for the acid-rain periodical that her boss is considering?

2. *Main types of information needed:*

   - Number of acid-rain magazines published
   - Types of articles now published
   - Types of government and nongovernment groups that now put out magazines containing acid-rain articles
   - Interest level and needs of potential readers

3. *Possible sources:*

   - Periodicals (indexed)
   - Newspapers (indexed)
   - Actual copies of environmental magazines
   - Questionnaire
   - Interviews with some potential readers, such as current clients

4. *Format for documentation:* Tanya will submit a short report to Rob McDuff, documenting her research using the author-date system of citing borrowed information (the same format used by McDuff engineers and scientists in their research reports).

With this basic plan in mind, Tanya can begin her work. She decides to start at a nearby university library, rather than the McDuff corporate library or the local library, because she knows about the university's extensive holdings in science and technology. The next section outlines major types of library research sources, along with examples from Tanya Grant's project. Then the following section examines primary research that Tanya will pursue.

SECONDARY RESEARCH: LOCATING LIBRARY SOURCES

The library can seem like an intimidating place when you first start a project. Once you learn a few basics, however, you will become comfortable and even confident about using this amazing resource. This section includes descriptions of seven types of research materials for technical projects: (1) books; (2) periodicals; (3) newspapers; (4) company directories; (5) dictionaries, encyclopedias, and other general references; (6) abstracts; and (7) computerized data bases. Though brief, the descriptions will help you "divide and conquer" your research task by learning about one type of resource at a time.

■ *Resource 1: Books*

Books provide well-supported, tested information about a topic, but, by definition, the information is often dated. Even a book just published has information that is one or two years old, given the time it takes to put a book-length manuscript into print. Keep this limitation in mind as you work in the library.

Despite its somewhat dated information, the book collection remains the most used part of the library. Most college libraries organize their collections by the Library of Congress (LC) system (see Figure 13−1). Some still offer the card catalog as the road map to the library collection; others have a catalog on microfiche; still others, in growing numbers, have computerized on-line catalogs that can be easily updated. In any case, you usually have three approaches, or means of access, to each book in the collection: author, title, and subject data. (See Figure 13−2 for the author, title, and subject cards for a book that Tanya Grant might have consulted on acid rain.) Here are some basic search guidelines, whether you are using catalog cards, a microfiche catalog, or on-line information:

1. If you already know specific titles of potentially useful books, consult the title entries for catalog numbers and other information.
2. If you know of authors who may have written books in the field, start with author entries.
3. If you do not know specific authors/titles or wish to view the library's range of books on a topic, consult the subject entries to build a list of available books.
4. If your library is automated, the on-line catalog will present you with ''help screens'' to assist you in using the three approaches previously described. (Note, however, that on-line catalogs may not be confined to book sources alone.)

In practice, defining your subject can cause some confusion. For example, your term for a subject might be different from the subject headings used in the catalog, or books on your topic might be grouped under several headings in the Library of Congress classification system. Fortunately, your library has a reference book to solve both these problems: the Library of Congress *Subject Headings*. Often kept near the library's catalog, this three-volume set of red books is your key to the LC system. Upon looking up your subject, you will discover (1) whether your wording of the topic is used in the LC catalog system and (2) if there is other phrasing under which information you need might be listed in the catalog.

Tanya Grant's McDuff Project

Tanya Grant, for example, rightly thinks that books will not be her main source of information about acid-rain periodicals, since the field has developed so recently. Yet she at least wants to see what range of sources the catalog offers, just as a starting point. Upon looking up ''acid rain'' in *Subject Headings,* she locates the information in Figure 13−3. Thus her quick inquiry has yielded several additional subject headings under which sources might be listed in the library catalog.

■ *Resource 2: Periodicals*

The *American Heritage Dictionary* defines periodical as a publication issued at ''regular intervals of more than one day.'' The term encompasses (1) popular magazines that take commercial advertising, such as *Time, Science,* and *National Geographic,* and (2) professional journals, such as *IEEE Transactions on Professional Communication.* Most library visitors are familiar with the section that houses current periodicals, either in alphabetical order or by subject area. Yet they are less familiar with the part of the library containing back issues. Libraries keep back issues of the

Library of Congress Cataloging System: A General Breakdown

A	Collections, Encyclopedias, Indexes, Yearbooks, Directories, etc.
B	Philosophy, Psychology, Religion
C	History of Civilization and Culture
D	History – General and Old World – Europe, etc.
E-F	American History
G-GF	Geography, Maps, Oceanography, Human Ecology
GN-GT	Anthropology and Related Subjects
GV	Recreation
H	Social Sciences – General
HA	Statistics
HB-HJ	Economics
HM-HX	Sociology
J	Political Science
K	Law
L	Education
M	Music
N	Art and Architecture (Architectural Design)
P	Language and Literature, Philology
PN-PX	Literary History, Literature
PR, PS, PZ	English and American Literature
Q	Science – General
QA	Math
QB	Astronomy
QC	Physics
QD	Chemistry
QE	Geology
QH	Natural History
QK	Botany
QL	Zoology
QM	Human Anatomy
QP	Physiology
QR	Microbiology
R	Medicine
S	Agriculture, Plant and Animal Culture, Hunting Sports
T	General Technology

FIGURE 13–1
General categories from LC system

General Engineering and Civil Engineering

TA	Engineering – General, Civil Engineering – General
TC	Hydraulic Engineering
TD	Environmental Technology, Sanitary Engineering
TE	Highway Engineering, Roads and Pavements
TF	Railroad Engineering and Operation
TG	Bridge Engineering
TH	Building Construction

Mechanical Group

TJ	Mechanical Engineering and Machinery
TK	Electrical Engineering, Electronics, Nuclear Engineering
TL	Motor Vehicles, Aeronautics, Astronautics

Chemical Group

TN	Mining Engineering, Metallurgy
TP	Chemical Technology
TR	Photography

Composite Group

TS	Manufactures
TT	Handicrafts, Arts and Crafts
TX	Home Economics
U	Military Science
V	Naval Science
Z	Bibliography

FIGURE 13–1, *continued*

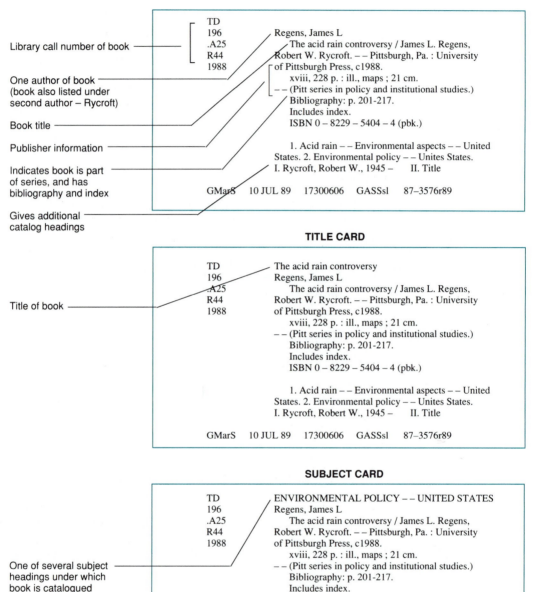

AUTHOR CARD

Library call number of book

TD
196
.A25
R44
1988

One author of book
(book also listed under
second author – Rycroft)

Book title

Publisher information

Indicates book is part
of series, and has
bibliography and index

Gives additional
catalog headings

Regens, James L
    The acid rain controversy / James L. Regens,
Robert W. Rycroft. – – Pittsburgh, Pa. : University
of Pittsburgh Press, c1988.
        xviii, 228 p. : ill., maps ; 21 cm.
    – – (Pitt series in policy and institutional studies.)
        Bibliography: p. 201-217.
        Includes index.
        ISBN 0 – 8229 – 5404 – 4 (pbk.)

        1. Acid rain – – Environmental aspects – – United
States. 2. Environmental policy – – Unites States.
I. Rycroft, Robert W., 1945 –   II. Title

GMarS   10 JUL 89   17300606   GASSsl   87–3576r89

TITLE CARD

Title of book

TD
196
.A25
R44
1988

The acid rain controversy
Regens, James L
        The acid rain controversy / James L. Regens,
Robert W. Rycroft. – – Pittsburgh, Pa. : University
of Pittsburgh Press, c1988.
        xviii, 228 p. : ill., maps ; 21 cm.
    – – (Pitt series in policy and institutional studies.)
        Bibliography: p. 201-217.
        Includes index.
        ISBN 0 – 8229 – 5404 – 4 (pbk.)

        1. Acid rain – – Environmental aspects – – United
States. 2. Environmental policy – – Unites States.
I. Rycroft, Robert W., 1945 –   II. Title

GMarS   10 JUL 89   17300606   GASSsl   87–3576r89

SUBJECT CARD

One of several subject
headings under which
book is catalogued

TD
196
.A25
R44
1988

ENVIRONMENTAL POLICY – – UNITED STATES
Regens, James L
        The acid rain controversy / James L. Regens,
Robert W. Rycroft. – – Pittsburgh, Pa. : University
of Pittsburgh Press, c1988.
        xviii, 228 p. : ill., maps ; 21 cm.
    – – (Pitt series in policy and institutional studies.)
        Bibliography: p. 201-217.
        Includes index.
        ISBN 0 – 8229 – 5404 – 4 (pbk.)

        1. Acid rain – – Environmental aspects – – United
States. 2. Environmental policy – – Unites States.
I. Rycroft, Robert W., 1945 –   II. Title

GMarS   10 JUL 89   17300606   GASSsl   87–3576r89

FIGURE 13–2
Author, title, and subject cards for same book

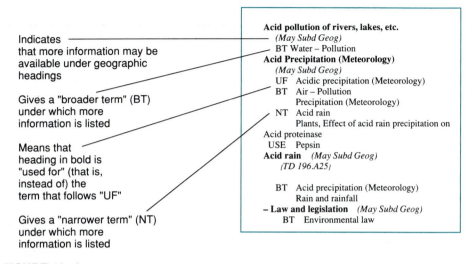

Indicates that more information may be available under geographic headings

Gives a "broader term" (BT) under which more information is listed

Means that heading in bold is "used for" (that is, instead of) the term that follows "UF"

Gives a "narrower term" (NT) under which more information is listed

Acid pollution of rivers, lakes, etc.
   (May Subd Geog)
   BT Water – Pollution
Acid Precipitation (Meteorology)
   (May Subd Geog)
   UF Acidic precipitation (Meteorology)
   BT Air – Pollution
       Precipitation (Meteorology)
   NT Acid rain
       Plants, Effect of acid rain precipitation on
Acid proteinase
   USE Pepsin
Acid rain   *(May Subd Geog)*
       [TD 196.A25]

   BT Acid precipitation (Meteorology)
       Rain and rainfall
 – Law and legislation   *(May Subd Geog)*
      BT Environmental law

FIGURE 13–3
Sample entries from subject headings (Library of Congress)

Source: Library of Congress, *Subject Headings*, 12th ed., vol. 1 (A–E) (Washington, D.C.: Cataloging Distribution Service, Library of Congress, 1989), 20.

periodicals considered most useful to the users. They may be in the form of hard copy (bound volumes of the actual issues) or on microfilm or some other reduced format.

Your key to information in periodicals is a periodical index. By looking up a subject area in an index, you can find articles listed that could provide the information you need. Some indexes, like the familiar *Readers' Guide to Periodical Literature*, deal with popular periodicals. Others, like the *Engineering Index*, deal with a broad range of technical information. Still others, like the *Society of Mechanical Engineers Technical Digest*, focus on periodicals, books, and papers in specialized technical fields. In all cases, periodicals covered in an index will be listed in the volume, along with the inclusive dates of the issues indexed in that volume.

When working in a library for the first time, ask a reference librarian about available indexes, for they may or may not be listed in the card catalog. Many libraries have handouts that list common indexes and their fields. Here are a few well-known technical indexes:

> *Applied Science and Technology Index*
> *Business Periodicals Index*
> *Computer Literature Index*
> *Energy Index*
> *Engineering Index*
> *Microcomputer Index*
> *Monthly Catalogue of U.S. Government Publications*

Also be aware that many libraries will give you access to electronic indexes. On-line data bases are becoming increasingly available in a wide variety of subject fields, but most libraries require that a librarian conduct the actual search (see Resource 7, which follows). Also, CD ROM technology is now coming into its own in the modern library. This service allows library users to conduct their own electronic searches at their own pace, with instant access to information. One common CD ROM system, *InfoTrac*, contains current periodical citations in business, the humanities, social sciences, and technology.

Tanya Grant's McDuff Project

Assume that in her search for information on acid-rain periodicals, Tanya decides to consult the *Applied Science and Technology Index* (ASTI) to see what has been published recently. This index pays special attention to periodical articles covering hands-on technical topics (as opposed to the *Engineering Index*, which deals more with research-related technical articles). One ASTI volume contains the information in Figure 13–4. Note that Tanya finds three related subjects in sequence, along with a "see also" reference that sends her to a topic a few pages away ("Atmospheric corrosion").

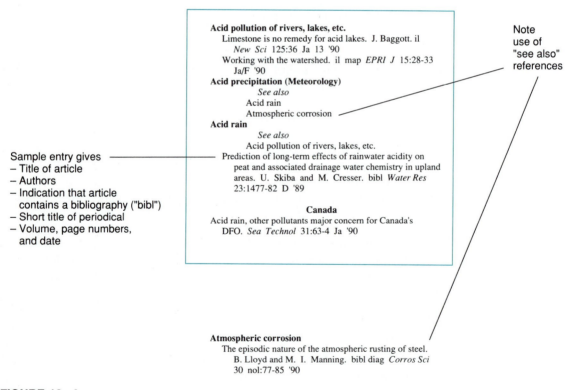

FIGURE 13–4

Sample entries from *Applied Science and Technology Index*

Source: *ASTI*, vol. 78, no. 4 (Bronx, N.Y.: H. W. Wilson Company, April 1990), 2, 19. *Applied Science and Technology Index*, 1990, Copyright © 1990 by the H. W. Wilson Company. Material reproduced by permission of the publisher.

Next Tanya searches for periodical indexes that deal just with environmental issues. A reference librarian refers her to *The Environment Index.* It indexes journal articles, conference papers, reports, and other documents. (Figure 13−5 shows some of the acid-rain entries contained in one yearly volume.) Tanya also discovers that the index includes a list of environmental periodicals. Her quick reading of the list finds no magazines or journals devoted exclusively to the study of acid rain.

■ *Resource 3: Newspapers*

If your research topic demands the most current information, newspapers provide an excellent source. One disadvantage is that newspaper information has not "stood the test of time"—that is, it is not validated to the same extent as information in journals and books. Another disadvantage is that many newspapers either are not indexed or are only indexed by local libraries or other agencies. Without an index, your only option would be to pore over each issue in search of information.

Fortunately, there are indexes—the *New York Times Index* and the *Wall Street Journal Index*—for two well-known newspapers. Consult the *New York Times* for general research topics; consult the *Wall Street Journal* for business-related topics, especially those dealing with specific corporations.

Tanya Grant's McDuff Project

As an aside in her research, Tanya decides it might be interesting to see what kinds of general articles on acid rain have appeared in the *New York Times.* In consulting one semimonthly volume of the *New York Times Index,* she locates the entry shown in Figure 13−6. Note that the *Index* provides brief summaries of the articles along with the bibliographic citations.

■ *Resource 4: Company Directories*

Often your research needs may require that you find detailed information about specific firms. For example, you could be completing research about a company that may hire you, or you may seek information about companies that compete with your own. Following are some useful directories. Consult them individually to discover each one's particular focus.

Corporate Technology Directory

Million Dollar Directory

Moody's Industrial Manual

Standard and Poor's Register of Corporations and Executives

Try Us: National Minority Business Directory

Ward's Directory of 51,000 U.S. Corporations

Ward's Directory of Major U.S. Private Companies

Who's Who in Engineering

Who's Who in Technology

World Business Directory

ACID RAIN

Atmospheric Deposition in Fenno-Scandia: Characteristics and Trends (Proceedings of the International Symposium on Acidic Precipitation, Muskoka, Ontario, September 15-20, 1985), Water Air & Soil Pollution, Sep 86, v30, n1-2, p5(12)   *01-87-00411

Pollutant Wet Deposition Mechanisms in Precipitation and Fog Water (Proceedings of the International Symposium on Acidic Precipitation, Muskoka, Ontario, September 15-20, 1985), Water Air & Soil Pollution, Sep 86, v30 , n1-2, p91(14)   *01-87-00413

Acidic Precipitation in Western North America: Trends, Sources, and Altitude Effects in New Mexico, 1979-1985 (Proceedings of the International Symposium on Acidic Precipitation, Muskoka, Ontario, September 15-20, 1985), Water Air & Soil Pollution, Sep 86, v30, n1-2, p125(9)   *01-87-00418

Alkaline Materials Flux from Unpaved Roads: Source Strength, Chemistry and Potential for Acid Rain Neutralization (Proceedings of the International Symposium on Acidic Precipitation, Muskoka, Ontario, September 15-20, 1985), Water Air & Soil Pollution, Sep 86, v30, n1-2, p285(9)   *01-87-00423

Overview of Historical and Paleoecological Studies of Acidic Air Pollution and its Effects (Proceedings of the International Symposium on Acidic Precipitation, Muskoka, Ontario, September 15-20, 1985) Water Air & Soil Pollution, Sep 86, v30, n1-2, p311(8)   *01-87-00425

A Review of the Chemical Record in Lake Sediment of Energy Related Air Pollution and its Effects on Lakes (Proceedings of the International Symposium on Acidic Precipitation, Muskoka, Ontario, September 15-20, 1985), Water Air & Soil Pollution, Sep 86, v30, n1-2, p331(15)   *02-87-00512

Application of Conceptual Environmental Model to Assess the Impact of Acid Precipitation on Drinking Water Quality II — Studies of Chemical Thermodynamic Equilbria in Water, Water Quality B, Jul 86, v11, n3, p152(9)   *19-87-00516

Bone Concentration of Manganese in White Sucker (Catostomus commersoni) from Acid, Circumneutral and Metal-Stressed Lakes (Proceedings of the International Symposium on Acidic Precipitation, Muskoka, Ontario, September 15-20, 1985), Water Air & Soil Pollution, Sep 86, v30, n1-2, p515(7)   *0287-00518

Issues in Environmental Science (Environmental Quality), CEQ Report 15, 1985, p433(44)   *07-87-00620

Gardsjon Project: Lake Acidification, Chemistry in Catchment Runoff, Lake Liming and Microcatchment Manipulations (Proceedings of the International Symposium on Acidic Precipitation, Muskoka, Ontario, September 15-20, 1985), Water Air & Soil Pollution, Sep 86, v30, n1-2, p31(16)   *19-87-00873

Changes in Fish Populations in Southernmost Norway During the Last Decade (Proceedings of the International Symposium on Acidic Precipitation, Muskoka, Ontario, September 15-20, 1985), Water Air & Soil Pollution, Sep 86, v30, n1-2, p381(6)   *19-87-00874

Evidence for Recent Acidification of Lentic Soft Waters in the Netherlands (Proceedings of the International Symposium on Acidic Precipitation, Muskoka, Ontario, September 15-20, 1985), Water Air & Soil Pollution, Sep 86, v30, n1-2, p387(6)   *19-87-00875

Ecological Effects of Acidification on Primary Producers in Aquatic Systems (Proceedings of the International Symposium on Acidic Precipitation, Muskoka, Ontario, September 15-20, 1985), Water Air & Soil Pollution, Sep 86, v30, n1-2, p421(18)   *19-87-00876

Ecological Effects of Acidification on Tertiary Consumers. Fish Population Responses (Proceedings of the International Symposium on Acidic Precipitation, Muskoka, Ontario, September 15-20, 1985), Water Air & Soil Pollution, Sep 86, v30, n1-2, p451(10)   *19-87-00877

FIGURE 13–5
"Acid rain" entries from *The Environment Index*

Source: *The Environment Index: 1987 in Retrospect,* vol. 17 (New York: EIC/Intelligence, Inc., 1988), 10.

Gives summary ────────
of article ─
along with
date, section,
and page
number.

ACID RAIN
Huge federal study finds acid rain has turned many lakes and streams acidic, not only in Northeast but also in other areas of East and parts of Midwest - - including northern Florida, Appalachians and New Jersey Pine Barrens - - but that overall acidification has been less extensive than was feared when research began a decade ago; findings in six areas across US discussed; map (M), Ja 16,C,4:1

Teachers and students at about a dozen schools in Connecticut and one in Massachusetts are taking part in project intended to involve students in science and provide much-needed information about acid rain in Connecticut; project is also pilot for Thames Science Center in New London, which is developing environmental science curriculum for state schools; photos (M), Ja 28,XII-CN,21:1

Editorial holds that preliminary results of National Acid Precipitation Assessment Program, 10-year Federal study on acid rain, lends urgency to proposals now before Congress that coal-burning utilities sharply cut sulfur dioxide emissions, Ja 29,A,22:1

FIGURE 13–6
"Acid rain" entry from the *New York Times Index*

Source: *New York Times Index,* vol. 78, no. 2, January 16–31, 1990 (semimonthly edition), 2–3. Copyright © 1990 by the New York Times Company. Reprinted by permission.

■ *Resource 5: Dictionaries, Encyclopedias, and Other General References*

Sometimes you may need some general information to help you get started on a research project. In this case, consult one of the specialized dictionaries or encyclopedias listed here. You can also use any number of general encyclopedias available in most libraries.

Cowles' Encyclopedia of Science, Industry, and Technology

Dictionary of Business and Economics

Encyclopedia of Business Information Sources

Encyclopedia of Engineering Materials and Processes

Encyclopedia of Physical Science and Technology

Engineering Encyclopedia

McGraw-Hill Dictionary of Science and Engineering

McGraw-Hill Encyclopedia of Engineering

McGraw-Hill Encyclopedia of Science and Technology

Van Nostrand's Scientific Encyclopedia

Of course, you should consider explaining the nature of your project to a librarian, who may refer you to a general reference book that could help you in the early stages of your research. Note the help that Tanya Grant received in her search.

Tanya Grant's McDuff Project

At this point, Tanya has spent a few hours on her own examining the library catalog (with the help of the *Subject Headings*), periodical indexes, and newspaper indexes. Although she suspects that there are no acid-rain periodicals published by environmental-science firms, she wants to pursue her research further.

Upon consulting a librarian at her local university library, Tanya learns about several specialized reference books that list magazines and journals from around the world. In one, called *Ulrich's International Periodicals Directory,* she turns to the subject heading of "Environmental Studies" and finds an alphabetical listing of periodicals. A quick look under "acid rain" reveals several publications associated with government and nonprofit agencies, such as *Acid Magazine, Acid Precipitation,* and *Acid Rain Update.* However, there are no newsletters, journals, magazines, or other periodicals published by corporations involved in environmental work.

A similar publication, the annual *Gale Directory of Publications and Broadcast Media,* lists broadcast stations and publications, such as newsletters, magazines, journals, and trade publications. There Tanya finds some of the same entries as in *Ulrich's* as well as at least one other, *Acid Precipitation Digest.* Nevertheless, she still does not find a newsletter of the kind being considered by McDuff—one published by a private firm in the environmental business.

■ *Resource 6: Abstracts*

Periodical indexes tell you where to find articles, but they do not indicate whether an article will provide the needed information. For this purpose, you can consult any number of periodical abstracts. Abstracts give you brief descriptions of articles so that you can decide whether or not an entire article is worth finding. They are especially useful when the article being summarized is not available in your library. The abstract can help you decide whether to (1) visit another library, (2) order an article through an interlibrary loan service, or (3) disregard an article altogether. Here are some common technical abstracts:

Abstracts of Health Care

Agricultural Index

Biological Abstracts

Chemical Abstracts

Energy Information Abstracts

Engineering Abstracts

Engineering Index

Geological Abstracts

Management Studies

Mineralogical Abstracts

Transportation Research Abstracts

World Textile Abstracts

The last section of this chapter will show you how to write abstracts yourself, for times when you may be called upon to summarize research items.

■ *Resource 7: Computerized Data Bases*

Traditionally, research has been a labor-intensive operation—with the emphasis on your labor. Now computers can reduce the time it takes you to collect information. The use of computerized data bases—or files of information—will lighten your load. Typically, a data base will provide a list of sources on your topic, perhaps along with brief descriptions of these sources. With this information you can then decide what sources from the computer-generated bibliography you want to collect in hard-copy form.

What are the advantages of using computer data bases? First, they are fast. You can avoid the laborious search through your library or the long wait to receive information requested from other libraries. Second, they give you access to the most current research on your topic, since data bases are often updated on a daily basis. There are, however, these four limitations to data bases.

1. Someone has to pay for the service—you, your company, or your library.
2. You must select the right "key words" that the person conducting the search can use to locate information. These words must be broad enough to encompass your research area and yet not so broad as to bring in sources too general for your purpose.
3. You must use the data base most appropriate for your field. Given the number available, talk with a research librarian before you proceed.
4. You do not learn much about the research process by using a data base. Thus the service is not recommended for someone wanting to learn how to use a library firsthand. Be sure you know how to complete the traditional research process before you rely on technology to do it for you.

PRIMARY RESEARCH: USING QUESTIONNAIRES AND INTERVIEWS

During research projects, libraries provide information collected or produced by others (called *secondary* information). Sometimes, however, your project may require collecting firsthand information yourself (called *primary* information). Such primary research most often includes questionnaires, personal interviews, or both. This section covers both questionnaires and interviews. You will learn how to (1) prepare, send out, and report the results of a questionnaire, and (2) prepare, conduct, and summarize a personal interview.

Tanya Grant's McDuff Project

You will recall that Tanya, who works in the Marketing Department at McDuff, has been asked by the company president to examine the feasibility of starting a company newsletter on acid rain. This newsletter will be used to inform general readers and prospective clients. Her library research has helped locate sources on acid rain that she may be able to use in newsletter issues. So far she has found no references to firms publishing the kind of newsletter her boss contemplates.

Now, before reporting her findings to Rob McDuff, she wants to find out for herself if readers would benefit from such a newsletter. She believes her best approach would be to (1) send a questionnaire to selected current clients of McDuff's services, and (2) personally interview three or four respondents who have completed the optional "your name" line on the survey. Tanya hopes that her primary research will reinforce whatever information she gathered from secondary sources in the library. In either case, she has rightly concluded that information personally gathered from clients will help her boss decide about the newsletter.

Questionnaires

Tanya Grant has the same challenge you would face in developing a questionnaire. Like you, she receives many questionnaires herself. Most of them she promptly tosses in the recycle bin because they don't warrant her time, are too long, or seem confusing. Now that the shoe is on the other foot, she wants to design a questionnaire that attracts the attention of her readers and entices them to complete it. To accomplish this feat, she goes through the following three-stage process suggested for your questionnaires.

■ *Step 1: Preparing the Questionnaire*

Let's start with an obvious but crucial point: Your questionnaire is useful only if sufficient numbers are completed and returned. Thus you must focus just as much on your readers' needs as you do on your own objectives. Before readers complete a questionnaire, they must perceive that (1) it benefits them personally or professionally and (2) it is easy to fill out and return. Keep these two points in mind as you design the form and the cover letter. Here are some specific guidelines for preparing a reader-focused document:

1. *Write a precise purpose statement.* As in other documents, a one-sentence statement of purpose provides a useful lead-in for your cover letter that accompanies the survey (see next section). For example, Tanya Grant has prepared the following purpose statement for her survey concerning the acid-rain newsletter: "The purpose of this survey is to find out ways in which a newsletter on acid rain might benefit our clients and the general public." As obvious as that statement sounds, it will help busy readers who don't have the time to wade through long, drawn-out rationales.

2. *Limit the number of questions.* Every question must serve to draw out information that relates to your purpose statement. For example, Tanya Grant knows that her questions must focus on the reader's interest in acquiring more data on acid rain. She must resist the temptation to clutter the questionnaire with irrelevant questions on other environmental topics, like wastewater treatment or asbestos abatement.

3. *Ask mostly objective questions.* You need to design your form so that (1) questions are easy to answer and (2) responses are easy to compile. Although open-ended questions yield more detailed information, the answers to them take time to write and are difficult to analyze statistically. Instead, your goal is breadth, not depth, of response. Perhaps with the exception of one or two open-ended questions at the end of your questionnaire, reserve long-answer responses for personal interviews you may conduct with a select audience. For example, Tanya Grant has decided to include an optional open-ended question at the end of her questionnaire. There she will encourage respondents to list sample topics they would like covered in an acid-rain newsletter.

Objective questions come in several forms. Four common types are described here, along with examples of each.

- *Either/Or Questions:* Such questions give the reader a choice between two, and only two, options, such as "yes" or "no." They are useful only when your questions present clear, obvious choices. EXAMPLE: "Would you or your technical staff find it useful to receive a technical newsletter on acid rain?" (followed by "yes" and "no" blocks), or "An acid-rain newsletter would be useful to your technical staff" (followed by "true" and "false" blocks).
- *Multiple-Choice Questions:* These questions expand the range of possibilities for your reader to three or more, requiring a longer response time but yielding you more detailed information than either/or questions. EXAMPLE: "If you answered 'yes' to the preceding question [a question asking if an acid-rain newsletter would be useful to the reader's firm], what publication schedule would best meet your needs: (a) semimonthly, (b) monthly, (c) quarterly, or (d) semiyearly?"
- *Graded-Scale Questions:* By permitting degrees of response, these questions help you gauge the relative strength of the reader's opinion. EXAMPLE: "Acid rain is an issue that has a strong impact on your day-to-day business: (a) strongly agree, (b) agree, (c) disagree, (d) strongly disagree, or (e) have no opinion."
- *Short-Answer Questions:* Use these questions when the possible short answers are too numerous to list on your form. EXAMPLE: "List any environmental newsletters you already receive that you find helpful in your business."

4. *Provide clear questions that are easy to answer.* Like other forms of technical writing, questionnaires can frustrate readers when individual questions are unclear. Four common problems are (1) bias in phrasing, (2) use of undefined terms, (3) use of more than one variable, and (4) questions that require too much homework. Here are some examples of right and wrong ways to phrase questions, along with a brief comment on each problem:

> **Biased Question:** Original question: "Is the state government's excessive taxation on business affecting your business?" (Words like *excessive* reflect a bias in the question, pushing a point of view and thus skewing the response.) Revised question: "Do you believe that the rate of state taxation on your own business is fair?"

> **Undefined Technical Terms:** Original question: "Are you familiar with EPA guidelines for using USTs?" (Your readers may not know that EPA is short for the Environmental Protection Agency or that UST stands for "underground storage tank." Thus some "no" answers may be generated by confusion about terminology.)

Revised question: "Are you familiar with Environmental Protection Agency guidelines for using underground storage tanks?"

Mixed Variables: Original question: "Were our technicians prompt and thorough in their work?" (There are two questions here, one dealing with promptness and the other with thoroughness.)
Revised question (two separate questions): "Were our technicians prompt? Were our technicians thorough in their work?"

Question That Requires Too Much Homework: Original question: "Besides the job just finished, how many times have you used our firm's services?" (This question may force readers to scout through company files for an accurate answer. If they do not have the time for that research, they may leave the answer blank or provide an inaccurate guess. In either case, you are not getting valid information.)
Revised question: "Other than the project just finished, have you used the services of our firm before?"

5. *Include precise and concise instructions at the top of the form.* Your instructions can be in the form of an easy-to-read list of points that start with action verbs, such as the following list:

- Answer Questions 1–20 by checking the correct box.
- Answer Questions 21–30 by completing the sentences in the blanks provided.
- Return the completed form in the envelope provided by October 15, 1993.

Or, if instructions are brief, they can be in the form of a short, action-centered paragraph, like this: "After completing this form, please return it in the enclosed stamped envelope by October 15, 1993."

6. *Apply principles of document design.* Although you must strive for economy of space when designing a questionnaire, use adequate white space and other design principles to make the document attractive to the eye.

7. *Test the questionnaire on a sample audience.* Some sort of "user test" is a must for every questionnaire. For example, after completing her acid-rain survey, Tanya decided to test it on three people. One is a fellow marketing colleague at McDuff, who has conducted several questionnaires for the firm. A second is a psychologist Tanya knows through a local professional association. A third is an actual McDuff client whom she knows well enough to ask for constructive criticism on the form. Thus her user test will solicit views from people with three quite different perspectives.

■ *Step 2: Conducting the Project*

After you have designed a good form, the next task is to distribute it. Following are guidelines for selecting a good sampling of potential respondents, introducing the questionnaire to your audience, and encouraging a quick response from a high percentage of readers.

1. *Choose an appropriate audience.* Selecting your audience depends on the purpose of your questionnaire. If, for example, you manage a 100-employee engineering firm and want to gauge customer satisfaction with recent construction jobs, you might send your questionnaire to all 156 clients you have served in the last two years. Restricting the mailing list would be unnecessary, since you have a small sample.

However, if you are in Tanya Grant's position at McDuff, with a client list totaling about 3200 over the last two years, you will need to select a random sample. Tanya's research suggests that she will receive about a 25 percent rate of return on her questionnaires. (Actually, this rate would be quite good for an anonymous questionnaire.) Given that she wants about 200 returned forms, she must send out about 800 questionnaires in expectation of the 25 percent return rate.

With a client list of 3200, she simply selects every fourth name from the alphabetized list to achieve a random list of 800 names. Note that the selection of client names from an alphabetized list preserves what is essential—that is, the random nature of the process.

Of course, you can create more sophisticated sampling techniques if necessary. For example, let's assume Tanya wants an equal sampling of clients from each of two years— 1992 (with 1200 clients) and 1993 (with 2000 clients). In other words, she wants to send an equal number of forms to each year's clients, even though the number of clients varies from year to year. In this case, first she would select 400 names—or every third name—from the 1200 alphabetized names for 1992. Then she would select the other 400 names—or every fifth name—from the 2000 alphabetized names for 1993. As a result, she has done all she can do to equalize the return rate for two years.

This strategy will help you choose the audience for simple questionnaire projects. You may want to consult a specialist in statistics if you face a sophisticated problem in developing an appropriate sampling.

2. *Introduce the questionnaire with a clear, concise cover letter.* In 15 or 20 seconds, your letter of transmittal must persuade readers that the questionnaire is worth their time. Toward this end, it should include three main sections (which correspond to the letter pattern presented in chapter 7):

- *Opening paragraph:* Here you precisely state the purpose of the questionnaire and perhaps indicate why this reader was selected.
- *Middle paragraph(s):* Here you state the importance of the project. Strive to emphasize ways that it may benefit the reader.
- *Concluding paragraph:* Specify when the questionnaire should be returned, even though this information will be included in the directions on the questionnaire itself.

3. *Encourage a quick response.* If your questionnaire is not anonymous, you may need to offer an incentive for respondents to submit the form by the due date. For example, you can offer to send them a report of survey results, a complimentary pamphlet or article related to their field, or even something more obviously commercial, when appropriate. Obviously, any incentive must be fitting for the context. Keep in mind also that some experts believe an incentive of any kind introduces a bias to the sample.

If the questionnaire is anonymous or if complimentary gifts are inappropriate or impractical, then you must encourage a quick response simply by making the form as easy as possible to complete. Clear instructions, frequent use of white space, a limited number of questions, and other design features mentioned earlier must be your selling points.

■ Step 3: Reporting the Results

After you have tabulated the results of the survey, you must return to the needs of your original audience—the persons who asked you to complete the questionnaire. They expect you to report the results of your work. Described next are the major features of such a report.

First, you want to show your audience that you did a competent job of preparing, distributing, and collecting the questionnaire. Thus the body of your report should give details about your procedures. Appendices may include a sample form, a list of respondents, your schedule, extensive tabulated data, and other supporting information.

Now, what about revealing the results of the survey? Here is where you need to be especially careful. You should present only those conclusions that flow clearly from the data. Choose a tone that is more one of suggesting than declaring. In this way, you give readers the chance to draw their own conclusions and to feel more involved in final decision making. Graphs are an especially useful way to present statistical information (see chapter 11, "Graphics").

Finally, remember that your report and the completed questionnaires may remain on file for later reference by employees who know nothing about your project. Be sure that your document is self-contained. Later readers who uncover the "time capsule" of your project should be able to understand its procedures and significance from the report you have written.

Interviews

Besides questionnaires, interviews are another common way to gather primary research. Often they are conducted after a questionnaire has been completed, as a follow-up activity with selected respondents. Interviews also may be done as primary research independent of questionnaires. In either case, you need to follow some common guidelines to achieve success in the interview. Here are a few basic pointers for preparing, conducting, and recording the results of your interviews:

■ Step 1: Preparing for the Interview

You should put at least as much effort into planning the interview as you do into conducting it. Good planning will put you at ease *and* show interviewees that you value their time. Specifically, follow these guidelines:

- *Develop a list of specific objectives for the interview.* Know exactly what you want to accomplish so that you can convey this significance to the person you interview.
- *Make clear your main objectives when you make contact for the interview.* This conversation should (1) stress the uniqueness of the person's contribution, (2) put him or her at ease with your goals and the general content of the proposed discussion, and (3) set a starting time and approximate length for the interview. If handled well, this preliminary conversation will serve as a prelude to the interview, giving direction to the next meeting.
- *Prepare an interview outline.* Persons you interview understand your need for written reference during the interview. Indeed, they will expect it of any well-prepared interviewer. A written outline should include (1) a sequential list of topics and subtopics you want to cover and (2) specific questions you plan to ask.
- *Show that you value your interviewee's time.* You can do this first by showing up a few minutes early so that the interview can begin on time. You also show this courtesy by staying on track and ending on time. Never go beyond your promised time limit unless it is absolutely clear that the person being interviewed wants to extend the conversation further than planned.

■ *Step 2: Conducting the Interview*

Your interview will be successful if you stay in control of it. Maintaining control has little to do with force of personality, so don't worry if you are not an especially assertive person. Instead, keep control by sticking to your outline and not letting time get away from you. If you find your speakers straying, for example, gently bring them back to the point with another question from your list. Here are additional pointers for conducting the interview:

- *Ask mostly "open" questions.* Open questions require your respondent to say something other than "yes," "no," or other short answers. They are useful to the speaker because they offer an opportunity to clarify an opinion or a fact. They are useful to you because you get the chance to listen to the speaker, to digest information, and to prepare for the next question.

    For example, McDuff's Tanya Grant may ask questions like these: "Could you describe two or three ways that a newsletter on acid rain might benefit your firm?" or "I've been told that members of your environmental group are concerned about levels of acid rain in the Denver area. What kinds of information do you think they would like to read in a newsletter on the topic of acid rain?"
- *Ask closed questions when you need to nail down an answer.* For example, Tanya Grant may ask persons she interviews, "Would you be willing to write a brief piece for our newsletter in an area of your expertise?" A "yes" or "perhaps" answer will give her an opening for calling the person several months later, when she is searching for news or opinion items for the newsletter. A closed question works when commitment is needed.
- *Use summaries throughout the interview.* Brief and frequent summaries serve as important resting points during the conversation. They give you the chance to make sure you understand the answers that have been given. They give your counterpart the chance to amplify or correct previous comments. For example,

Tanya Grant may comment to her speaker, "So, in other words, you're saying that an acid-rain newsletter would be most useful to your members if it includes three main types of information: (1) data on acid-rain levels in the Appalachians, (2) views of forestry experts on the possible effects on tree die-offs in high elevations, and (3) new technology that can serve to scrub smokestacks to lower acid levels at the plants." This summary will elicit either a "yes" or a clarification, either of which will help Tanya record the interview accurately.

■ *Step 3: Recording the Results*

Throughout the interview you will have taken notes. The actual mechanics of this process may influence the accuracy of your note taking. Here are three possible approaches:

- **Option 1: Number reference.** Using this approach, you will begin the interview with a list of numbered questions on your outline page. Then when you are taking notes, simply list the number of the question, followed by your notes. This approach gives you as much space as you want to write questions, but it does require that you move back and forth between your numbered question list and note page.
- **Option 2: Combined question-and-answer page.** Here you place a major question or two on each page, leaving the rest of the page to record answers to these and related questions that may be discussed. While this strategy requires considerably more paper and separates your prepared list of questions, it does help you focus quickly on each specific question and answer.
- **Option 3: Split page.** Some interviewers prefer to split each page lengthwise, writing questions in the left column and corresponding answers in the right column. Some questions may have been prepared ahead of time, as in Option 2. Others may be written as they are asked. In either case, you have a clear visual break between questions on one side and answers on the other. The advantage over Option 2 is that you have a visual map that shows you your progress during the conversation. Questions and answers are woven together into the fabric of your interview.

USING BORROWED INFORMATION CORRECTLY

Most errors in research papers occur in transferring borrowed information with accuracy and correct form to your report, paper, or other document. This section has two goals: (1) to explain why you must acknowledge sources you have used, and (2) to outline a research process from the point at which you identify possible sources of information to the point at which you have written the first draft.

Avoiding Plagiarism

One basic rule underlies the mechanical steps described in the rest of this chapter:

With the exception of "common knowledge," you should cite sources for ALL borrowed information used in your final document. This includes quotations, paraphrases, and summaries.

"Common knowledge" is information generally available from basic sources in the field. In the case of Tanya Grant's research project, common knowledge would be a definition for acid rain. When you are uncertain whether or not a piece of borrowed information is common knowledge, go ahead and cite the source. It is better to err on the side of excessive documentation than to leave out a citation and risk a charge of *plagiarism* (the intentional or unintentional use of the ideas of others as your own). Here are three main reasons for documenting sources thoroughly and accurately:

1. **Ethics:** You have an *ethical* obligation to show your reader where your ideas stop and those of another person begin. Otherwise you would be parading the ideas of others as your own.
2. **Law:** You have a *legal* obligation to acknowledge information borrowed from a copyrighted source. In fact, you should seek written permission for the use of borrowed information that is copyrighted when you plan to publish your document or when you are using your document to bring in profit to your firm (as in a proposal or report). If you need more specific information about copyright laws or about the legalities of documentation, see a research librarian.
3. **Courtesy:** You owe readers the *courtesy* of citing sources where they can seek additional information on the subject. Presumably, sources for quotations, paraphrases, and summaries would provide such a reference point.

Certainly some plagiarism occurs when unscrupulous writers intentionally copy the writing of others without acknowledging sources. However, most plagiarism results from sloppy work during the research and writing process. Described here are two common types of unintentional plagiarism. Though the errors are unintentional—that is, the writer did not intend to "cheat"—nevertheless both result in the inappropriate use of another person's work. That's plagiarism.

Mike Pierson, a supervisor at McDuff's Cleveland office, has been asked to deliver a presentation at an upcoming conference on acid rain. In his last-minute rush to complete the presentation—which will be published in a collection of papers from the meeting—Mike is taking notes from a source in the company library. He hurriedly writes a note card on a source but fails to indicate the source. Later, when he is writing the paper draft, he finds the card and does not know whether it contains information that was borrowed from a source or ideas that came to him during the research process. If he incorporates the passage into his paper without a source, he will have committed plagiarism.

In our second case, Mike transfers a direct quotation from a source onto a note card. He remembers to place source data on the card, but he forgets to include quotation marks. If later he were to incorporate the quotation into his presentation *with* the source citation but *without* quotation marks, he would have plagiarized.

Why? Because he would be parading the exact words of another writer as his own paraphrase. The passage would give the appearance of being his own words that are supported by the ideas of another, when in fact the passage would be a direct quote. Again, remember that the test for plagiarism is not one's intent; it is the result.

The next section shows you how to avoid plagiarism by completing the research process carefully. In particular, it focuses on a methodical process that involves (1) bibliography cards, (2) a rough outline, (3) note cards of three main kinds, (4) a final outline, and (5) drafts.

Following the Research Process

The research process is no different from most other technical tasks. If you carefully attend to this five-step procedure, you can avoid problems at the end.

■ Step 1: Write Complete Bibliography Cards

Using 3″ × 5″ index cards, you should write a complete bibliography card for each source that may help you later. Called a "working bibliography," this batch of cards becomes the foundation for the rest of your research. (Figure 13−7 shows a bibliography card that Tanya Grant might have collected in her research on acid rain.)

Errors made at this stage—in transferring information from sources to your cards—can easily work their way into a final document. In writing your cards, therefore, be sure to take these precautions:

1. *Include all information needed for the final-copy citation in your paper.* Use the exact wording for titles and publication information. Common errors are to leave off articles (*a, an, the*) and to abbreviate words in titles, with the writer thinking

FIGURE 13−7
Sample bibliography card

there will be time later to double-check the original source. In fact, that final check often does not occur, leading to errors in the final citation.

2. *Save space at the bottom of the card for a reminder to yourself about the usefulness of the source.* For example, a notation such as ''includes excellent chapter on water resources'' may help you later as you begin your research. While they are mainly to record source information, bibliography cards can also provide some guidance in the next stage of note taking.

3. *Include information in the exact format as it will appear in your final bibliography, down to the indenting, punctuation, and capitalization.* Tanya Grant's card in Figure 13–7, for example, follows the capitalization guidelines described on pages 434–435 for the author-date system. Again, do not assume that you will have the time later to transcribe every card into another format. That time will not be there. Also, using the same format ensures that you will make sure to take down all information that will be needed later for the source page.

■ *Step 2: Develop a Rough Outline*

This outline is *not* the one you will use to write the first draft. Instead, it is essentially a list of topics in the approximate sequence they will be covered in the paper. It serves to direct your writing of note cards during the next step.

■ *Step 3: Take Careful Notes on Large Note Cards*

Most plagiarism results from sloppy note taking. This important stage requires that you attend to detail and follow a rigorous procedure. The procedure suggested here divides note cards into three types: summary, paraphrase, and quotation. Figure 13–8 gives examples of all three types.

- **Summary cards** are written in your words and reduce a good deal of borrowed information to a few sentences. They are best written by reading a section of source material, looking away from the source, and summarizing the passage in your own words. In this way, you can later use any of this information with confidence in your paper, without worrying about the absence of quotation marks. Of course, a summary card may contain a few quoted passages, but the card's main purpose is to reduce considerable source material to a short summary.

- **Paraphrase cards** include a close rephrasing of material from your sources. Unlike summary cards, which condense a considerable amount of information, paraphrase cards usually include more of the original text. Thus they demand even more attention than summary cards to the problem of plagiarism. Like summary cards, they are best written by looking away from the source for a moment and then rewriting the passage in your own words. You can use a few key words from the passage, but do *not* duplicate exact phrasing or sentence structure. Using the ''look away'' technique will help avoid creating a paraphrase that too closely resembles the original.

- **Quotation cards** include only words taken directly from the source. Your main concern should be the care with which you transfer sentences from source to card—with absolute accuracy. Even include grammatical or spelling errors that

Quotation card records words exactly as they occur in source. Ellipses (. . .) indicate some words are deleted. Hash marks (//) indicate a page change. "Sic" indicates a possible grammar error ("were" instead of "was") present in the source.

Quotation Card

Regeus, pp. 5-6

"For at least the past two millennia, air pollution has been looked upon as a nuisance. As early as A.D. 61, the philosopher Seneca noted Rome's polluted vistas. Almost a thousand years later, the pollution associated with wood burning at Tutbury Castle in Nottingham was considered unendurable by Eleanor of Aquitaine, the wife of King Henry II of England, forcing her to move. Moreover, starting as early as 1273, a series of royal decrees were [sic] issued barring the combustion of coal in London in a futile attempt to address that city's burgeoning air quality problem. Such illustrations . . . //p6 underscore the enduring nature of air pollution as a public concern."

Paraphrase card rephrases quoted passage above, including most of the main ideas.

Paraphrase Card

Regeus, pp. 5-6

Air pollution has plagued mankind for at least 2000 years. For example, Seneca commented on the polluted air of ancient Rome in A.D. 61. Also, Eleanor of Aquitaine, (wife of Henry II) was so bothered by the pollution at Tutbury Castle that she moved out. Although there were royal decrees about coal burning in London, the city continued to have major problems with dirty air.

Summary card presents only a brief overview of quoted passage.

Summary Card

Regeus, pp. 5-6

Air pollution has been a problem for a long time. In fact, it was noted by Seneca in A.D. 61 regarding ancient Rome.

FIGURE 13-8
Sample note cards on same passage

the original source may contain (and, in so doing, use the word *sic*—see Handbook). Also, as shown in Figure 13–8, you should use ellipses (spaced dots) when you leave out words that you deem unnecessary. When using ellipses, however, be sure not to alter the meaning of the passage you are quoting.

■ Step 4: Organize Research in an Outline

With notes cards in hand, you are now ready to render order from chaos—to create an outline that flows from the technical research related in your note cards. (See outline suggestions in chapter 1.) Key your note cards to the outline by placing one or both of these items at the top of the card: (1) the wording of the related topic, as it appears on the outline, or (2) the letter and/or number that exists on the outline for the related topic. Then place your cards in the order that you will use them in the writing of your first draft.

■ Step 5: Write the Draft From the Outline and the Cards

This step poses the greatest challenge. Here you must incorporate borrowed information with your own ideas to create fluid prose. Your goal should be to demonstrate (a) a smooth transition between your ideas and those you have borrowed and (b) absolute clarity about when borrowed ideas and quotations start and end.

SELECTING AND FOLLOWING A DOCUMENTATION SYSTEM

"Documentation" refers to the mechanical system you use to cite sources from which you borrow information. This section first highlights the main approaches to documentation and then gives details about one common approach.

Varieties of Documentation

There are almost as many systems as there are professional organizations. Yet all have the same goal of showing readers the sources from which you gathered information. Here are just a few documentation manuals commonly used in business, industry, and the professions:

> *The Chicago Manual of Style*
>
> *Council of Biology Editors Style Manual*
>
> *MLA Handbook for Writers of Research Papers*
>
> *Publication Manual of the American Psychological Association*
>
> *Style Manual for Engineering Authors and Editors*
>
> *U.S. Government Printing Office Style Manual*

These and other approaches to documentation can be grouped into three main categories, with these features:

1. Footnotes at the bottom of pages, with an alphabetical list of sources at the end of the document
2. Endnotes collected on a page at the end of the document, followed by an alphabetical list of all sources
3. Parenthetical references in the text of the paper, with an alphabetical list of sources at the end of the document

Most professions have adopted some form of the third system—parenthetical documentation. It is the simplest one for writers to use and the easiest one for readers to decipher. Source information is given right within the text, rather than at the bottom of the page or the end of the document. Then a list of works cited occurs at the end of the document. The next section gives guidelines for one such form of parenthetical documentation.

Author-Date System With Pages

Often used in scientific and technical fields, the author-date system closely resembles systems recommended by the Modern Language Association (for the humanities), the American Psychological Association (for behavioral sciences), and the *Chicago Manual of Style* (for a variety of fields). The version described here includes the page number as well as author and date, giving it enough versatility to be used in many professional fields.

Citations Within the Text. Follow these six guidelines for acknowledging borrowed information within the text. Remember that you must note *any* quotations, paraphrases, or summaries that are not considered common knowledge.

1. *Immediately after the borrowed information occurs in the text, place the author's last name, publication date, and page number in parentheses.* As always, your main goal is to consider the reader's needs. EXAMPLE: "Aerial photos made at this site in 1965 revealed geomorphic features indicative of fault activity (Spears 1966, 76–77)." Note that the parenthetical citation is placed *inside* the punctuation for the sentence and that there is no comma between author and year.
2. *To show where borrowed information starts, or just to vary your style, you also have the option of placing the author's name at the beginning of the text passage.* In this case, remove the author's name from the parenthetical citation. EXAMPLE: "As Spears has shown, aerial photos made at the site in 1965 revealed geomorphic features indicative of fault activity (1966, 76–77)."
3. *When a source has no author or editor, use a short form of the title.* EXAMPLE: (Faults today 1981, 12). Make sure that the first word of your short title is also the first word under which the title is alphabetized in the list of references.
4. *For works with two authors, list both.* EXAMPLE: (Hobbs and Smith 1989, 52). For works with three or more authors, list the first author's name followed by *et al.*, Latin for "and others." EXAMPLE: (Munson et al. 1967).
5. *When your references contain two or more works with the same author and with the same date, distinguish them from each other by placing letters after the date.* EXAM-

PLE: (Jones 1987a) versus (Jones 1987b). Then you will make this same distinction in the list of references (Works Cited).

6. *If you need to mention several sources in the same citation, separate them by semicolons.* EXAMPLE: (Barns 1945, 34; Timm 1956, 12).

List of References. Like other forms of parenthetical documentation, the author-date system requires that you place a list of references at the end of the document. This alphabetized listing includes all sources you cited in the document text. However, it does *not* include references you may have consulted but did not cite. View the list as the spot to which readers turn for complete bibliographical information on sources mentioned in the text, in case they want to find the sources themselves in a library. A sample list of references is shown in Figure 13−9. When assembling such a list, follow these basic rules:

1. *Arrange the sources alphabetically according to author.* When there is no author, list a source by its title.

2. *For books, give (a) the author (last name first, followed by initials or by first name and middle initial), (b) year, (c) book title (underlined or in italic with only the first word and proper nouns in initial caps), (d) edition (if any), (e) city of publication (along with state if city is not well known), and (f) publisher.* Note punctuation of examples in Figure 13−9. For additional authors, place names in last-name-last order, after the first author's name.

3. *For articles, give the (a) author, (b) year, (c) article title (no quotation marks, and with only first word and proper nouns in caps), (d) periodical title (underlined or in italic with main words in initial caps), (e) volume number in arabic numbers, and (f) inclusive page numbers on which the article appears.*

Figure 13−9 lists additional details about books and article citations and about other specialized sources.

WRITING RESEARCH ABSTRACTS

The term *abstract* has been used throughout this book to describe the summary component of any technical document. As the first part of the ABC pattern, it gives decision-makers the most important information they need. However, here we use abstract for a narrower purpose. It is a stand-alone summary that provides readers with a capsule version of a piece of research, such as an article or a book. This section (1) describes the two main types of research abstracts, with examples of each, and (2) gives five guidelines for writing research abstracts.

Types of Abstracts

There are two types of abstracts: informational and descriptive. As the definitions below indicate, informational abstracts include more detail than descriptive abstracts:

Book with two or more authors: Note that the second author's name is given in normal order, with a comma between the two authors' names.

Andrews, T. S., and L. Kelly. 1965a. *Geography of central Oregon.* 4th ed. New York: Jones and Caliber Press.

Book by same authors: Note that the blank line indicates another source by same authors. The "b" is used because the same authors published two cited books in the same year. Also, the state abbreviation is used for clarity, since Hiram is a small town.

_____. 1965b. *Geography today.* Hiram, OH.: Pixie Publishers.

Journal article: Note that there are no spaces between the first page number of this journal article and the colon that follows the volume number. Also, there is only a space, but no punctuation, between the journal title and the volume number.

Cranberg, E. V. 1986. Fossils are fun: The life of a geologist in the 1980s. *Geology Issues* 34:233–344.

Article in collection: This entry is for an article that appears in a collection, with a general editor. Note the comma and "ed." (for "edited by") after the collection title. This same format would be used in other cases where you were referring to a piece from a collection, such as a paper in a conference proceedings.

Fandell, C. N., L. Guest, H. M. Smith, and Z. H. Taylor. 1976. Achieving purity in your sampling techniques. In *Geotechnical Engineering Practices,* ed. J. Schwartz, 23–67. Cleveland: Hapsburg Press.

Interview: Refer to yourself in the third person as the "author."

Iris, J. G. 1988. Resident of Summer Hills Subdivision. Interview with the author at site of toxic-waste dump, 23 March.

Newspaper article: This reference includes the day, section, and page number of the article. If no author had been listed, the entry would have begun with the article title.

Mongo, G. P. 1989. Sinkhole psychology: The ground is falling, Chicken Little. *Dayton Gazette,* 12 July, sec. D, 5.

Article from popular magazine: This reference is handled like a journal article except that the date of the particular issue is also included, after the volume number. Note that there is no extra spacing around the parentheses.

Runyon, D. G., and L. P. Goss. 1967. Sinkholes are coming to your area soon. *Timely News* 123(15 July):65–66.

FIGURE 13–9
Sample list of references

Informational Abstract.

- **Format:** This type of abstract includes the major points from the original document.
- **Purpose:** Given their level of detail, informational abstracts give readers enough information to grasp the main findings, conclusions, and recommendations of the original document.
- **Length:** Though longer than descriptive abstracts, informational abstracts are still best kept to one to three paragraphs.
- **Example:** A sentence from such an abstract might read, "The article notes that functional resumes should include a career objective, academic experience, and a list of the applicant's skills." (See corresponding example in definition of a descriptive abstract.)

Descriptive Abstract.

- **Format:** This type of abstract gives only main topics of the document, without supplying supporting details such as findings, conclusions, or recommendations.
- **Purpose:** Given their lack of detail, descriptive abstracts can only help readers decide whether they want to read the original document.
- **Length:** Their lack of detail usually ensures that descriptive abstracts are no more than one paragraph.
- **Example:** A sentence from such an abstract might read, "The article lists the main parts of the functional resume." (See corresponding example in definition of an informational abstract.)

You may wonder when you'll need to write abstracts during your career. First, your boss may ask you to summarize some research, perhaps because he or she lacks your technical background. Second, you may want to collect abstracts as part of your own research project. In either case, you need to write abstracts that accurately reflect the tone and content of the original document.

Assume, for example, that your McDuff supervisor has asked you to read some current research on strategies for negotiating. Later your boss plans to use your abstracts to get an overview of the field and to decide which, if any, of the original full-length documents should be read in full. The examples that follow show both informational and descriptive abstracts of the section of chapter 14 that covers negotiating (pp. 457–462). Note that the informational abstract actually lists the guidelines contained in the chapter, whereas the descriptive abstract notes only that the article includes the guidelines.

Informational Abstract: "Guidelines for Negotiating"

This article suggests that modern negotiations should replace "I win, you lose" thinking with a "we can both win" attitude. To achieve this change, these six main guidelines are prescribed: (1) think long-term, (2) explore many options, (3) find the shared interests, (4) listen carefully, (5) be patient, and (6) DO look back. Although this strategy applies to all types of negotiation, this article focuses on a business context. It includes an extended example that involves establishing an appropriate entry salary for a job applicant in computer systems engineering.

Descriptive Abstract: "Guidelines for Negotiating"
This article describes six main guidelines that apply to all types of negotiations. The emphasis is on strategies to be used in the context of business. All the suggestions in the article support the need for a "we both win" attitude in negotiating, rather than an "I win, you lose" approach.

Guidelines for Writing Research Abstracts

The guidelines given here will help you (1) locate the important information in a document written by you or someone else and (2) present it with clarity and precision in an abstract. In every case, you must present a capsule version of the document in language the reader can understand. The ultimate goal is to save the readers time.

■ Abstracting Guideline 1: Highlight the Main Points

This guideline applies whether you are abstracting a document written by you or someone else. To extract information that will be used in your abstract, follow these steps:

- Find a purpose statement in the first few paragraphs.
- Skim the entire piece quickly, getting a sense of its organization.
- Read the piece more carefully, underlining main points and placing comments in margins.
- Pay special attention to information gained from headings, first sentences of paragraphs, listings, graphics, and beginning and ending sections.

■ Abstracting Guideline 2: Sketch Out an Outline

From the notes and marginal comments gathered in Abstracting Guideline 1, write a brief outline that contains the main points of the piece. If you are dealing with a well-organized piece of writing, it will be an easy task. If you are not, it will be a challenge. Here is an outline for the negotiation section of chapter 14, as abstracted in the previous examples.

Outline for "Guidelines for Negotiating"

Purpose: to provide rules that help readers adopt a "we both win" strategy

  I. Think long term
    A. Focus on building mutual trust
    B. Project the long-term attitude into every part of the negotiation process
  II. Explore many options
    A. Get away from thinking there are only two choices
    B. Put diverse options on the table early in the negotiation
  III. Find the shared interests
    A. Stress points of agreement, rather than conflict
    B. Use mutual concerns to defuse contentious issues

IV. Listen carefully
  A. Ask questions and listen, rather than talk
  B. Use probing questions to move discussion along in salary discussion
    1. Break out of attack/counterattack cycle
    2. Uncover motivations
    3. Expose careless logic and unsupported demands
    4. Move both sides closer to objective standards
 V. Be patient
  A. Avoid the mistakes that come from hasty decisions
  B. Avoid the bad feeling that results when people feel pressured
VI. DO look back
  A. Keep a journal in which you reflect on your negotiations
  B. Analyze the degree to which you followed the previous five guidelines

■ *Abstracting Guideline 3: Begin With a Short Purpose Statement*

Both descriptive and informational abstracts should start with a concise overview sentence. This sentence acquaints the reader with the document's main purpose. Stylistically, it should include an action verb and a clear subject. Here are three options that can be adapted to any abstract:

- The article "Recycle Now!" states that Georgia must intensify its effort to recycle all types of waste.
- In "Recycle Now!" Laurie Hellman claims that Georgia must intensify its effort to recycle all types of waste.
- According to "Recycle Now!" Georgians must intensify their efforts to recycle all types of waste.

■ *Abstracting Guideline 4: Maintain a Fluid Style*

One potential hazard of the abstracting process is that you may produce disjointed, awkward paragraphs. You can reduce the possibility of this stylistic flaw by following these steps:

- Writing in complete sentences, without deleting articles (*a, an, the*)
- Using transitional words and phrases between sentences
- Following the natural logic and flow of the original document itself

■ *Abstracting Guideline 5: Avoid Technical Terms Readers May Not Know*

Another potential hazard is that the abstract writer, in pursuit of brevity, will use terms unfamiliar to the readers of the abstract. This flaw is especially bothersome to readers who do not have access to the original document. As a general rule, use no technical terms that may be unclear to your intended audience. If a term or two are needed, provide a brief definition in the abstract itself.

Note, also, that abstracts that might become separated from the original document should include a bibliographical citation (see the previous examples).

CHAPTER SUMMARY

This chapter highlights the process of conducting technical research and writing about the results. Much of the information is presented within the context of a research project conducted at McDuff.

Before starting your search for information, you need to decide what main question you are trying to answer. Also, think about the types of information you need (secondary and/or primary), the types of sources that would be useful, and the format required for the final document. Once in the library, you have many sources available to you: books, periodicals, newspapers, company directories, general references such as dictionaries and encyclopedias, abstracts, and computerized data bases. Also, you can gather information from primary sources such as questionnaires and interviews.

As you begin to locate sources, follow this five-step research process: (1) write complete bibliography cards, (2) develop a rough outline, (3) take careful notes on large note cards, (4) organize research in an outline, and (5) write the draft from the outline and the cards. For the final paper, choose a documentation system appropriate for your field or organization. Today the preferred approach is to use some form of parenthetical citations, such as the author-date system detailed in this chapter.

Another research skill is writing research abstracts (summaries) of articles, books, or other sources of information. Abstracts can be either descriptive (quite brief) or informational (somewhat more detailed).

ASSIGNMENTS

If your instructor considers it appropriate, use a copy of the Planning Form at the end of the book for completing these assignments.

1. **General Research Paper.** Using a topic approved by your instructor, follow the procedure suggested in this chapter for writing a paper that results from some technical research. Be sure that your topic (1) relates to a technical field in which you have an interest, by virtue of your career or academic experience, and (2) is in a field about which you can find information in nearby libraries.

2. **Research Paper—Your Major Field.** Write a research paper on your major field. Consider some of these questions in arriving at your thesis for the paper: What is the history of your major? What types of jobs do majors in your discipline pursue? Do their job responsibilities change after 5, 10, or 15 years in the field? What kinds of professional organizations exist to support your field?

3. **Research Paper—McDuff.** As a McDuff engineer or scientist, you have been asked to write a research paper for McDuff's upper management. Choose your topic from one of the technical fields listed here. Assume that your readers are gathering information about the topic because they may want to conduct consulting work for companies or government agencies involved in these fields. Focus on advantages and disadvantages associated with the particular technology you choose. Follow the procedure outlined in this chapter.

   - Artificial intelligence
   - Chemical hazards in the home
   - Fiber optics
   - Forestry management
   - Geothermal energy
   - Human-powered vehicles
   - Lignite-coal mining
   - Organic farming
   - Satellite surveying
   - Solar power
   - Wind power

4. **Abstract.**
   Option A. Visit your college library and find a magazine or journal in a technical area, perhaps your major field. Then photocopy a short article (about five pages) that does not already contain a separate abstract or summary at the beginning of the article. Using the guidelines in this chapter, write both an informative and a descriptive abstract for a nontechnical audience. Submit the two abstracts, along with the copy of the article.
   Option B. Follow the instructions in Option A, but use a short article that has been selected or provided by your instructor.

5. **Questionnaire—Analysis.** Using the guidelines in this chapter for questionnaires, point out problems posed by the following questions:
   a. Is the poor economy affecting your opinion about the current Congress?
   b. Do you think the company's severe morale problem is being caused by excessive layoffs?
   c. Was the response of our salespeople both courteous and efficient?
   d. Of all the computer consultants you have used in the last 15 years, which category most accurately reflects your ranking of our firm: (a) the top 5%, (b) the top 10%, (c) the top 25%, (d) the top 50%, or (e) the bottom 50%?
   e. In choosing your next writing consultant, would you consider seeking the advice of a professional association such as the STC or the CPTSC?
   f. Besides the position just filled, how many job openings at your firm have been handled by Dowry Personnel Services?

6. **Questionnaire—Writing.** Design a brief questionnaire to be completed by students on your campus. Select a topic of general interest, such as the special needs of evening students or the level of satisfaction with certain college facilities or services. Administer the questionnaire to at least 20 individuals (in classes, at the student union, in dormitories, etc.). After you analyze the results, write a brief report that summarizes your findings.
   NOTE: Before completing this exercise, make sure that you gain any necessary approvals of college officials, if required.

7. **Interviews—Simulation and Analysis.** Divide into groups of three or four students, as your instructor directs. Two members of the group will take part in a simulated

interview between a placement specialist at your school and a personnel representative from an area company. Assume that the firm might have a number of openings for your school's graduates in the next few years.

As a group, create some questions that would be useful during the simulated interview. Then have the two members perform the role-playing exercise for 15-20 minutes. Finally, as a group, critique the interview according to the suggestions in this chapter and share your findings with the entire class.

8. **Interview—Real-World.** Select a simple research project that would benefit from information gained from an interview. (Your project may or may not be associated with a written assignment in this course.) Using the suggestions in this chapter, conduct the interview with the appropriate official.

9. **Library Search.** Use your library skills to find the answers to the following questions. Be sure that you can explain exactly how you found the answers.

   a. Largest corporate employer in your state
   b. Number of companies on the New York Stock Exchange
   c. Number of doctors in the American Medical Association
   d. Largest university in New Zealand
   e. Average pay of secondary schoolteachers in your state last year
   f. Percentage of children below the poverty line in the United States
   g. Year that *Huckleberry Finn* was first published
   h. Last names of the nine justices of the U.S. Supreme Court
   i. Average starting salary of chemical engineers in 1991
   j. Number of nuclear power plants in operation
   k. Address of the STC (Society for Technical Communication)
   l. Contrast of ABS and Kevlar, two materials for constructing canoes
   m. Current head of the USGS (United States Geological Survey)
   n. Median age of a new mother in 1991
   o. Month and year that the Berlin Wall came down
   p. Last 10 presidents of the United States
   q. Number of members in the president's cabinet
   r. Greek island that is largest in land mass
   s. Number of novels published by Isaac Asimov
   t. Year in which Fidel Castro came to power in Cuba
   u. Names of two "grammar-checker" computer packages
   v. Brief biography of Gunning (of *Gunning Fog Index* fame)
   w. Last five locations of the summer Olympics
   x. Three most popular dog breeds in a recent year of your choice
   y. Brief description of what a micropaleontologist does
   z. Name of the organization in Utah that supplies genealogical information

14 | *The Job Search*

The job search tests *all* your communication skills: researching career opportunities, writing letters and resumes, interviewing with potential employers, and negotiating for a contract.

*T*his chapter offers suggestions for all major steps for landing a job in your profession. You'll find information on these main activities:

- Researching occupations and companies
- Writing job letters and resumes
- Succeeding in job interviews
- Negotiating on the job

RESEARCHING OCCUPATIONS AND COMPANIES

Before writing a job letter and resume, you may need information about (1) career fields that interest you (if you have not already chosen one) and (2) specific companies that hire graduates in your field. Here are some pointers for finding both types of information:

■ *Do Basic Research in Your College Library or Placement Office*

Libraries and placement centers offer the best starting point for getting information about professions. Following are a few well-known handbooks and bibliographies found in reference collections. They either give information about occupations or provide names of other books that supply such information:

Career Choices Encyclopedia: Guide to Entry-Level Jobs

Dictionary of Occupational Titles

Directory of Career Training and Development Programs

Encyclopedia of Business Information Sources

Encyclopedia of Careers and Vocational Guidance

High-Technology Careers

Occupational Outlook Handbook

Professional Careers Sourcebook: An Information Guide for Career Planning

■ *Interview Someone in Your Field of Interest*

To get the most current information, arrange an interview with someone working in an occupation that interests you. This abundant source of information often goes untapped by college students, who mistakenly think such interviews are difficult to arrange. In fact, usually you can locate people to interview through your college placement office or through your own network of family and friends. Another possibility is to call a reputable firm in the field and explain that you wish to interview someone in a certain occupation. Make it clear, however, that you are not looking for a job—only information about a profession.

Once you set up the interview, prepare well by listing your questions in a notebook or on a clipboard that you take with you to the interview. This preparation will keep you on track and show persons being interviewed that you value their time and information. Here are some questions to ask:

■ How did you prepare for the career or position you now have?
■ What college course work or other training was most useful?
■ What types of activities fill your typical working day?
■ What features of your career do you like the most? The least?
■ What personality characteristics are most useful to someone in your career?
■ How would you describe the long-term outlook of your field?
■ How do you expect your career to develop in the next 5 years, 10 years, or 15 years?
■ Do you know any books or periodicals that might help me find out more about your field?
■ Do you know any individuals who, like you, might permit themselves to be interviewed about their choice of a profession?

■ *Find Information on Companies in Your Field*

With a profession selected, you need to begin screening companies that employ people in the field you have chosen. First, determine the types of information you want to find. Examples might include location, net worth, number of employees, number of workers in your specific field, number of divisions, types of products or services, financial rating, and names and titles of company officers. The following are some sources that might include such information. They can be found in the reference sections of libraries and in college placement centers:

Business Bankings Annual contains ranked lists related to business and industry, such as "Largest Data Communication Companies."

The Career Guide gives overviews of many American companies and includes information such as types of employees hired, training opportunities, and fringe benefits.

Corporate Technology Directory profiles high-tech firms and covers topics such as sales figures, number of employees, locations, and names of executives.

CPC Annual: A Guide to Employment Opportunities for College Students includes four volumes that give general information on career planning (Vol. I) and specific data about work in nontechnical fields (Vol. II), technical fields (Vol. III), and health care (Vol. IV).

Dun's Million Dollar Directory: America's Leading Public and Private Companies lists information about 160,000 U.S. businesses with net worth over $500,000.

Facts on File Directory of Major Public Corporations gives essential information on 5700 of the largest U.S. companies listed on major stock exchanges.

Peterson's Business and Management Jobs provides background on employers of business, management, and liberal-arts graduates.

Peterson's Engineering, Science and Computer Jobs provides background information on employers of technical graduates.

Standard and Poor's Register of Corporations, Directors, and Executives lists names and titles of officials at 55,000 public and private U.S. corporations.

■ *Do Intensive Research on a Selected List of Potential Employers*

The previous steps will help get you started finding information on occupations and firms. Ultimately, you will develop a selected list of firms that interest you. Your research may have led you to these companies, or your college placement office may have told you that openings exist there. Now you need to conduct an intensive search to learn as much as you can about the firms. Here are a few sources of information, along with the kinds of questions each source will help to answer.

- **Annual reports** (often available in your library or placement office):
  How does the firm describe its year's activities to stockholders?
- **Media or press kits** (available from public relations offices):
  How does the firm portray itself to the public?
- **Personnel manuals and other policy guidelines:**
  What are actual features of the firm's "corporate culture"?
- **Graduates of your college or university now working for the firm:**
  What sort of reputation does your school have among decision-makers at the firm?
- **Company newsletters and in-house magazines:**
  How open and informative is the firm's internal communication?
- **Business sections of newspapers and magazines:**
  What kind of news gets generated about the firm?
- **Professional organizations or associations:**
  Is the firm active within its profession?

- **Stock reports:**
  Is the firm having a good year financially?
- **Accrediting agencies or organizations:**
  How has the firm fared during peer evaluations?
- **Former employees of the company:**
  Why have people left the firm?
- **Current employees of the company:**
  What do employees like, or dislike, about the company?

In other words, you should thoroughly examine an organization from the outside. The information you gather will help you decide where to apply and, if you later receive a job offer, where to begin or continue your career.

JOB CORRESPONDENCE

Job letters present a special challenge in your career. You must attract the attention of a reader who may spend only 20–30 seconds deciding whether your letter and resume deserve further consideration. This section gives you the tools to write a successful letter and resume. "Successful," of course, means a letter and resume that will get you an interview. After that, your interpersonal skills will help you land the job. Remember: The letter and resume aim to get you only to the next step—the personal interview.

This section assumes you already have a good idea of the kind of job you want and the preparation needed to get it. It also assumes you are in a competitive market, that your potential employer can choose from a number of people with qualifications similar to yours. In that context, your letter and resume really count and must call attention to you.

Job Letters

In preparing to write a job letter, first try to take the point of view of the persons to whom you are writing. What criteria will they use to evaluate your credentials? How much or how little do they want in the letter? What main points will they be hunting for as they scan your resume? Accordingly, this section first examines the needs of these readers and then gives guidelines for you, the writer. Models 14–1 and 14–2 on pages 464–467 include sample job letters and resumes.

The Readers' Needs.   You probably will not know personally the readers of your job letter, so you must think hard about what they may want. Your task is complicated by the fact that often there are several readers of your letter and resume, who may have quite different backgrounds.

Here is one possible scenario: The letter may go first to the personnel office, where a staff member specializing in employment selects letters and resumes that meet the criteria stated in the position announcement. (In some large employers, letters and resumes may even be stored in a computer where they are scanned for key words that relate to specific jobs.) Applications that pass this screening are sent to the department manager who will supervise the employee that is hired. This manager will interview applicants and ultimately hire the employee. The manager may then select a group to be interviewed. One variation of this process has the personnel department doing an initial interview as well as a screening of the letters and resumes—before the department manager even hears about any applications.

This dual audience—personnel staff members and the manager who will hire you—complicates the business of writing a job letter and resume. Yet most readers, whatever their professional background, usually have these five characteristics in common:

■ *Feature 1: They Read Job Letters in Stacks*

Most search-and-screen processes are such that letters get filed until there are many to evaluate. Your reader faces this intimidating pile of paper, from which you want your letter to emerge as victor.

■ *Feature 2: They Are Tired*

Some employment specialists may save job letters for their fresher moments, but many people who do the hiring get to job letters at the end of a busy day or at home in the evening. So they have even less patience than usual for flowery wording or hard-to-read print.

■ *Feature 3: They Are Impatient*

Your readers expect major points to jump right out at them. In most cases, they will not dig for information that cannot be found quickly.

■ *Feature 4: They Become Picky Grammarians*

Readers of all professional and academic backgrounds expect good writing when they read job letters. There is an unspoken assumption that a letter asking for a chance at a career should reflect solid use of the language. Furthermore, it should have no typographical errors. If the letter does contain a typo or grammar error, the reader may wonder about the quality of writing you will produce on the job.

■ *Feature 5: They Want Attention-Grabbers but Not Slickness*

You walk a fine line in deciding on matters such as format, paper, and fonts. That fine line separates what readers consider a professional appearance, on the one hand, from a showy, inappropriate performance, on the other. Of course, exact likes and dislikes about format vary from field to field and from person to person. An advertising director, who works all day with graphics, would probably want a bolder format design than an engineering manager, who works with documents that are less flashy. If you cannot decide, it is best to err in the direction of a conservative format and style.

The Letter's Organization.   The job-letter guidelines that follow relate to the features mentioned about readers. Remember that your one and only goal with the letter and accompanying resume is to tantalize the reader enough to want to interview you. That is all. With that goal and the reader's needs in mind, your job letter should follow this ABC format:

ABC Format: Job Letters
Abstract

- Apply for a specific job
- Refer to ad, mutual friend, or other source of information about the job
- (Optional) Briefly state how you can meet the main need of your potential employer

Body

- Specify your understanding of the reader's main needs
- Provide main qualifications that satisfy these needs (but only *highlight* points from resume—do NOT simply repeat all resume information)
- Avoid mentioning weak points or deficiencies
- Keep body paragraphs to six or fewer lines
- Use a bulleted or numbered list if it helps draw attention to three or four main points
- Maintain the "you" attitude throughout (see Chapter 7, p. 176)

Conclusion

- Tie the letter together with one main theme or selling point, as you would a sales letter
- Refer to your resume
- Explain how and when the reader can contact you for an interview

This pattern gives you a starting point, but it is not the whole story. There is one feature of application letters that cannot be placed easily in a formula about writing letters. That feature is style. You need to work hard with your draft to develop a smoothness and a facility of thought that, by itself, will set you apart from the crowd. Your attention-grabber will engage interest. But the clarity of your prose will keep readers attentive and persuade them that you are an applicant who should be interviewed. And, of course, you need to do all of this on *one page*.

Resumes

Resumes almost always accompany application letters. Three points make writing resumes a challenge:

1. **Emphasis:** You should select just a *few major points of emphasis* from your personal and professional life. Avoid the tendency to include college and employment details best left for the interview.
2. **Length:** You usually should use only *one page*. For those with extensive experience, a two-page resume is acceptable—if it is arranged evenly over both pages.

3. **Arrangement:** You should arrange information so that it is *pleasing to the eye and easy to scan.* (Prospective employers spend less than a minute assessing your application.) You may even want to include an appropriate, simple illustration. (See the innovative format of Model 14−7 on pages 474−475.)

There is no easy formula for writing excellent resumes. Stylistic preferences vary greatly. This section distills the best qualities of many formats into three basic patterns: (1) the chronological resume, which emphasizes employment history, (2) the functional resume, which emphasizes the skills you have developed, and (3) the combined resume, which merges features of both the chronological and functional formats. See the "Experience" section that follows to learn when to use each format. Choose the pattern that best demonstrates your strengths.

The following paragraphs describe the main parts of the resume. The "Experience" section explains the differences between chronological, functional, and combined resumes. Refer to the following models on pages 464−476 for resume examples:

Model 14−1: Job Letter and Chronological Resume

Model 14−2: Job Letter and Chronological Resume

Model 14−3: Job Letter and Functional Resume

Model 14−4: Job Letter and Functional Resume

Model 14−5: Combined Resume

Model 14−6: Combined Resume

Model 14−7: Resume with Innovative Format

Model 14−8: Resume with Innovative Format

Objective.   Personnel directors and other people in the employment cycle often sort resumes by the "Objective" statement. Writing a good one is hard work, especially for new graduates, who often just want a chance to start working at a firm at any level. Despite this eagerness to please, do not make the mistake of writing an all-encompassing statement such as "Seeking challenging position in innovative firm in civil-engineering field." Your reader will find such a general statement of little use in sorting your application. It gives the impression that you have not set clear professional goals.

Most objectives should be short, preferably one sentence. Also, they should be detailed enough to show that you have prepared for, and are interested in, a specific career, yet open-ended enough to reflect a degree of flexibility. If you have several quite different career options, you might want to design a different resume for each job description, rather than trying to write a job objective that takes in too much territory. Word processing allows you to tailor resume objectives to the particular employer to whom you are writing.

Education.   Whether you choose to follow the objective with the "Education" or "Experience" section depends on the answer to one main question: Which topic is most important to the reader? Most recent college graduates lead off with "Education," particularly if the completion of the degree prompted the job search.

This section seems simple at the outset. Obligatory information includes your school, school location, degree, and date of graduation. It is what you include beyond the bare details, however, that most interests prospective employers. Here are some possibilities:

- **Grade point average:** Include it if you are proud of it; do not if it fails to help your case.
- **Honors:** List anything that sets you apart from the crowd—such as dean's list or individual awards in your major department. If you have many, include a separate "Recognitions" heading toward the end of the resume.
- **Minors:** Highlight any minors or degree options, whether they are inside or outside your major field. Employers place value on this specialized training, even if (and sometimes *especially* if) it is outside your major field.
- **Key courses:** When there is room, provide a short list of courses you consider most appropriate for the kind of position you are seeking. Because the employer probably will not look at your transcripts until a later stage of the hiring process, use this brief listing as an attention-grabber.

Experience.   This section poses a problem for many applicants just graduating from college. Students often comment that experience is what they are *looking* for, *not* what they have. Depending on the amount of work experience you have gained, consider three options for completing this section of the resume: (1) emphasize specific positions you have held (chronological resume), (2) emphasize specific skills you have developed in your experience (functional resume), or (3) emphasize both experience and skills (combined resume).

Option 1: Chronological Format.   This option works best if your job experience has led logically toward the job you now seek. Follow these guidelines:

- List relevant full-time or part-time experience, including co-op work, in *reverse* chronological order.
- Be specific about your job responsibilities, while still being brief.
- Be selective if you have had more jobs than can fit on a one-page resume.
- Include nonprofessional tasks (such as working on the campus custodial staff) *if* it will help your case (for example, the employer might want to know that you worked your way through college).
- Remember that if you leave out some jobs, the interview will give you the chance to elaborate upon your work experience.
- Select a readable format with appropriate white space and good use of print options.
- Use action verb phrases and lists to emphasize what you did or what you learned at these jobs—for example, "Provided telephone support to users of System/23." Use parallel form in each list.

Option 2: Functional Format.   This approach works best if (1) you wish to emphasize the skills and strengths you have developed in your career, rather than specific jobs you have had, or (2) you have had "gaps" in your work history, which would be obvious if you used the chronological format. Although it is sometimes used by those whose job experience is not a selling point, this is not always the case. Sometimes your skills built up over time may be the best argument for your being considered for a position, even if your job experience also is strong. For example, you may have five years' experience in responsible positions at four different retailers. You then decide to write a functional resume focused on the three skill areas you developed: sales, inventory control, and management.

If you write a functional resume that stresses skills, you may still want to follow this section with a brief employment history (see Option 3). Most potential employers want to know where and when you worked, even though this issue is not a high priority. If you decide to leave out the history, at least be sure to bring it with you to the interview on a separate sheet.

Option 3: Combined Format.   The combined format uses features of both the chronological and the functional formats. This format works best when you want to emphasize the skills you have developed, while still giving limited information on the chronology of your employment.

Models 14–5 and 14–6 on pages 472–473 show two variations of the combined format. In Model 14–5, the experience section looks exactly as it would in a functional resume, with subheadings giving the names of skills. However, the writer adds a brief skeleton work history near the end of the page; he believes the reader will want some sort of chronological work history, even if it is not the writer's strength. Model 14–6 integrates chronological information into the skills section. The positions held may not be prestigious, but together they show that the applicant has considerable experience developing the two sets of skills listed: Editing/Writing and Teaching/Research.

Activities, Recognitions, Interests.   Most resumes use one or two of these headings to provide the reader with additional background information. The choice of which ones, if any, to use depends on what you think will best support your job objective. Here are some possibilities:

- **Activities:** selected items that show your involvement in your college or your community or both.
- **Recognitions:** awards and other specific honors that set you apart from other applicants. (Do not include awards that might appear obscure, meaningless, or dated to the reader, such as most high-school honors.)
- **Interests:** hobbies or other interests that give the reader a brief look at the "other" you.

However you handle these sections, they should be fairly brief and should not detract from the longer, more significant sections described previously.

References.   Your resume opens the door to the job interview and later stages of the job process, when references will be called. There are two main approaches to the reference section of the resume:

1. Writing "Available upon request" at the end of the page
2. Listing names, addresses, and phone numbers at the end of the resume

The first approach assumes that the reader prefers the intermediate step of contacting you before references are sent or solicited. The second approach assumes that the reader prefers to call or write references directly, without having to contact you first. Use the format most commonly used in your field or, most important, the one most likely to meet the needs of a particular employer. As always, be ready to tailor your letter and resume each time you put it in the mail.

Your goal is to write an honest resume that emphasizes your good points and minimizes your deficiencies. To repeat a point made at the outset, you want your resume and job letter to open the door for later stages of the application process. Look upon this writing task as your greatest persuasive challenge. Indeed, it is the ultimate sales letter, for what you are selling is the potential you offer to change an organization and, perhaps, the world as well. Considering such heady possibilities, make sure to spend the time necessary to produce first-rate results.

JOB INTERVIEWS

Your job letter and resume have only one purpose: to secure a personal interview by the personnel director or other official who screens applicants for a position. Much has been written about job interviews. Fortunately, most of the good advice about interviewing goes back to just plain common sense about dealing with people. Following are some suggestions to show you how to prepare for a job interview, perform at your best, and send a follow-up letter.

Preparation

■ *Do Your Homework on the Organization*

You have learned how to locate data about specific companies. Once you have been selected for an interview, review whatever information you have already gathered about the employer. Then go one step further by searching for the *most current* information you can find. Your last source may be someone you know at the organization, or a "friend of a friend."

When you don't have personal contacts, use your research skills again. For large firms, locate recent periodical or newspaper articles by consulting general indexes—such as the *Business Periodicals Index, Wall Street Journal Index, Readers' Guide to Periodicals, New York Times Index*, or the index for any newspaper in a large metropolitan area. For smaller firms, consult recent issues of local newspapers for announcements about the company. Being aware of current company issues will demonstrate your initiative and show your interest in the firm.

■ *Write Out Answers to the Questions You Consider Likely*

You probably would not take written answers with you to the interview. But writing them out will give you a level of confidence unmatched by candidates who only ponder possible questions that might come their way. This technique resembles the manner in which some people prepare for oral presentations: First they write out a speech, then they commit it to notes, and finally they give an extemporaneous presentation that reflects confidence in themselves and knowledge of the material. This degree of preparation will place you ahead of the competition.

There are few, if any, original questions asked in job interviews. Most interviewers simply select from some standard questions to help them find out more about you and your background. Here are some typical questions, along with tips for responses in parentheses:

1. **Tell me a little about yourself.** (Keep your answer brief and relate it to the position and company—do *not* wander off into unrelated issues, like hobbies, unless asked to do so.)

2. **Why did you choose your college or university?** (Be sure your main reason relates to academics—for example, the academic standing of the department, the reputation of the faculty, or the job placement statistics in your field.)

3. **What are your strengths?** (Focus on two or three qualities that would directly or indirectly lead to success in the position for which you are applying.)

4. **What are your weaknesses?** (Choose weaknesses that, if viewed from another perspective, could be considered strengths—for example, your perfectionism or overattention to detail.)

5. **Why do you think you would fit into this company?** (Using your research on the firm, cite several points about the company that correspond to your own professional interests—for example, the firm may offer services in three fields that relate to your academic or work experience.)

6. **What jobs have you held?** (Use this question as a way to show that each previous position, no matter how modest, has helped prepare you for this position—for example, part-time employment in a fast-food restaurant developed teamwork and interpersonal skills.)

7. **What are your long-term goals?** (Be ready to give a 5- or 10-year plan that, preferably, fits within the corporate goals and structure of the firm to which you are applying—for example, you may want to move from the position of technical field engineer into the role of a project manager, to develop your management skills.)

8. **What salary range are you considering?** (Avoid discussing salary if you can. Instead, note that you are most interested in criteria such as job satisfaction and professional growth. If pushed, give a salary range that is in line with the research you did on the career field in general and this company in particular; see the last section of this chapter on negotiating.)

9. **Do you have any questions of me?** (*Always* be ready with questions that reinforce your interest in the organization and your knowledge of the position—for example, "Given the recent opening of your Tucson warehouse, do you plan other expansions in the Southwest?" or "What types of in-house or off-site training do you offer new engineers who are moving toward project management?" Other questions can concern issues such as (a) benefits, (b) promotions, (c) availability of personal computers, and (d) travel requirements.)

■ *Do a Dress Rehearsal*

You can improve your chances considerably by practicing for your job interview. One of the easiest and best techniques is role-playing. Ask a friend to serve as the interviewer, and give him or her a list of questions from which to choose. Also, inform that person about the company so that he or she can improvise during the session. In this way you will be better prepared for the real thing.

You can get additional information about your interviewing abilities by videotaping your role-playing session. Reviewing the videotape will help you highlight (1) questions that pose special problems for you and for which you need further preparation and (2) mannerisms that need correction.

■ *Be Physically Prepared for the Interview*

Like oral presentations, job interviews work best when you are physically at your best. Thus all the old standbys apply:

- Get a good night's rest before the interview.
- Avoid caffeine or other stimulants.
- Eat about an hour beforehand so that you are not distracted by hunger pangs during the session.
- Take a brisk walk to dispel nervous energy.

Performance

Good planning is your best assurance of a successful interview. Of course, there are always surprises that may catch you. Remember, however, that most interviewers are seriously interested in your application and want you to succeed. Help them by selling *yourself* and thus giving them a reason to hire you. Here are some guidelines for the interview.

■ *Dress Appropriately*

Much has been written on the topic of appropriate attire for interviews. Here are some practical suggestions that are often emphasized:

- Dress conservatively and thus avoid drawing attention to your dress—for example, do not use the interview as an opportunity to break in a garment in the newest style.
- Consider the organization—for example, a brokerage-firm interview may require a dark suit for a man and a tailored suit for a woman, whereas an interview at a construction firm may require less formal attire.
- Avoid excessive jewelry.
- Pay attention to the fine points—for example, wear shined shoes and carry a tasteful briefcase or notebook.

■ *Take an Assertive Approach*

Either directly or indirectly, use everything you say to make the case for your hiring. Be positive, direct, and unflappable. Use every question as a springboard to show your capabilities and interest, rather than waiting for pointblank questions about your qualifications. To be sure, the degree to which you assert yourself partly depends on your interpretations of the interviewer's preference and style. Although you do not want to appear "pushy," you should take the right opportunities to sell yourself and your abilities.

■ Use the First Few Minutes to Set the Tone

What you have heard about first impressions is true: Interviewers draw conclusions quickly. Having given many interviews, they are looking for an applicant who injects vitality into the interview and makes their job easier. Within a minute or two, establish the themes and the tone that will be reinforced throughout the conversation—that is, your relevant background, your promising future, and your eagerness (*not* pushiness). In this sense, the interview subscribes to the Preacher's Maxim mentioned in chapter 12: "First you tell 'em what you're going to tell 'em, then you tell 'em, and then you tell 'em what you told 'em."

■ Maintain Eye Contact While You Speak

Although you may want to look away occasionally, much of the time your eyes should remain fixed on the person interviewing you. In this way you show interest in what she or he is saying.

If you are being interviewed by several people, make eye contact with *all* of them throughout the interview. No one should feel ignored. You are never quite certain exactly who may be the decision-maker in your case.

■ Be Specific in the Body of the Interview

In every question you should see the opportunity to say something specific about you and your background. For example, rather than simply stating that your degree program in computer science prepared you for the open position, cite three specific courses and briefly summarize their relevance to the job.

■ Do Not Hesitate

A job interview is no time to hesitate, unless you are convinced the job is not for you. If the interviewer notes that the position involves 40 percent travel, quickly respond that the prospect of working around the country excites you. The question is this: Do you want the job or not? If you do, then accept the requirements of the position and show excitement about the possibilities. You can always turn down the job if you receive an offer and decide later that some restrictions, like travel, are too demanding.

■ Reinforce Main Points

The interviewer has no text for the session other than your resume. Therefore, you should drive home main points by injecting short summaries into the conversation. After a five-minute discussion of your recent work experience, take 15 seconds to present a capsule version of relevant employment. Similarly, orchestrate the end of the interview so that you have the chance to summarize your interest in the position and your qualifications. Here is your chance to follow through on the "tell 'em what you told 'em" part of the Preacher's Maxim.

Follow-Up Letters

Follow *every* personal contact with a letter to the person with whom you spoke. Send it within 24 hours of the interview or meeting so that it immediately reinforces the person's recollection of you. This simple strategy gives you a powerful tool for showing interest in a job.

Follow-up letters abide by the same basic letter pattern discussed in chapter 7. In particular, follow these guidelines:

- Write no more than one page.
- Use a short first paragraph to express appreciation for the interview.
- Use the middle paragraph(s) to (a) reinforce a few reasons why you would be the right choice for the position or (b) express interest in something specific about the organization.
- Use a short last paragraph to restate your interest in the job and to provide a hopeful closing.

See chapter 7 for the various formats appropriate for all types of business letters. Here is sample text of a thank-you letter:

Dear Ms. Ferguson:

I enjoyed meeting with you yesterday about the career possibilities at Klub Kola's district headquarters. The growth that you are experiencing makes Klub an especially exciting company to join.

As I mentioned, I believe my marketing background at Seville College has prepared me for the challenge of working in your new Business Development Department. Several courses last semester focused specifically on sales strategies for consumer goods. In addition, my internship this semester has given me the chance to try out marketing strategies in the context of a local firm.

Again, thank you for the chance to learn about your firm's promising future. I remain very interested in joining the Klub Kola team.

Sincerely,

Marcia B. Mahoney

When your audience might appreciate a less formal response, consider writing your interviewer a personal note instead of a typed letter. This sort of note is most appropriate when you plan a short message.

NEGOTIATING

All of us negotiate every day of our lives. Both on the job and in our personal lives, we constantly find ourselves in give-and-take discussions to negotiate issues as diverse as those that follow:

- Major and minor purchases
- Relationships with spouses and friends
- Performance evaluations—with bosses and with subordinates
- Salaries—with those to whom we report and with those who report to us

Because negotiating will become an important part of your career, it receives attention in this final section. After some brief background information, the chapter focuses on six guidelines that will steer you toward successful negotiations—when you are hired and also at other points in your career. The main example used in this section is a salary negotiation for an entry-level position.

How has the art of negotiating changed recently? In the past, the process was often characterized by words like *trickery, intimidation,* and *manipulation.* In this game's lexicon there were "winners" and "losers" and lots of warlike imagery. Participants, seen as battlefield adversaries, took up extreme positions, defended and attacked each other's flanks, finally agreed reluctantly to some middle ground, and then departed wounded and usually uncertain of who had won the battle.

Today, however, the trend is away from this war-zone approach that demands an "I win, you lose" mentality. As a negotiator, you must enter the process searching for common ground for a very practical reason: Long-term relationships are at stake. In later negotiations, you are much more likely to achieve success if the present negotiation helps both parties. This goal—"we both win"—requires a new set of practices at the negotiation table.

Specifically, six guidelines should drive the negotiation process. All of them embody the viewpoint that successful negotiations involve honest communication wherein both parties benefit. Try to weave these six guidelines into the style of negotiating that you develop.

■ *Negotiating Guideline 1: Think Long Term*

Enter every negotiation with a long-term strategy for success. You need to establish and nurture a continuing relationship with the person on the other side of the table. Later dealings might depend on mutual understandings and goodwill that result from your first meeting. First impressions *do* count.

How might such long-term thinking apply to actual contract discussions for jobs, especially for your first position after graduating? If you are fortunate enough to be in demand in the job market, you will have the leverage to discuss salary expectations and other benefits during an interview. Such discussions often are characterized by you and the employer sharing details about your expectations and the employer's offer. You should enter such sessions with a realistic idea of what you can command in the marketplace. Neither sell yourself short nor harbor inflated ideas of your worth. Your college or university placement office should be able to provide information about salary ranges and benefit options for graduates in your field and organizations in your region.

Of course, the "real world" of the job hunt is such that the supply of new talent may overshadow the demand. You may be so glad to receive a good offer that you hesitate to jeopardize it by attempting to negotiate. Yet, ironically, you can damage your long-term interests in an organization by being overly timid before accepting an offer. Even if there is little or no room for salary negotiation, you should engage in a wide-ranging discussion that allows you to explore options for your contract and learn about features of the position. This dialogue helps you learn about the organization. It also gives the employer a healthy respect for your ability to ask serious questions about your career.

Whatever your bargaining position, therefore, take advantage of the opportunity to discuss features of your job and the organization. Questions like those that follow may yield important information for you *and* show your interest in developing a long-term relationship with the employer:

- What philosophy underlies the firm's approach to management?
- What is the general timetable for career advancement?
- Where will your specific job lead?
- What opportunities exist for company-sponsored training?
- How will you be evaluated and how often?

Employers respect applicants who have done enough homework to ask informed questions about the firm's employment practices. Both parties benefit from a frank, detailed discussion. You get what you need to make an informed decision about the firm; and your potential employer can showcase the organization and observe your ability to ask perceptive questions.

■ Negotiating Guideline 2: Explore Many Options

The negotiation process sometimes begins with only two options—your salary objective and the employer's offer—with seemingly little room for movement. You can escape this "either/or" trap by working to explore many options in the early stages of contract negotiation. This technique opens both parties to a variety of possible solutions and keeps the discussion rolling.

For example, assume that McDuff recently decided to add a new computer systems engineer to the staff at the corporate office in Baltimore. As a college senior about to graduate with a degree in computer science, you have applied for the job and have had a good first interview. The next week you are called back for a second interview and are offered a job, with a starting salary of $26,000. You are told this firm offer reflects the standard salary for new engineers with no experience. However, your research suggests that entry-level jobs in your field should pay closer to $29,000 a year, a full $3,000 more than the McDuff offer. While this difference concerns you, you have heard good things about the working environment at McDuff and would like to join the firm.

If you immediately were to state your need for a $29,000 starting salary, the negotiation might be thrown into the "either/or" trap that leaves little room for agreement. Instead, you should keep the conversation going by putting additional options on the table and asking open-ended questions (that is, questions that require more than a "yes" or "no" answer). For example, you could temporarily put aside your salary objective and ask how McDuff arrived at the offer figure. While giving the McDuff representative the chance to get facts on the table, this strategy also gives you opportunities to develop and then offer alternatives other than the two salary figures. The discussion might lead to options like these: (1) starting at $26,000 but moving to a higher figure after a successful 90-day trial period, (2) starting at the $26,000 figure but receiving an enhanced stock-option package upon being hired, or (3) starting at $28,000 but giving up the standard $2,000 moving allowance offered to entry-level employees.

The point is that you must be careful to avoid rigidity. Consider possibilities other than the two ideal goals both parties brought to the negotiation table.

■ Negotiating Guideline 3: Find the Shared Interests

If you succeed in keeping options open during the negotiation process, you inevitably begin to discover points on which you agree. Psychologically, it is to your advantage to draw attention to these points rather than to points of conflict. Finding such shared interests helps establish a friendship that, in turn, makes your counterpart more willing to compromise.

Let's go back to the preceding McDuff example. Assume you are continuing to discuss a number of salary options but have reached no agreement. Chances for closure may increase if you temporarily disengage from a discussion of salary and instead search for points, however minor, upon which you do agree. For example, you could ask about the specific job tasks in the position. When you learn that new engineers spend about 25 percent of their workday writing reports, you comment that your college training included two advanced electives in technical and business writing, along with a senior-level research report. The McDuff representative praises the extra effort you made to prepare for the communication tasks in a technical profession.

This discussion about writing, though brief, has highlighted information that may have been missed during McDuff's early reviews of your application. The company's interest in clear, concise writing overlaps with the extra effort you gave to this discipline in your collegiate curriculum. That one shared interest may motivate your counterpart to offer a salary figure closer to what you desire. At the very least, you will have reinforced the decision that McDuff officials made to offer you the job over three other finalists.

■ Negotiating Guideline 4: Listen Carefully

Despite multiple options and shared interests, negotiations often return to basic differences. An effective technique at this point is to seek information on the rationale behind your counterpart's views. It furthers the negotiation and, in fact, your own case to ask questions and then listen carefully to the answers coming from the other side of the table.

How are we helped by asking questions? Returning to the McDuff example, acknowledge it when you are confronted with the salary offer, and then ask how McDuff arrived at that figure. Your questions may uncover what is really behind the offer. Did McDuff recently make similar offers to other applicants? Is McDuff aware of the national salary surveys that tend to support your request? Asking such probing questions benefits both you and the entire negotiation process in four ways:

- You give your counterparts the opportunity to explain their views (thus breaking out of the attack/counterattack cycle).
- You discover what motivates them (making it more likely that you will find an appropriate response and reach consensus).
- You expose careless logic and unsupported demands.
- You move closer to objective standards on which to base negotiations.

From your persistent questioning, careful listening, and occasional responses, information may emerge that would otherwise have remained buried. You may discover, for example, that McDuff is basing its salary offer on data pertaining to another part of the country, where both salaries and costs of living are lower. That would give you the opportunity to argue for a higher starting salary, on the basis of regional differences in compensation.

■ *Negotiation Guideline 5: Be Patient*

In the old hard-sell negotiations, participants frequently pushed for quick decisions, often to the regret of at least one of the parties. The better approach is to slow down the process. For example, you might want to delay agreement on a final salary figure until a later meeting, giving both you and your counterpart the chance to digest the conversation and consider options.

The main benefit of slowing down the process is to prevent basing decisions on the emotionalism of the moment. When objectivity takes a backseat to emotions in any negotiation—with an applicant, a client, a spouse, or a vendor—it is always best to put on the brakes, for two reasons:

- Good negotiated settlements should stand the test of time. When one party feels pressured, mistakes are made.
- Well-thought-out decisions are more likely to produce better long-term relationships, a major goal of your negotiations.

■ *Negotiating Guideline 6: Do Look Back*

Conventional wisdom has it that once you have negotiated an agreement, you should not look back to second-guess yourself, since it will only make you less satisfied with what cannot be changed. That kind of thinking assumes that negotiations are spontaneous phenomena that cannot be analyzed, which is not true. If you have conducted your negotiations methodically, you will have much to gain from postmortems—particularly if they are in writing. Keep a negotiation journal to review before every major negotiation starts. Besides reminders, this journal should contain a short summary of previous negotiations. Make these entries immediately after a session ends, being sure to answer these questions:

- What options were explored before a decision was made?
- What shared interests were discovered?
- Did you emphasize these shared interests?
- What questions did you ask?
- How did you show that you were listening to responses?

So do look back. Analyze every negotiation to discover what went right and what went wrong during the proceedings. Like other communication skills, such as writing and speaking, the ability to negotiate improves with use. With a few basic guidelines in mind and a journal upon which to reflect, you will discover the power of friendly persuasion.

CHAPTER SUMMARY

This chapter surveys the entire process of searching for jobs, from performing your initial research to negotiating a contract. As a first step in the process, use the library and other sources of information to learn about occupations and specific employers that interest you. Second, write letters and resumes that get attention and respond to specific needs of employers. You can choose from chronological, functional, combined, or more innovative resume formats, using the patterns of organization and style that best highlight your background. Third, prepare carefully for your job interview, especially in anticipating the questions that may be asked. Then perform with confidence. And do not forget to send a thank-you letter soon after the interview. Finally, use the negotiation phase of the job-search process to begin building a long-term relationship with your future employer, rather than focusing only on short-term gains of your first contract.

ASSIGNMENTS

1. **Job Letter and Resume.** Search the newspapers or visit your college placement office for a job advertisement that matches your qualifications now or will match them after you complete the academic program upon which you are now working. Write a job letter and resume that respond to the ad. Submit the letter, resume, and written advertisement to your instructor.

   If useful for this assignment and if permitted by your instructor, you may fictionalize part of your resume so that it lists a completed degree program and other experience not yet acquired. In this way, the letter and resume will reflect the background you would have if you were applying for the job. Choose the resume format that best fits your credentials.

2. **Job Interview.** Pair up with another classmate for this assignment. First, exchange the letters, resumes, and job ads referred to in Assignment 1. Discuss the job ads so that you are familiar with the job being sought by your counterpart, and vice versa. Then perform a role-playing exercise during which you act out the two interviews, one person as applicant and the other as interviewer.

   Option: Include a third member in your group. Have this person serve as a recorder, providing an oral critique of each interview at the end of the exercise. Then collaborate among the three of you in producing a written critique of the role-playing exercise. Specifically, explain what the exercise taught you about the main challenges of the job interview.

3. **Follow-Up Letter.** Write a follow-up letter to the interview that resulted from Assignment 2.

4. **Negotiation for Entry-Level Job.** As in Assignment 2, pair up with another student. Assume that the letters, resumes, interviews, and follow-up letters from the preceding assignments have resulted in a second interview for one of you. (That is, select one of the positions, with one of you acting as applicant and the other as interviewer.)

   The topic of this second interview is the position being offered to you. After talking with your team member about the context of this simulated interview, conduct a negotiation session wherein the two of you discuss one or more aspects of the position being offered (salary, benefits, travel schedule, employee orientation, training arrangement, career development, etc.).

5. **Negotiation With McDuff Client.** In this exercise, you and a classmate will simulate a negotiation session between Sharon Gibbon, a McDuff training manager at the Cleveland office, and Bernard Claxton, training director of Cleveland's Mercy Hospital. Study the following details before beginning your 10- or 15-minute discussion.

   Option: Collaborate with your teammate in writing an evaluation of this role-playing exercise. Explain the major obstacles encountered by both Gibbon and Claxton, and describe the techniques attempted by both parties to overcome these obstacles.

   General Background: Gibbon recently submitted a proposal to Claxton, offering to have McDuff conduct three hazardous-waste seminars for the plant staff at Mercy Hospital. Claxton calls Gibbon to say that he wants to go ahead with the seminars, contingent on some final negotiations between the two. Claxton and Gibbon agree to meet in a few days, presumably to iron out a final agreement. Gibbon wants the contract, and Claxton's staff needs the training. Yet the deal won't be sealed until they have their discussion and resolve several issues. Following are their respective points of view.

   Gibbon's Viewpoint: Sharon Gibbon has offered to have McDuff teach five one-day seminars for a fee of $15,000 ($3,000 per seminar). Each seminar will be team-taught by two of McDuff's certified industrial hygienists, Tom Rusher and Susan Sontack. They are expert trainers with much field experience in the identification and safe use of hazardous chemicals and other wastes. The $3,000 course fee is standard for McDuff's hazardous waste seminars, though the company has on rare occasions given 10 percent discounts for any of the same seminars after the first one for the same client. Gibbon is interested in picking up Mercy Hospital as a client, but she also recognizes that the two instructors she has committed for the seminars may be needed for jobs the company has not yet scheduled. She is leery of cutting fees for Mercy Hospital when there may be other full-fee work right around the corner.

   Claxton's Viewpoint. Claxton knows that the hospital staff must have hazardous-waste training to conform to new county and city regulations, and he has heard from other hospitals that McDuff has the best training in the business. Yet he has real problems spending $15,000 on five seminars. Proposals from other firms were in the $9,000 to $12,000 range for the five seminars, and Claxton's training budget is modest. Though the other firms that submitted proposals did not share McDuff's reputation, they too offered team-taught seminars by certified industrial hygienists. Although Claxton would prefer to hire McDuff and although he knows that good training is worth the money, he is hoping to get Gibbon to lower McDuff's fee when they meet for their negotiation session. He knows that one benefit he can offer McDuff is continued training contracts from the hospital, since the high employee turnover will necessitate frequent training in hazardous waste. In addition, there may be other training opportunities for McDuff at the hospital, once Claxton completes his upcoming needs assessment of the staff training program.

201 Edge Drive
Norcross, PA 17001
May 4, 1994

Mr. James Vernon, Personnel Director
McDuff, Inc.
105 Halsey Street
Baltimore, MD 21212

Dear Mr. Vernon:

My academic advisor, Professor Sam Singleton, informed me about an electrical-engineering opening at McDuff where he worked until last year. I am writing to apply for the job.

I understand that McDuff is making a major effort to build a full-scale equipment-development laboratory. That prospect interests me greatly, because of my academic background in electrical engineering technology. At Northern Tech, I took courses in several subjects that might be useful in the lab's work—for example, microprocessor applications, artificial intelligence, and fiber optics.

Also, related work at two firms has given me experience building and developing new electronics systems. In particular, more than two years' work as an assembler taught me the importance of precision and quality control. I'd like the opportunity to apply this knowledge at McDuff.

Personal business will take me to Baltimore June 5-8. Could you meet with me on one of those days to discuss how McDuff might use my skills? Please let me know if an interview would be convenient at that time.

Enclosed is a resume that highlights my credentials. I hope to be talking with you in June.

Sincerely,

Donald Vizano

Donald Vizano

Enclosure: Resume

MODEL 14–1
Job letter and chronological resume

Donald Vizano
201 Edge Drive
Norcross, PA 17001
(300) 555-7861

OBJECTIVE: A full-time position in electrical engineering, with emphasis on designing new equipment in automation and microprocessing

EDUCATION: 1987-1993   Bachelor of Science in Electrical Engineering (June1993)
Northern College of Technology, Shipley, PA
3.5 GPA (out of 4.0 scale)

Major Courses:

Fiber Optics	Artificial Machine Intelligence
Robotic Systems	Communication Control Systems
Microprocessor Control	Microcomputer Applications
Microcomputer Systems	Digital Control Systems
	Semiconductor Circuits & Devices

Related Courses:

BASIC Programming	FORTRAN
Business Communication	Engineering Economy
Industrial Psychology	Technical Communication

ACTIVITIES
AND HONORS: Institute of Electrical and Electronic Engineering (IEEE)
Dean's List, 8 quarters.

EMPLOYMENT:

1989-1993	Electronic Assembler (part-time) Jones Energy & Automation, Inc. Banner, PA
1988-1989	Lab Monitor (part-time) Computer Services Northern College of Technology Shipley, PA
1987-1988	Electronic Assembler (part-time) Jones Energy & Automation, Inc. Banner, PA
1985-1987	Electronic Assembler (full-time) Jones Energy & Automation, Inc. Banner, PA

PERSONAL: Willing to travel, fluent in Spanish

REFERENCES: Available upon request

MODEL 14–1, *continued*

1523 River Lane
Worthville, OH 43804
August 6, 1994

Mr. Willard Yancy
Director, Automotive Systems
XYZ Motor Company, Product Development Division
Charlotte, NC 28202

Dear Mr. Yancy:

Recently I have been researching the leading national companies in automotive computer systems. Your job ad in the July 6 *National Business Employment Weekly* caught my eye because of XYZ's recent innovations in computer controlled safety systems. I would like to apply for the automotive computer engineer job.

Your advertisement notes that experience in computer systems for machinery or robotic systems would be a plus. I have had extensive experience in the military with computer systems, ranging from a digital communications computer to an air traffic control training simulator. In addition, my college experience includes courses in computer engineering that have broadened my experience. I am eager to apply what I have learned to your company.

My mechanical knowledge was gained from growing up on my family's dairy farm. After watching and learning from my father, I learned to repair internal combustion engines, diesel engines, and hydraulic systems. Then for five years I managed the entire dairy operation.

With my training and hands-on experience, I believe I can contribute to your company. Please contact me at 614/882-2731 if you wish to arrange an interview.

Sincerely,

James M. Sistrunk

James M. Sistrunk

Enclosure: Resume

MODEL 14-2
Job letter and chronological resume

James M. Sistrunk
1523 River Lane
Worthville, OH 43804
(614) 882-2731

Professional Objective:

To contribute to the research, design, and development of automotive computer control systems

Education:

B.S., Computer Engineering, 1991-present
Major concentration in Control Systems with minor in Industrial Engineering. Courses included Microcomputer Systems, Digital Control Systems, and several different programming courses.
Columbus College, Columbus, Ohio

Computer Repair Technician Certification Training, 1988-1989
General Computer Systems Option with emphasis on mainframe computers. Student leader in charge of processing and orientation for new students from basic training.
U.S. Air Force Technical Training Center, Keesler Air Force Base, Biloxi, MS.

Career Development:

Computer Repair Technician, U.S. Air Force, 1988-1991
Secret Clearance

Responsibilities and duties included:
- Repair of computer systems
- Documentation of work accomplished
- Preventative maintenance inspections
- Diagnostics and troubleshooting of equipment

Accomplishments include:
- "Excellent" score during skills evaluation
- Award of an Air Force Specialty Code "5" skill level

Assistant Manager, Spring Farm, Wootan, Ohio, 1982-1987
Responsible for dairy operations on this 500-acre farm. Developed the management and technical skills, learned to repair sophisticated farm equipment.

Special Skills:

Macintosh desk-top publishing
IBM - MS DOS
Assembly Language
C++ Programming

References:

Available upon request

MODEL 14–2, *continued*

456 Cantor Way, #245
Gallop, Minnesota 55002
September 3, 1994

Ms. Judith R. Gonzalez
American Hospital Systems
3023 Center Avenue
Randolf, Minnesota 55440

Dear Ms. Gonzalez:

My placement center recently informed me about the Management Trainee opening with Mercy Hospital. As a business major with experience working in hospitals, I wish to apply for the position.

Your job advertisement notes that you seek candidates with a broad academic background in business and an interest in hospital management. At Central State College, I've taken extensive coursework in three major areas in business: finance, marketing, and personnel management. This broad-based academic curriculum has provided a solid foundation for a wide variety of management tasks at Mercy Hospital.

My summer and part-time employment also matches the needs of your position. While attending Central State, I've worked part-time and summers as an assistant in the Business Office at Grady Hospital. That experience has acquainted me with the basics of business management within the context of a mid-sized hospital, much like Mercy.

The enclosed resume highlights the skills that match your Management Trainee opening. I would like the opportunity to talk with you in person and can be reached at 612-111-1111 for an interview.

Sincerely,

Denise Ware Sanborn

Denise Ware Sanborn

MODEL 14–3
Job letter and functional resume

Denise Ware Sanborn
456 Cantor Way, #245
Gallop, Minnesota 55002
612-111-1111

Objective

Entry-level management position in the health care industry. Seek position that includes exposure to a wide variety of management and business-related tasks.

Education

Bachelor of Arts Degree, June 1994
Central State College
Gallop, Minnesota

Major: Business Administration
Grade Point Average: 3.26 of possible 4.0, with 3.56 in all major courses
All college expenses financed by part-time and summer work at Grady Hospital in
St. Paul, Minnesota.

Skills and Experience

Finance
    Helped with research for three fiscal year budgets
    Developed new spreadsheet for monthly budget reports
    Wrote accounts payable correspondence
    Called on collections from insurance companies
Marketing
    Solicited copy from managers for new brochure
    Designed and edited new brochure
    Participated in team visits to ten area physicians
    Wrote copy for one-page flyer
Personnel
    Designed new performance appraisal form for secretarial staff
    Interviewed applicants for Maintenance Department jobs
    Coordinated annual training program for nursing staff

Awards

1993 Arden Award for best senior project in the Business Administration Department (paper that examined latest developments in Total Quality Management)

Dean's list for six semesters.

References

Academic and work references available upon request.

MODEL 14–3, *continued*

2389 Jenson Court
Gulfton, MS 39200
(601) 111-1111
February 17, 1994

Mr. Nigel Pierce, Personnel Director
Structural Systems, Inc.
105 Paisley Way
Jackson, MS 39236

Dear Mr. Pierce:

I am writing in response to your ad for a technical representative in the February 13 (Sunday) edition of the *Jackson Journal*. I believe my experience and education make me an excellent candidate for this position.

I am very familiar with your products for the wood construction market. The laminated beams and floor joists your company manufactures were specified by many of the architects I have worked with during my co-op experience at Mississippi College. Work I have done in the residential and small commercial construction industry convinced me of the advantages of your products over nominal lumber.

Enclosed is my resume, which focuses on the skills gained from my co-op work that would transfer to your firm. I look forward to meeting you and discussing my future with your company.

Sincerely,

Todd L. Fisher

Todd L. Fisher

Enclosure: Resume

MODEL 14–4
Job letter and functional resume

Todd L. Fisher
2389 Jenson Court
Gulfton, MS 39200
(601) 111-1111

PROFESSIONAL OBJECTIVE	Use my education in civil engineering and my construction experience to assume a technical advisory position.
EDUCATION	Mississippi College Hart, Mississippi; Bachelor of Science, Civil Engineering Technology June 1993, GPA: 3.00 (out of 4.00)
PROFESSIONAL EXPERIENCE	Financed education by working as co-op student for two Jackson construction firms for 18 months.
Design skills	Assisted with the layout and design of wall panels for Ridge Development condominium project.
	Created layout and design for complete roof and floor systems for numerous churches and small commercial projects.
Computer skills	Introduced computerization to the design offices of a major construction company (HP hardware in HPbasic operating system).
	Designed trusses on Sun workstations in the UNIX operating system. Operated as the system administrator for the office.
	Learned DOS operating system and the Windows environment (on IBM hardware and its clones).
Leadership skills	Instructed new CAD (Computer-assisted design) operators on the operation of design software for panel layout and design.
	Designed and implemented management system for tracking jobs in plant.
INTERESTS	Family, gardening, sailing, travel
REFERENCES	References and transcripts available upon request.

MODEL 14–4, *continued*

SUSAN A. MARTIN

PRESENT ADDRESS	**PERMANENT ADDRESS**
540 Wood Drive	30 Avon Place
Bama, CA 90012	Atlas, CA 90000
(901) 666-2222	(901) 555-6074

PROFESSIONAL OBJECTIVE: Analyze and solve problems involving natural and pollution control systems as an Environmental Scientist.

EDUCATION:

Pierce College, Bama, California
Bachelor of Science, Environmental Science
May 1991, GPA: 3.15 (out of 4.00)

Pleasant Valley College, Barnes, Nevada
Associate in Applied Science, Engineering Science
May 1989, GPA: 3.15 (out of 4.00)

PROFESSIONAL EXPERIENCE:

Research Skills:
- Worked as lab assistant in a research project to analyze the effect of acid rain on frog reproduction in Lake Lane.
- Designed Pierce College computer program to analyze data on ozone depletion.

Leadership Skills:
- Taught inventory procedures to new employees of Zane's Office Supply.
- Helped incoming freshmen and transfer students adjust to Pierce College (as dormitory resident assistant).

Organizational Skills:
- Maintained academic department files as student assistant in Environmental Science Department
- Organized field trips for Pierce College Mountaineering Club.

HONORS AND ACTIVITIES: Dean's List (five semesters)
President of Cycling Club

INTERESTS: Photography, camping, biking, traveling

EMPLOYMENT HISTORY: *Dormitory Resident Assistant*, Pierce College, Bama, CA, 1990–1991
Trainer, Zane's Office Supply, Bama, CA, 1989–1990

REFERENCES: References and transcripts available upon request.

MODEL 14–5
Combined resume

Karen S. Patel
300 Park Drive
Burtingdale, New York 20092

Home: (210) 400-2112   **Messages:** (210) 400-0111

OBJECTIVE	Position as in-house technical writer and as trainer in communication skills
EDUCATION	**Sumpter College, Marist, Vermont** Master of Science in Technical Communication, GPA: 4.0 December 1993
	Warren College, Aurora, New York M.A. in English, Cum Laude, June 1990
	University of Bombay, India B.A. in English, First Class Honors, June 1987
EMPLOYMENT *Editing/* *Writing*	**Public Relations Office, Sumpter College, 1993–present** Administrative Assistant: Write press releases and conduct interviews. Publish new stories in local newspapers and in *Sumpter Express*. Edit daily campus newsletter.
	Hawk Newspapers, Albany, New York, 1988–1989 Wasseu College Internship: Covered and reported special events; conducted interviews; assisted with proofreading, layout, headline count. Scanned newspapers for current events; conducted research for stories. Published feature stories.
Teaching/ *Research*	**Sumpter College, Marist, Vermont, 1992–1993** Teaching assistant: Tutored English at the Writing Center, answered "Grammar Hotline" phone questions, edited and critiqued student papers, taught English to non-English speakers and helped students prepare for Regents exams.
	Warren College, Aurora, New York, 1989–1990 Teaching assistant: Taught business writing, supervised peer editing and in-class discussions, held student conferences, and graded student papers.
	Research Assistant: Verified material by checking facts, wrote brief reports related to research, researched information and bibliographies.
COMPUTER SKILLS	Wordperfect, Microsoft Word, Pagemaker, Unix, Excel
REFERENCES	Available upon request

MODEL 14–6
Combined resume

EXPERIENCE

12/91 to Present
DataCorp, Atlanta, Georgia
Administrative/Document Manager
Responsibilities consist of creating, editing and organizing marketing and sales documents, manuals and proposals. Also responsible for hiring and supervising sales support staff.

3/91 – 12/91
Sunvie Corporation, Atlanta, Georgia
Director of Client Support Services
Responsibilities included formatting, editing and final production of various reports, graphics, and publications developed from research data. Sourced and indexed information for a reference library. Managed office activities and functions.

12/89 – 2/91
ComKing, Atlanta, Georgia
National Accounts Coordinator
Set up account-tracking system. Responsible for initiating all paperwork involved with opening a new account and the follow up, including invoicing. Handled client requests. Proofed and edited print materials before being sent to press. Researched new markets. Developed charts and data for presentations.

9/89 – 12/89
Self-Employed Consultant
Editorial Assistant/Administrative Assistant
Proofed and edited materials for brochures and booklets.

EXPERIENCE

8/88 – 8/89
Lidgate Press Ltd, Craftrends Magazine,
Norcross, Georgia
Advertising Coordinator
Responsibilities included calling advertisers monthly for ad material. Generated reader service and sales reports. Proofed boards and chromalins before they went to the printers. Assisted advertisers in writing their ads. Arranged photography shoots for advertising material. Billed our advertising monthly.

3/84 – 7/88
Capstone Channels, Inc.
Atlanta, Georgia
Marketing Assistant
Coordinated research projects that involved acquisition of new magazines, tabulated in-house and outside studies, and ran competitive analysis and individual research projects for company personnel. Provided reports for various departments and outside firms. Generated a company directory and managed some in-house promotions.

10/80 – 3/84
Genuine Parts, Inc.,
Charlotte, North Carolina
Administrative Assistant
Responsible for maintaining sales records and creating sales charts. Coordinated sales activities and general support duties. Typed manuscripts and correspondence for 16 editors. Assisted promotion director in writing and organizing promotions.

EDUCATION

Currently working on *Master of Science, Technical Communication*
Southern College of Technology, Marietta, Georgia
Estimated completion date–January 1993.

1/79 – 8/80
Bachelor of Arts, Journalism
Univeristy of Georgia, Athens, Georgia
School of Journalism

9/76 – 8/78
Associate in Journalism
Abraham Baldwin Junior College, Tifton, Georgia

SKILLS

Writing, editing, proofreading.
Working knowledge of Windows 3.0, Ventura Windows Version, PageMaker IBM version, Excel 3.0, Charisma 2.1, Omnipage, Scan Gallery, HP Paintbrush, Hijaak Conversion software, Zybuild/ZyFind, Lotus 2.2 & 2.3, Excel, WordPerfect 5.1, Microsoft Word, Multimate, Grammatik, Q & A database, Paradox database, Filepro database, Harvard Graphics 3.0, Charisma, Freelance 3.0, the HP Plotter, HP Laserjet III, IBM ExecJet and the HP ScanJet.

INTERESTS

Writing, computer graphics, desktop publishing, racquetball, guitar, oil painting.

MODEL 14–7
Resume with innovative format

RESUME
of
Becky Dacnell

RESUME

Becky Dacnell
39 Rock Drive
Marietta, Georgia
30062
(404) 000-6000

Leslie Highland
997 Simmons Drive
Boise, Idaho 88822

OBJECTIVE:	A full time position in architectural design with emphasis on model-making and renderings for future buildings.
EDUCATION:	**Boise Architectural College** Boise, Idaho Bachelor of Science Architectural Engineering Technology June 1993
	Harvard University Cambridge, Massachusetts Certificate in Advance Architectural Delineation August 1987
ACTIVITIES AND HONORS:	**Boise Architectural College** Winner of Senior Design Project Architectural Engineering Technology Charter Member of American Society of Architectural Perspectives
EMPLOYMENT: 1987–1993	**Architectural Designer and Delineator** Dorsey-Hudson, Architects Boise, Idaho
1985–1987	**Architectural Designer and Renderer** Windsor and Associates, Architects St. Lake, Utah
1982–1985	**Architectural Renderer and Draftsman** Sanders and Associates, Architects Provo, Utah
1980–1982	**Architectural Draftsman** Brown Engineering St. Lake, Utah
REFERENCES:	References and portfolio available upon request.

MODEL 14–8

15 | *Style in Technical Writing*

Alone with his thoughts and his computer, this McDuff manager puts finishing touches on a report written by his group.

*T*his chapter, as well as the Handbook that follows, focuses on the last stage of the writing process—revising. As you may already have discovered, revision sometimes gets short shrift during the rush to finish documents on time. That's a big mistake. Your writing must be clear, concise, and correct if you expect the reader to pay attention to your message. Toward that end, this chapter offers a few basic guidelines on style. The Handbook contains alphabetized entries on grammar and mechanics.

After defining style and its importance, this chapter offers suggestions for achieving five main stylistic goals:

- Writing clear sentences
- Being concise
- Being accurate in wording
- Using the active voice
- Using nonsexist language

STYLE OVERVIEW

Just as all writers have distinct personalities, they also display distinct features in their writing. Writing style can be defined in this way:

> **Style:** the features of one's writing that show its individuality, separating it from the writing of another. Style results from the conscious and subconscious decisions each writer makes in matters like word choice, word order, sentence length, and active and passive voice. These decisions are different from the "right and wrong" matters of grammar and mechanics (see the Handbook). Instead, they comprise choices writers make in deciding how to transmit ideas to others.

Style is largely a series of personal decisions you make when you write. As noted in chapter 1, however, much writing is being done these days by teams of two or more writers in organizations. Collaborative writing forces individual writers to combine their efforts to produce a consensus style, usually a compromise of stylistic preferences of the individuals involved. Thus, personal style can become absorbed into a jointly produced product.

Similarly, many companies tend to develop a company style that is fairly recognizable in documents like reports and proposals. The reports at McDuff, Inc., for example, tend to be formal, objective, and "scientific" in tone. In the last few years, the company has tried to change its overly technical and scientific style to one that is more informal, readable, and reflective of the speech of average readers. You might very well be hired by an organization that is making this same shift toward a more readable style.

Despite the need to make style conform to group or company guidelines, each individual remains the final arbiter of her or his own style in technical writing. Most of us will be forced to be our own stylists, even in firms in which in-house editors help "clean up" writing errors. This chapter aims to help such writers deal with everyday decisions of sentence arrangement, word choice, and the like. But in seeing style as a personal statement, you should not presume that "anything goes." Certain fundamentals are part of all good technical style in the professional world. Let's take a look at these basics.

WRITING CLEAR SENTENCES

Each writer has his or her own approach to sentence style. Yet every one of us has the same tools with which to work: words, phrases, and clauses. This section first defines some basic terminology in sentence structure. Then it provides some simple stylistic guidelines for writing clear sentences.

Sentence Terms

The most important parts of the sentence are the subject and the verb. The *subject* names the person doing the action or the thing being discussed (*He* completed the study/ The *figure* shows that); the *verb* conveys action or state of being (She *visited* the site/ He *was* the manager).

Whether they are subjects, verbs, or other parts of speech, words are used in two main units: phrases and clauses. A *phrase* lacks a subject or verb or both and it thus must always relate to or modify another part of the sentence (She went *to the office./As project manager,* he had to write the report). A *clause,* on the other hand, has both a subject and a verb. Either it stands by itself as a *main clause* (He talked to the group) or it relies on another part of the sentence for its meaning and is thus a *dependent clause* (After she left the site, she went home.)

Beyond these basic terms for sentence parts, you also should be aware of the four main types of sentences. A *simple sentence* contains one main clause (He completed his work). A *compound sentence* contains two or more main clauses connected by conjunctions (He completed his work, but she stayed at the office to begin another job). A *complex sentence* includes one main clause and at least one dependent clause (After he finished the project, he headed for home). Finally, a *compound-complex sentence* contains at least two main clauses and at least one dependent clause (After they studied the maps, they left the fault line, but they were unable to travel much farther that night).

Guidelines for Sentence Style

Knowing the basic terms of sentence structure makes it easier to apply some stylistic guidelines. Here are a few fundamental ones that should form the underpinnings for all good technical writing. As you review and edit your own writing or that of others, try to put these principles into practice.

■ Guideline 1: Place the Main Point Near the Beginning

One way to satisfy this criterion for good sentence style is to avoid excessive use of the passive voice (see "Using the Active Voice" on pages 488–489). Another way is to avoid lengthy introductory phrases or clauses at the beginnings of sentences. Remember that the reader usually wants the most important information first.

Original:	"After reviewing the growth of the Cleveland office, it was decided by the corporate staff that an additional lab should be constructed at the Cleveland location."
Revision:	"The corporate staff decided to build a new lab in Cleveland after reviewing the growth of the office there."

■ Guideline 2: Focus on One Main Clause in Each Sentence

When you start stringing together too many clauses with "and" or "but," you dilute the meaning of your text. However, an occasional compound or compound-complex sentence is acceptable, just for variety.

Original:	"The McDuff hiring committee planned to interview Jim Steinway today, but bad weather delayed his plane departure, and the committee had to reschedule the interview for tomorrow."

Revision: "The McDuff hiring committee had to change Jim Steinway's interview from today to tomorrow because bad weather delayed his flight."

■ *Guideline 3: Vary Sentence Length but Seek an Average Length of 15–18 Words*

Of course, do not inhibit your writing process by actually counting words while you write. Instead, analyze one of your previous reports to see how you fare. If your sentences run too long, start making an effort to shorten them, such as by making two sentences out of one compound sentence connected by an "and" or a "but."

Yet you should also remember the importance of varying the length of sentences. Such variety keeps your reader's attention engaged. Make an effort to place important points in short, emphatic sentences. Reserve longer sentences for supporting main points.

Original: "Our field trip for the project required that we conduct research on Cumberland Island, a national wilderness area off the Georgia Coast, where we observed a number of species that we had not seen on previous field trips. Armadillos were common in the campgrounds, along with raccoons that were so aggressive that they would come out toward the camp fire for a handout while we were still eating. We saw the wild horses that are fairly common on the island and were introduced there by explorers centuries ago, as well as a few bobcats that were introduced fairly recently in hopes of checking the expanding population of armadillos."

Revision: "Our field trip required that we complete research on Cumberland Island, a wilderness area off the Georgia Coast. There we observed many species we had not seen on previous field trips. Both armadillos and raccoons were common in the campgrounds. Whereas the armadillos were docile, the raccoons were quite aggressive. They would approach the camp fire for a handout while we were still eating. We also encountered Cumberland's famous wild horses, introduced centuries ago by explorers. Another interesting sighting was a pair of bobcats. They were brought to the island recently to check the expanding armadillo population."

BEING CONCISE

Some experts believe that careful attention to conciseness would shorten technical documents by 10 percent to 15 percent. As a result, reports and proposals would take less time to read and cost less to produce. This section on conciseness offers several techniques for reducing verbiage without changing meaning.

■ Guideline 1: Replace Abstract Nouns With Verbs

Concise writing depends more on verbs than it does on nouns. Sentences that contain abstract nouns, especially ones with more than two syllables, can be shortened by focusing on strong verbs instead. By converting abstract nouns to action verbs, you can eliminate wordiness, as the following sentences illustrate:

Wordy:	"The *acquisition* of the property was accomplished through long and hard negotiations."
Concise:	"The property was *acquired* through long and hard negotiations."
Wordy:	"*Confirmation* of the contract occurred yesterday."
Concise:	"The contract was *confirmed* yesterday."
Wordy:	"*Exploration* of the region had to be effected before the end of the year."
Concise:	"The region had to be *explored* before the end of the year."
Wordy:	"*Replacement* of the transmission was achieved only three hours before the race."
Concise:	"The transmission was *replaced* only three hours before the race."

As the examples show, abstract nouns often end with "-tion" or "-ment" and are often followed by the preposition "of." These words are not always "bad" words; they cause problems only when they replace action verbs from which they are derived. The following examples show some noun phrases along with the preferred verb substitutes:

assessment of	assess
classification of	classify
computation of	compute
delegation of	delegate
development of	develop
disbursement of	disburse
documentation of	document
elimination of	eliminate
establishment of	establish
negotiation of	negotiate
observation of	observe
requirement of	require
verification of	verify

■ Guideline 2: Shorten Wordy Phrases

Many wordy phrases have worked their way into business and technical writing. Weighty expressions add unnecessary words and rob prose of clarity. Here are some of the culprits, along with their concise substitutes:

afford an opportunity to	permit
along the lines of	like
an additional	another
at a later date	later
at this point in time	now
by means of	by
come to an end	end
due to the fact that	because
during the course of	during
for the purpose of	for
give consideration to	consider
in advance of	before
in the amount of	of
in the event that	if
in the final analysis	finally
in the proximity of	near
prior to	before
subsequent to	after
with regard to	about

■ *Guideline 3: Replace Long Words With Short Ones*

In grade school, most students are taught to experiment with multisyllabic words. Although this effort helps build vocabularies, it also can lead to a lifelong tendency to use long words when short ones will do. Of course, sometimes you want to use longer words just for variety—for example, using an occasional "approximately" for the preferred "about." As a rule, however, the following long words in the left column routinely should be replaced by the short words in the right column:

advantageous	helpful
alleviate	lessen, lighten
approximately	about
cognizant	aware
commence	start, begin
demonstrate	show
discontinue	end, stop
endeavor	try
finalize	end, complete

implement	carry out
initiate	start, begin
inquire	ask
modification	change
prioritize	rank, rate
procure	buy
terminate	end, fire
transport	move
undertake	try, attempt
utilize	use

■ Guideline 4: Leave out Clichés

Clichés are worn-out expressions that add words to your writing. Though they once were fresh descriptive phrases, they became clichés when they no longer conveyed their original meaning. You can make writing more concise by replacing clichés with a good adjective or two. Here are some clichés to avoid:

as plain as day

ballpark figure

efficient and effective

few and far between

last but not least

leaps and bounds

needless to say

reinvent the wheel

skyrocketing costs

step in the right direction

■ Guideline 5: Make Writing More Direct by Reading It Aloud

Much wordiness results from a tendency to talk around a topic. Sometimes called "circumlocution," this stylistic flaw arises from an inclination to write indirectly. It can be avoided by reading passages aloud. Hearing the sound of the words makes problems of wordiness quite apparent. It helps condense all kinds of inflated language, including the wordy expressions mentioned earlier. Remember, however, that direct writing must also retain a tactful, diplomatic tone when it conveys negative or sensitive information.

Indirect: "We would like to suggest that you consider directing your attention toward completing the project before the commencement of the seasonal monsoon rains in the region of the project area."

Direct: "We suggest you complete the project before the monsoons begin."
Indirect: "At the close of the last phase of the project, a bill for your services should be expedited to our central office for payment."
Direct: "After the project ends, please send your bill immediately to our central office."
Indirect: "It is possible that the well-water samples collected during our investigation of the well on the site of the subdivision could possibly contain some chemicals in concentrations higher than is allowable according to the state laws now in effect."
Direct: "Our samples from the subdivision's well might contain chemical concentrations beyond those permitted by the state."

■ *Guideline 6: Avoid "There Are," "It Is," and Similar Constructions*

"There are" and "it is" should not be substituted for concrete subjects and action verbs, which are preferable in good writing. Such constructions delay the delivery of information about who or what is doing something. Thus they tend to make your writing lifeless and abstract. Avoid them by creating (1) main subjects that are concrete nouns and (2) main verbs that are action words. Note that the following revised passages give readers a clear idea of who is doing what in the subject and verb positions.

Original: "There are many McDuff projects that could be considered for design awards."
Revision: "Many McDuff projects could be considered for design awards."
Original: "It is clear to the hiring committee that writing skills are an important criterion for every technical position."
Revision: "The hiring committee believes that writing skills are an important criterion for every technical position."
Original: "There were 15 people who attended the meeting at the client's office in Charlotte."
Revision: "Fifteen people attended the meeting at the client's office in Charlotte."

■ *Guideline 7: Cut Out Extra Words*

This guideline covers all wordiness errors not mentioned earlier. You need to keep a vigilant eye for *any* extra words or redundant phrasing in your work. Sometimes the problem comes in the form of needless connecting words, like *to be* or *that*. Other times it appears as redundant points—that is, those that have been made earlier in a sentence, paragraph, or section and do not need repeating. Delete extra words when their use (1) does not add a necessary transition between ideas or (2) does not provide new information to the reader. (One important exception is the intentional repetition of main points for emphasis, as in repeating important conclusions in different parts of a report.) The following examples display a variety of wordy or redundant writing, with corrections made in longhand:

Example 1: Preparing the client's final bill involves ~~the~~ checking ~~of~~ all [project] invoices ~~for the project.~~

Example 2: The report examined what the McDuff project manager considered ~~to be~~ a technically acceptable risk.

Example 3: During ~~the course of~~ its field work, the McDuff team will ~~be engaged in the process of~~ reviewing all ~~of the~~ notes ~~that have been~~ accumulated in previous studies.

Example 4: ~~Because of his position~~ as head of ~~the~~ [McDuff] public relations group ~~at McDuff~~, he planned ~~such that he would be able~~ to attend the meeting.

Example 5: She believed ~~that the~~ recruiting ~~of~~ more minorities for the technical staff is essential.

Example 6: The department must determine its ~~aims and~~ goals so that they can be included in ~~the annual~~ [McDuff's 1995] strategic plan ~~produced by McDuff for the year of 1995.~~

Example 7: Most McDuff managers ~~generally~~ agree that all ~~of the~~ company~~'s~~ employees ~~at all the offices~~ deserve ~~at least~~ some ~~degree of~~ training each year ~~that they work for the firm.~~

BEING ACCURATE IN WORDING

Good technical writing style also demands accuracy in phrasing. Indeed, many technical professionals place their reputations and even their financial lives on the line with every document that goes out their door. That fact makes clear the importance of taking your time on any editing pass that deals with the accuracy of phrasing. Accuracy often demands more words, not fewer. The main rule is never to sacrifice clarity for conciseness.

Careful writing helps to limit the liability that your organization may incur. Your goal is really very simple: Make sure words convey the meaning you intend—no more, no less. Here are some basic guidelines to follow:

■ Guideline 1: Distinguish Facts From Opinions

In practice, this guideline means that you must always identify your opinions and judgments as such by using phrases like "we recommend," "we believe," "we suggest," or "in our opinion." EXAMPLE: "In our opinion, spread footings would be an acceptable foundation for the building you plan at the site." If you want to avoid repetitious use of such phrases, group your opinions into listings or report sections. Thus a single lead-in can indicate to the reader that opinions, not facts, are forthcoming. EXAMPLE: "On the basis of our site visit and our experience at similar sites, we believe that (1) _____ , (2) _____ , and (3) _____ ."

■ Guideline 2: Include Obvious Qualifying Statements When Needed

This guideline does not mean that you have to be overly defensive in every part of the report. It does mean, however, that you must be wary of possible misinterpretations. EXAMPLE: "Our summary of soil conditions is based only on information obtained during a brief visit to the site. We did not drill any soil borings."

■ Guideline 3: Use Absolute Words Carefully

In particular, avoid words that convey an absolute meaning or that convey a stronger meaning than you intend. One notable example is "minimize," which means to reduce to the lowest possible level or amount. If a report claims that a particular piece of equipment will "minimize" breakdowns on the assembly line, the passage could be interpreted as an absolute commitment. In theory, the reader could consider any breakdown at all to be a violation of the report's implications. If instead the writer had used the verb "limit" or "reduce," the wording would have been more accurate and less open to misunderstanding.

USING THE ACTIVE VOICE

Striving to use the active voice can greatly improve your technical writing style. This section defines the active and passive voices and then gives examples of each. It also lists some practical guidelines for using both voices.

What Do Active and Passive Mean?

Active-voice sentences emphasize the person (or thing) performing the action—that is, somebody (or something) does something ("Matt completed the field study yesterday"). Passive-voice sentences emphasize the recipient of the action itself—that is, something is being done to something by somebody ("The field study was completed [by Matt] yesterday"). Here are some other examples of the same thoughts being expressed in first the active and then the passive voice:

- EXAMPLES: Active-Voice Sentences:

  1. "We *reviewed* aerial photographs in our initial assessment of possible fault activity at the site."
  2. "The study *revealed* that three underground storage tanks had leaked unleaded gasoline into the soil."
  3. "We *recommend* that you use a minimum concrete thickness of 6 in. for residential subdivision streets."

- EXAMPLES: Passive-Voice Sentences:

  1. "Aerial photographs *were reviewed* [by us] in our initial assessment of possible fault activity at the site."
  2. "The fact that three underground storage tanks had been leaking unleaded gasoline into the soil *was revealed* in the study."
  3. "*It is recommended* that you use a minimum concrete thickness of 6 in. for residential subdivision streets."

Just reading through these examples gives the sense that passive constructions are wordier than active ones. Also, passives tend to leave out the person or thing doing the action. Although occasionally this impersonal approach is appropriate, often the reader becomes frustrated by writing that fails to say who or what is doing something.

When Should Actives and Passives Be Used?

Both the active and passive voices have a place in your writing. Knowing when to use each is the key. Here are a few guidelines that will help:

- *Use the active voice when you want to:*

  1. Emphasize who is responsible for an action ("*We recommend* that you consider . . .")
  2. Stress the name of a company, whether yours or the reader's ("*PineBluff Contracting has expressed* interest in receiving bids to perform work at . . .")

3. Rewrite a top-heavy sentence so that the person or thing doing the action is up front (*"Figure 1 shows* the approximate locations of . . ."*)
4. Pare down the verbiage in your writing, since the active voice is usually a shorter construction

- *Use the passive voice when you want to:*

1. Emphasize the receiver of the action rather than the person performing the action (*"Samples will be sent* directly from the site to our laboratory in Sacramento"*)
2. Avoid the kind of egocentric tone that results from repetitious use of "I," "we," and the name of your company (*"The project will be directed* by two programmers from our Boston office"*)
3. Break the monotony of writing that relies too heavily on active-voice sentences

Although the passive voice has its place, it is far too common in business and technical writing. This stylistic error results in part from the common misperception that passive writing is more objective. In fact, excessive use of the passive voice only makes writing more tedious to read. In modern business and technical writing, strive to use the active voice.

USING NONSEXIST LANGUAGE

Language usually *follows* changes in culture, rather than *anticipating* such changes. A case in point is today's shift away from sexist language in business and technical writing—indeed, in all writing and speaking. The change reflects the increasing number of women entering previously male-dominated professions such as engineering, management, medicine, and law. It also reflects the fact that many men have taken previously female-dominated positions as nurses and flight attendants.

This section on style defines sexist and nonsexist language. Then it suggests ways to avoid using gender-offensive language in your writing.

Sexism and Language

Sexist language is the use of wording, especially masculine pronouns like "he" or "him," to represent positions or individuals who could be either men or women. For many years, it was perfectly appropriate to use "he," "his," "him," or other masculine words in sentences such as these:

- "The operations specialist should check page 5 of his manual before flipping the switch."
- "Every physician was asked to renew his membership in the medical association before next month."
- "Each new student at the military academy was asked to leave most of his personal possessions in the front hallway of the administration building."

The masculine pronoun was understood to be an indefinite representation for any person—male or female. That seems simple enough, and many people still wonder why such usage came under criticism. Yet there are two good reasons for attacks on such usage:

1. As already mentioned, the entry of many more women into formerly male-dominated professions has called attention to the inappropriate generic use of masculine pronouns.
2. Many people believe that the use of masculine pronouns in a context that could include both genders constrains women from achieving equal status in the professions and, generally, in the culture. That is, the use of masculine pronouns becomes a subtle but pervasive way of continuing sexism in society as a whole.

Either point supplies a good enough reason to avoid drawing attention to your style with sexist language. There are many women in positions of responsibility who may be reading your on-the-job writing. If you fail to rid your writing of sexist language, you take the risk of drawing negative attention toward your style and away from your ideas. Common sense argues for following some basic style techniques to avoid sexist language.

Techniques for Nonsexist Language

This section offers techniques to help you shift from sexist to nonsexist language. Not all these strategies will suit your taste in writing style; choose the ones that will work for you.

Technique 1: Avoid Personal Pronouns Altogether

One of the easiest ways to avoid problems with sexist language is to delete unnecessary pronouns from your writing:

Example:

Sexist Language: "During his first day on the job, any new employee in the toxic-waste laboratory must report to the company doctor for his employment physical."

Nonsexist Language: "During the first day on the job, each new employee in the toxic-waste laboratory must report to the company doctor for a physical."

Technique 2: Use Plural Pronouns Instead of Singular

In most contexts you can shift from singular to plural pronouns without altering your meaning. The plural usage avoids the problem of using masculine pronouns for mixed usage.

Example:

Sexist Language: "Each geologist should submit his time sheet by noon on the Thursday before checks are issued."

> **Nonsexist Language:** "All geologists should submit their time sheets on the Thursday before checks are issued."

Interestingly, you may encounter sexist language that uses generic female pronouns inappropriately. For example, "Each nurse should make every effort to complete her rounds each hour." As in the preceding case, a shift to plural pronouns is appropriate: "Nurses should make every effort to complete their rounds each hour."

■ Technique 3: Alternate Masculine and Feminine Pronouns

Some writers, who prefer to use singular gender pronouns, avoid sexist use by alternating "he" and "him" with "she" and "her." Writers using this technique usually avoid the unsettling practice of switching pronoun use within too-brief a passage, such as a paragraph or page. Instead, they may switch every few pages, or every section or chapter. Although this technique is not yet in common usage, its appeal is growing. It gives writers the linguistic flexibility to continue to use masculine and feminine pronouns in a generic fashion.

One problem is that the alternating use of masculine and feminine pronouns tends to draw attention to itself. Also, the writer must work to balance the use of masculine and feminine pronouns, in a sense to give "equal treatment." Perhaps such obstacles will seem less onerous as the use of alternating pronouns grows.

■ Technique 4: Use Forms Like "He or She," "Hers or His," and "Him or Her"

This solution simply requires the writer to include both genders of pronouns in contexts that are not gender-specific.

Example:

> **Sexist Language:** "The president made it clear that each McDuff branch manager will be responsible for the balance sheet of his respective office."
>
> **Nonsexist Language:** "The president made it clear that each McDuff branch manager will be responsible for the balance sheet of his or her respective office."

You need to know, however, that this stylistic correction of sexist language is bothersome to some readers. They feel that the doublet structure, "her or his," is wordy and awkward. Many readers are bothered even more by the formations "he/she," "his/her," and "her/him." Avoid this usage.

■ Technique 5: Shift to Second-Person Pronouns

Consider shifting to the use of "you" and "your," words without any sexual bias. This technique is effective only with documents in which it is appropriate to use an instructions-related "command" tone associated with the use of "you."

Example:

> **Sexist Language:** "After selecting her insurance option in the benefit plan, each new nurse should submit her paperwork to the Human Resources Department."

> **Nonsexist Language:** "Submit your paperwork to the Human Resources Department after selecting your insurance option in the benefit plan."

■ *Technique 6: Be Especially Careful of Titles and Letter Salutations*

When you do not know how a particular woman prefers to be addressed, always use "Ms." Even better, call the person's employer and ask if the recipient goes by "Miss," "Mrs.," "Ms.," or some other title. (When calling, also check on the correct spelling of the person's name and her current job title.) Receptionists and secretaries are used to receiving such inquiries.

When you do not know who will read your letter, never use "Dear Sir" or "Gentlemen" as a generic greeting. Such a mistake may offend women reading the letter and may even cost you some business. "Dear Sir or Madam" is also inappropriate. It shows you do not know your audience, and it includes the archaic form "madam." Instead, call the organization and get the name of a particular person to whom you can direct your letter. If you must write to a group of people, replace the generic greeting with an "Attention" line that denotes the name of the group.

Examples:

> **Sexist Language:** Dear Miss Finnegan: [to a single woman for whom you can determine no title preference]

> **Nonsexist Language:** Dear Ms. Finnegan:

> **Sexist Language:** Dear Sir: [to a collective audience]

> **Nonsexist Language:** Attention: Admissions Committee

No doubt the coming years will bring additional suggestions for solving the problem of sexist language. Whatever the culture finally settles on, it is clear that good technical writing style will no longer tolerate the use of such language.

CHAPTER SUMMARY

Style is an important part of technical writing. During the editing process, each writer makes the kinds of changes that place his or her personal stamp on a document. Style can also be shaped (1) by a group, in that writing done collaboratively can acquire features of its diverse contributors, or (2) by an organization, in that an organization may encourage (or require) writers to adopt a particular writing style. Yet the decision-making process of individual writers remains the most important influence on the style of technical documents.

This chapter offers five basic suggestions for achieving good technical writing style. First, sentences should be clear, with main ideas at the beginning and with one main clause in most sentences. Although sentences should average only 15–18 words, you should vary sentence patterns in every document. Second, technical writing should be concise. You can achieve this goal by reading prose aloud as you rewrite and edit. Third, wording should be accurate. Fourth, the active voice should be dominant, though the passive voice also has its place in good technical writing. And fifth, the language of technical documents should be free of sexual bias.

ASSIGNMENTS

The first exercise includes a variety of stylistic errors covered in the chapter. The second exercise deals exclusively with conciseness. In revising the passages, your goal is to improve style without altering meaning.

1. **General Rules of Style.** Revise the following sentences by applying the guidelines mentioned in this chapter. When you change passive verbs to active, you may need to make some assumptions about the agent of the action, since the sentences are taken out of context.

   a. Based on our review of the available records, conversations with the various agencies involved, including the Fire Department and the Police Department, and a thorough survey of the site where the spill occurred, it was determined that the site contained chemicals that were hazardous to human health.

   b. After seven hours at the negotiation table, the union representatives and management decided that the issues they were discussing could not be resolved that evening, so they met the next day at the hotel complex, at which point they agreed on a new contract that would increase job security and benefits.

   c. It is recommended by us that your mainframe computer system be replaced immediately by a newer model.

   d. After the study was completed by the research team and the results were published in the company newsletter the following month, the president decided to call a meeting of all senior-level managers to discuss strategies for addressing problems highlighted by the research team.

   e. Our project activities can be generally described in this way. The samples were retrieved from the site and then were transported to the testing lab in the containers made especially for this project, and at the lab they were tested to determine their soil properties; the data were analyzed by all the members of the team before findings and conclusions were arrived at.

   f. First the old asbestos tile was removed. Then the black adhesive was scraped off. Later the floor was sanded smooth. The wood arrived shortly. Then the floor was installed.

   g. The figures on the firm's profit margins in July and August, along with sales commissions for the last six months of the previous year and the top 10 salespersons in the firm, are included in the Appendix.

   h. It was suggested by the team that the company needs to invest in modern equipment.

   i. It is the opinion of this writer that the company's health plan is adequate.

   j. Shortly after the last change in leadership, and during the time that the board of directors was expressing strong views about the direction that the company was taking, it became clear to me and other members of the senior staff that the company was in trouble.

   k. Each manager should complete and submit his monthly report by the second Tuesday of every month.

   l. After completing our engineering analysis, it is clear that metal fatigue caused the structure to fail.

   m. Upon hearing the captain's signal, each flight attendant should complete her checklist of preflight procedures.

   n. Our weed-spraying procedure will have minimal impact on shrubbery that surrounds the building site.

   o. It was reported today from the corporate headquarters that the health-care plan has been approved by the president.

2. **Conciseness.** The following sentences contain more words than necessary. Rewrite each passage more concisely, without changing the meaning. If appropriate, make two sentences out of one.

   a. The disbursement of the funds from the estate will occur on the day that the proceedings concerning the estate are finalized in court.

   b. During the course of the project that we conducted for Acme Pipe, several members of our project team were in the unfortunate position of having to perform their fieldwork at the same time that torrential rains hit the area, totaling three inches of rain in one afternoon.

   c. At a later date we plan to begin the process of prioritizing our responsibilities on the project so that we will have a clear idea of which activities deserve the most attention from the project personnel.

   d. Needless to say, we do not plan to add our participation to the project if we conclude that the skyrocketing costs of the project will prohibit our earning what could be considered to be a fair profit from the venture.

   e. The government at this point in time plans to discontinue its testing of every item but will undertake to implement testing again in approximately five months.

   f. Hazerd, Inc., will endeavor to finalize the modifications of the blueprints for a ballpark figure of about $850.

   g. For us to supply the additional supplies that the client wishes to procure from us, the client will have to initiate a change order that permits additional funds to be transferred into the project account.

   h. Upon further analysis of the many and varied options that we are cognizant of at this time, it is our opinion that the long-term interests of our firm would be best served by reducing the size of the production staff by 300 workers.

   i. Prior to the implementation of the state law with regard to the use of asbestos as a building material, it was common practice to utilize this naturally occurring mineral in all kinds of facilities, some of which became health hazards subsequently.

   j. In the event that we are given permission to undertake the research, be sure to make certain to perform an efficient and effective search of available literature in a research facility so that we do not end up, in the final analysis, reinventing the wheel with regard to knowledge of superconductors.

3. **Editing Paper of Classmate.** For this assignment, exchange papers with a member of your class. Use either the draft of a current assignment *or* a paper that was completed earlier in the term. Edit your classmate's work in accordance with this chapter's guidelines on style. Then explain your changes to the writer.

4. **Editing Sample Memo.** Using the guidelines in this chapter, edit the following memorandum.

DATE: January 12, 1994
TO: All Employees of Denver Branch
FROM: Leonard Schwartz, Branch Manager
SUBJECT: New Loss-Prevention System

As you may have recently heard, lately we received news from the corporate headquarters of the company that it would be in the best interest of the entire company to pay more attention to matters of preventing accidents and any other safety-related measures that affect the workplace, including both office and field activities related to all types of jobs that we complete. Every single employee in each office at every branch needs to be ever mindful in this regard so that he is most efficient and effective in the daily performance of his everyday tasks that relate to his job responsibilities such that safety is always of paramount concern.

With this goal of safety ever present in our minds, I believe the bottom line of the emphasis on safety could be considered to be the training that each of us receives in his first, initial weeks on the job as well as the training provided on a regular basis throughout each year of our employment with McDuff, so that we are always aware of how to operate in a safe manner. The training vehicle gives the company the mechanism to provide each of you with the means to become aware of the elements of safety that relate to the specific needs and requirements of your own particular job. Therefore, at this point in time I have come to the conclusion in the process of contemplating the relevance of the new corporate emphasis on safety to our particular branch that we need, as a branch, to give much greater scrutiny and analysis to the way we can prevent accidents and emphasize the concern of safety at every stage of our operation for every employee. Toward this end, I have asked the training coordinator, Kendra Jones, to assemble a written training program that will involve every single employee and that can be implemented beginning no later than June of this year. When the plan has been written and approved at the various levels within the office, I will conduct a meeting with every department in order to emphasize the major and minor components of this upcoming safety program.

It is my great pleasure to announce to all of you that effective in the next month (February) I will give a monthly safety award of $100 to the individual branch employee at any level of the branch who comes up with the best, most useful suggestion related to safety in any part of the branch activities. Today I will take the action of placing a suggestion box on the wall of the lunchroom so that all of you will have easy access to a way to get your suggestions for safety into the pipeline and to be considered. As an attachment to the memo you are now reading from me I have provided you with a copy of the form that you are to use in making any suggestions that are then to be placed in the suggestion box. On the last day of each month that we work, the box will be emptied of the completed forms for that month, and before the end of the following week a winner will be selected by me for the previous month's suggestion program and an announcement will be placed by me to that effect on the bulletin board in the company workroom.

If you have any questions in regard to the corporate safety program as it affects our branch or about the suggestion program that is being implemented here at the Denver office at McDuff, please do not hesitate to make your comments known either in memorandum form or by way of telephonic response to this memorandum.

HANDBOOK

*T*his handbook includes entries on the basics of writing. Here you will find three main types of information:

1. **Grammar:** the rules by which we edit sentence elements. Examples include rules for the placement of punctuation, the agreement of subjects and verbs, and the placement of modifiers.
2. **Mechanics:** the rules by which we make final proofreading changes. Examples include the rules for abbreviations and the use of numbers. This handbook also includes a list of commonly misspelled words.
3. **Usage:** information on the correct use of particular words, especially pairs of words that are often confused. Examples include problem words like "affect/ effect," "complement/compliment," and "who/whom."

Another editing concern, technical style, is the topic of chapter 15. In that chapter you will find guidelines for sentence structure, conciseness, accuracy of wording, active and passive voice, and nonsexist language. Together, chapter 15 and this handbook will help you turn unedited drafts into final revised documents.

This handbook is presented in alphabetized fashion for easy reference during the editing process. A table of contents follows. Grammar and mechanics entries are in caps; usage entries are in lowercase. Several exercises follow the entries.

a/an
a lot/alot
ABBREVIATIONS
affect/effect
agree to/agree with
all together/altogether
alternately/alternatively
amount/number
anticipate/expect
assure/ensure/insure
augment/supplement
awhile/a while
between/among
CAPITALIZATION
complement/compliment
compose/comprise
continuous/continual
data/datum
definite/definitive
discrete/discreet/discretion
disinterested/uninterested

A/An

These two words are different forms of the same article. "A" occurs before words that start with consonants or consonant sounds. EXAMPLES:

- a three-pronged plug
- a once-in-a-lifetime job ("once" begins with the consonant sound of *w*)
- a historic moment (many speakers and some writers mistakenly use "an" before "historic")

"An" occurs before words that begin with vowels or vowel sounds. EXAMPLES:

- an eager new employee
- an hour before closing

A lot/Alot

The correct form is the two-word phrase "a lot." Though acceptable in informal discourse, "a lot" usually should be replaced by more formal diction in technical writing. EXAMPLE: "They retrieved many [*not* a lot of] soil samples from the construction site."

Abbreviations

Technical writing uses many abbreviations. Without this shorthand form, you would end up writing much longer reports and proposals—without any additional content. Follow these seven basic rules in your use of abbreviations, paying special attention to the first three:

■ Rule 1: Do Not Use Abbreviations When Confusion May Result

When you want to use a term just once or twice and you are not certain your readers will understand an abbreviation, write out the term rather than abbreviating it. EX-AMPLE: "They were required to remove creosote from the site, according to the directive from the Environmental Protection Agency." Even though "EPA" is the accepted abbreviation for this government agency, you should write out the name in full *if* you are using the term only once to an audience that may not understand it.

■ Rule 2: Use Parentheses for Clarity

When you use a term *more* than twice and are not certain that your readers will understand it, write out the term the first time it is used and place the abbreviation in parentheses. Then use the abbreviation in the rest of the document. In long reports or proposals, however, you may need to repeat the full term in key places. EXAMPLE: "According to the directive from the Environmental Protection Agency (EPA), they were required to remove the creosote from the construction site. Furthermore, the directive indicated that the builders could expect to be visited by EPA inspectors every other week."

■ Rule 3: Include a Glossary When There Are Many Abbreviations

When your document contains many abbreviations that may not be understood by all readers, include a well-marked glossary at the beginning or end of the document. A glossary simply collects all the terms and abbreviations and places them in one location, for easy reference.

■ Rule 4: Use Abbreviations for Units of Measure

Most technical documents use abbreviations for units of measure. Do not include a period unless the abbreviation could be confused with a word. EXAMPLES: mi, ft, oz, gal., in. and lb. Note that units-of-measurement abbreviations have the same form for both singular and plural amounts. EXAMPLES: 1/2 in., 1 in., 5 in.

■ Rule 5: Avoid Spacing and Periods

Avoid internal spacing and internal periods in most abbreviations that contain all capital letters. EXAMPLES: ASTM, EPA, ASEE. Exceptions include professional titles and degrees such as P.E., B.S., and B.A.

■ Rule 6: Be Careful With Company Names

Abbreviate a company or other organizational name only when you are sure that officials from the organization consider the abbreviation appropriate. IBM (for the company) and UCLA (for the university) are examples of commonly accepted organizational abbreviations. When in doubt, follow Rule 2 — write the name in full the first time it is used, followed by the abbreviation in parentheses.

■ Rule 7: Use These Common Abbreviations

These common abbreviations are appropriate for most writing in your technical or business career. They are placed into three main categories of measurements, locations, and titles:

Measurements: Use these abbreviations only when you place numbers before the measurement.

ac:	alternating current	db:	decibel
amp:	ampere	dc:	direct current
bbl:	barrel	dm:	decimeter
Btu:	British thermal unit	doz *or* dz:	dozen
bu:	bushel	F:	Fahrenheit
C:	Celsius	f:	farad
cal:	calorie	fbm:	foot board measure
cc:	cubic centimeter	fig.:	figure
circ:	circumference	fl oz:	fluid ounce
cm:	centimeter	FM:	frequency modulation
cos:	cosine	fp:	foot pound
cot:	cotangent	ft:	foot (feet)
cps:	cycles per second	g:	gram
cu ft:	cubic feet	gal.:	gallon

gpm:	gallons per minute	oz:	ounce
hp:	horsepower	ppm:	parts per million
hr:	hour	psf:	pounds per square foot
Hz:	hertz	psi:	pounds per square inch
in.:	inch	pt:	pint
j:	joule	qt:	quart
K:	Kelvin	rev:	revolution
ke:	kinetic energy	rpm:	revolutions per minute
kg:	kilogram	sec:	second
km:	kilometer	sq:	square
kw:	kilowatt	sq ft:	square foot (feet)
kwh:	kilowatt-hour	T:	ton
l:	liter	tan.:	tangent
lb:	pound	v:	volt
lin:	linear	va:	volt-ampere
lm:	lumen	w:	watt
log.:	logarithm	wk:	week
m:	meter	wl:	wavelength
mm:	millimeter	yd:	yard
min:	minute	yr:	year

Locations: Use these common abbreviations for addresses (on envelopes and letters, for example), but write out the words in full in other contexts.

AL:	Alabama	ID:	Idaho
AK:	Alaska	IL:	Illinois
AS:	American Samoa	IN:	Indiana
AZ:	Arizona	IA:	Iowa
AR:	Arkansas	KS:	Kansas
CA:	California	KY:	Kentucky
CZ:	Canal Zone	LA:	Louisiana
CO:	Colorado	ME:	Maine
CT:	Connecticut	MD:	Maryland
DE:	Delaware	MA:	Massachusetts
DC:	District of Columbia	MI:	Michigan
FL:	Florida	MN:	Minnesota
GA:	Georgia	MS:	Mississippi
GU:	Guam	MO:	Missouri
HI:	Hawaii	MT:	Montana

NE:	Nebraska	VT:	Vermont
NV:	Nevada	VI:	Virgin Islands
NH:	New Hampshire	VA:	Virginia
NJ:	New Jersey	WA:	Washington
NM:	New Mexico	WV:	West Virginia
NY:	New York	WI:	Wisconsin
NC:	North Carolina	WY:	Wyoming
ND:	North Dakota	Alta.:	Alberta
OH:	Ohio	B.C.:	British Columbia
OK:	Oklahoma	Man.:	Manitoba
OR:	Oregon	N.B.:	New Brunswick
PA:	Pennsylvania	Nfld.:	Newfoundland
PR:	Puerto Rico	N.W.T.:	Northwest Territories
RI:	Rhode Island	N.S.:	Nova Scotia
SC:	South Carolina	Ont.:	Ontario
SD:	South Dakota	P.E.I.:	Prince Edward Island
TN:	Tennessee	P.Q.:	Quebec
TX:	Texas	Sask.:	Saskatchewan
UT:	Utah	Yuk.:	Yukon

Titles: Some of these abbreviations go before the name (such as Dr., Ms., and Messrs.), while others go after the name (such as college degrees, Jr., and Sr.).

Atty.:	Attorney
B.A.:	Bachelor of Arts
B.S.:	Bachelor of Science
D.D.:	Doctor of Divinity
Dr.:	Doctor (used mainly with medical and dental degrees but also with other doctorates)
Drs.:	plural of Dr.
D.V.M.:	Doctor of Veterinary Medicine
Hon.:	Honorable
Jr.:	Junior
LL.D.:	Doctor of Laws
M.A.:	Master of Arts
M.S.:	Master of Science
M.D.:	Doctor of Medicine
Messrs.:	Plural of Mr.
Mr.:	Mister

Mrs.:	used to designate married, widowed, or divorced women
Ms.:	used increasingly for all women, especially when one is uncertain about a woman's marital status
Ph.D.:	Doctor of Philosophy
Sr.:	Senior

Affect/Effect

These two words generate untold grief among many writers. The key to using them correctly is remembering two simple sentences: (1) "affect" with an "a" is a verb meaning "to influence"; (2) "effect" with an "e" is a noun meaning "result." There are some exceptions, however, such as the following: "Effect" can be a verb that means "to bring about," as in "He effected considerable change when he became a manager." EXAMPLES:

- "His progressive leadership greatly *affected* the company's future."
- "One *effect* of securing the large government contract was the hiring of several more accountants."
- "The president's belief in the future of microcomputers *effected* change in the company's approach to office management." (For a less wordy alternative, substitute "changed" for "effected change in.")

Agree to/Agree With

In correct usage, "agree to" means that you have *consented to* an arrangement, offer, proposal, etc. "Agree with" is less constraining and only suggests that you are *in harmony with* a certain statement, idea, person, etc. EXAMPLES:

- "Representatives from McDuff *agreed to* alter the contract to reflect the new scope of work."
- "We *agree with* you that more study may be needed before the nuclear power plant is built."

All Together/Altogether

"All together" is used when items or people are being considered in a group or are working in concert. "Altogether" is a synonym for "utterly" or "completely." EXAMPLES:

- "The three firms were *all together* in their support of the agency's plan."
- "There were *altogether* too many pedestrians walking near the dangerous intersection."

Alternately/Alternatively

Because many readers are aware of the distinction between these two words, any misuse can cause embarrassment or even misunderstanding. Follow these guidelines for correct use.

Alternately. As a derivative of "alternate," "alternately" is best reserved for events or actions that occur "in turns." EXAMPLE: "While digging the trench, he used a backhoe and a hand shovel *alternately* throughout the day."

Alternatively. A derivative of "alternative," "alternatively" should be used in contexts where two or more choices are being considered. EXAMPLE: "We suggest that you use deep foundations at the site. *Alternatively,* you could consider spread footings that were carefully installed."

Amount/Number

"Amount" is used in reference to items that *cannot* be counted, whereas "number" is used to indicate items than *can* be counted. EXAMPLES:

- "In the last year, we have greatly increased the *amount* of computer paper ordered for the Boston office."
- "The last year has seen a huge increase in the *number* [*not* amount] of boxes of computer paper ordered for the Boston office."

Anticipate/Expect

These two words are *not* synonyms. In fact, their meanings are distinctly different. "Anticipate" is used when you mean to suggest or state that steps have been taken beforehand to prepare for a situation. "Expect" only means you consider something likely to occur. EXAMPLES:

- "*Anticipating* that the contract will be successfully negotiated, Jones Engineering is hiring three new hydrologists."
- "We *expect* [*not* anticipate] that you will encounter semicohesive and cohesive soils in your excavations at the Park Avenue site."

Assure/Ensure/Insure

"Assure" is a verb that can mean "to promise." It is used in reference to people, as in "We want to assure you that our crews will strive to complete the project on time." In fact, "assure" and its derivatives (like "assurance") should be used with care in technical contexts, for these words can be viewed as a guarantee.

The synonyms "ensure" and "insure" are verbs meaning "to make certain." Like "assure," they imply a level of certainty that is not always appropriate in engineering or the sciences. When their use is deemed appropriate, the preferred word is "ensure"; reserve "insure" for sentences in which the context is insurance. EXAMPLES:

- "Be *assured* that our representatives will be on site to answer questions that the subcontractor may have."
- "To *ensure* that the project stays within schedule, we are building in 10 extra days for bad weather." (An alternative: "So that the project stays within schedule, we are building in 10 extra days for bad weather.")

Augment/Supplement

"Augment" is a verb that means to increase in size, weight, number, or importance. "Supplement" is either (1) a verb that means "to add to" something to make it complete or to make up for a deficiency or (2) a noun that means "the thing that has been added." EXAMPLES:

- "The power company supervisor decided to *augment* the line crews in five counties."
- "He *supplemented* the audit report by adding the three accounting statements."
- "The three accounting *supplements* helped support the conclusions of the audit report."

Awhile/A While

Though similar in meaning, this pair is used differently. "Awhile" means "for a short time." Because "for" is already a part of its definition, it cannot be preceded by the preposition "for." The noun "while," however, can be preceded by the two words "for a," giving it essentially the same meaning as "awhile." EXAMPLES:

- "Kirk waited *awhile* before trying to restart the generator."
- "Kirk waited *for a while* before trying to restart the generator."

Between/Among

The distinction between these two words has become somewhat blurred. However, many readers still prefer to see "between" used with reference to only two items, reserving "among" for three or more items. EXAMPLES:

- "The agreement was just *between* my supervisor and me. No one else in the group knew about it."
- "The proposal was circulated *among* all members of the writing team."
- "*Among* Sallie, Todd, and Fran, there was little agreement about the long-term benefits of the project."

Capitalization

As a rule, you should capitalize *specific* names of people, places, and things—sometimes called "proper nouns." For example, capitalize specific streets, towns, trademarks, geological eras, planets, groups of stars, days of the week, months of the year, names of organizations, holidays, and colleges. However, remember that excessive capitalization—as in titles of positions in a company—is inappropriate in technical writing and can appear somewhat pompous.

The following rules cover some frequent uses of capitals:

1. Major words in titles of books and articles. Only capitalize prepositions and articles when they appear as the first word in titles. EXAMPLES:
- *For Whom the Bell Tolls*
- *In Search of Excellence*
- *The Power of Positive Thinking*

2. Names of places and geographical locations. EXAMPLES:
   - Washington Monument
   - Cleveland Stadium
   - Dallas, Texas
   - Cobb County
3. Names of aircraft and ships. EXAMPLES:
   - *Air Force One*
   - SS *Arizona*
   - *Nina, Pinta,* and *Santa Maria*
4. Names of specific departments and offices within an organization. EXAMPLES:
   - Humanities Department
   - Personnel Department
   - International Division
5. Political, corporate, and other titles that come before names. EXAMPLES:
   - Chancellor Hairston
   - Councilwoman Jones
   - Professor Gainesberg
   - Congressman Buffett

   Note, however, that general practice does not call for capitalizing most titles when they are used by themselves or when they follow a person's name. EXAMPLES:
   - Jane Cannon, a professor in the Business Department.
   - Rob McDuff, president of McDuff, Inc.
   - Chris Presley, secretary of the Oil Rig Division.

Complement/Compliment

Both words can be nouns and verbs, and both have adjective forms (complementary, complimentary).

Complement. This word is used as a noun to mean "that which has made something whole or complete," as a verb to mean "to make whole, to make complete," or as an adjective. You may find it easier to remember the word by recalling its mathematical definition: Two complementary angles must always equal 90 degrees. EXAMPLES:

- (As noun): "The *complement* of five technicians brought our crew strength up to 100 percent."
- (As verb): "The firm in Canada served to *complement* ours in that together we won a joint contract."
- (As adjective): "Seeing that project manager and her secretary work so well together made clear their *complementary* relationship in getting the office work done."

Compliment. This word is used as a noun to mean "an act of praise, flattery, or admiration," as a verb to mean "to praise, to flatter," or as an adjective to mean "related to praise or flattery, or without charge." EXAMPLES:

- (As noun): "He appreciated the verbal *compliments,* but he also hoped they would result in a substantial raise."
- (As verb): "Howard *complimented* the crew for finishing the job on time and within budget."
- (As adjective): "We were fortunate to receive several *complimentary* copies of the new software from the publisher."

Compose/Comprise

These are both acceptable words, with an inverse relationship to each other. "Compose" means "to make up or be included in," whereas "comprise" means "to include or consist of." The easiest way to remember this relationship is to memorize one sentence: "The parts compose the whole, but the whole comprises the parts." One more point to remember: The common phrase "is comprised of" is a substandard, unacceptable replacement for "comprise" or "is composed of." Careful writers do not use it. EXAMPLES:

- "Seven quite discrete layers *compose* the soils that were uncovered at the site."
- "The borings revealed a stratigraphy that *comprises* [*not* is comprised of] seven quite discrete layers."

Continuous/Continual

The technical accuracy of some reports may depend on your understanding of the difference between these two words. "Continuous" and "continuously" should be used in reference to uninterrupted, unceasing activities. However, "continual" and "continually" should be used with activities that are intermittent, or repeated at intervals. If you think your reader may not understand the difference, you should either (1) use synonyms that will be clearer (such as "uninterrupted" for "continuous," and "intermittent" for "continual") or (2) define each word at the point you first use it in the document. EXAMPLES:

- "We *continually* checked the water pressure for three hours before the equipment arrived, while also using the time to set up the next day's tests."
- "Because it rained *continuously* from 10:00 A.M. until noon, we were unable to move our equipment onto the utility easement."

Data/Datum

Coming as it does from the Latin, the word "data" is the plural form of "datum." Although many writers now accept "data" as singular or plural, traditionalists in the technical and scientific community still consider "data" exclusively a plural form. Therefore, you should maintain the plural usage. EXAMPLES:

- "These *data* show that there is a strong case for building the dam at the other location."
- "This particular *datum* shows that we need to reconsider recommendations put forth in the original report."

If you consider the traditional singular form of "datum" to be awkward, use substitutes such as "This item in the data shows" or "One of the data shows that." Singular subjects like "one" or "item" allow you to keep your original meaning without using the word "datum."

Definite/Definitive

Though similar in meaning, these words have slightly different contexts. "Definite" refers to that which is precise, explicit, or final. "Definitive" has the more restrictive meaning of "authoritative" or "final." EXAMPLES:

- "It is now *definite* that he will be assigned to the London office for six months."
- "He received the *definitive* study on the effect of the oil spill on the marine ecology."

Discrete/Discreet/Discretion

The adjective "discrete" suggests something that is separate, or something that is made up of many separate parts. The adjective "discreet" is associated with actions that require caution, modesty, or reserve. The noun "discretion" refers to the quality of being "discreet" or the freedom a person has to act on her or his own. EXAMPLES:

- "The orientation program at McDuff includes a writing seminar, which is a *discrete* training unit offered for one full day."
- "The orientation program at McDuff includes five *discrete* units."
- "As a counselor in McDuff's Human Resources Office, Sharon was *discreet* in her handling of personal information about employees."
- "Every employee in the Human Resources Office was instructed to show *discretion* in handling personal information about employees."
- "By starting a flextime program, McDuff, Inc., will give employees a good deal of *discretion* in selecting the time to start and end their workday."

Disinterested/Uninterested

In contemporary business use, these words have quite different meanings. Because errors can cause confusion for the reader, make sure not to use the words as synonyms. "Disinterested" means "without prejudice or bias," whereas "uninterested" means "showing no interest." EXAMPLES:

- "The agency sought a *disinterested* observer who had no stake in the outcome of the trial."
- "They spent several days talking to officials from Iceland, but they still remain *uninterested* in performing work in that country."

Due to/Because of

Besides irritating those who expect proper English, mixing these two phrases can also cause confusion. "Due to" is an adjective phrase meaning "attributable to" and almost always follows a "to be" verb (such as "is," "was," or "were"). It should not be used in place of prepositional phrases such as "because of," "owing to," or "as a result of." EXAMPLES:

- "The cracked walls were *due to* the lack of proper foundation fill being used during construction."
- "We won the contract *because of* [not *due to*] our thorough understanding of the client's needs."

e.g./i.e.

The abbreviation "e.g." means "for example," whereas "i.e." means "that is." These two Latin abbreviations are often confused, a fact that should give you pause before using them. Many writers prefer to write them out, rather than risk confusion on the part of the reader. EXAMPLES:

- "During the trip, he visited 12 cities where McDuff is considering opening offices—*e.g.*, [or, preferably, *for example*] Kansas City, New Orleans, and Seattle.
- "A spot along the Zayante Fault was the earthquake's epicenter—*i.e.*, [or, preferably, *that is*] the focal point for seismic activity."

Fewer/Less

The adjective "fewer" is used before items that can be counted, whereas the adjective "less" is used before mass quantities. When errors occur, they usually result from "less" being used with countable items, as in this *incorrect* sentence: "We can complete the job with less men at the site." EXAMPLES:

- "The newly certified industrial hygienist signed with us because the other firm in which he was interested offered *fewer* [not *less*] benefits."
- "There was *less* sand in the sample taken from 15 ft than in the one taken from 10 ft."

Flammable/Inflammable/Nonflammable

Given the importance of these words in avoiding injury and death, make sure to use them correctly—especially in instructions. "Flammable" means "capable of burning quickly" and is acceptable usage. "Inflammable" has the same meaning, but it is *not* acceptable usage for this reason: Some readers confuse it with "nonflammable." The word "nonflammable," then, means "not capable of burning" and is accepted usage. EXAMPLES:

- "They marked the package *flammable* because its contents could be easily ignited by a spark." (Note that *flammable* is preferred here over its synonym, *inflammable*.)
- "The foreman felt comfortable placing the crates near the heating unit, since all the crates' contents were *nonflammable*."

Fortuitous/Fortunate

The word "fortuitous" is an adjective that refers to an unexpected action, without regard to whether it is desirable or not. The word "fortunate" is an adjective that indicates an action that is clearly desired. The common usage error with this pair is the wrong assumption that "fortuitous" events must also be "fortunate." EXAMPLES:

- "Seeing McDuff's London manager at the conference was quite *fortuitous,* since I had not been told that he also was attending."
- "It was indeed *fortunate* that I encountered the London manager, for it gave us the chance to talk about an upcoming project involving both our offices."

Generally/Typically/Usually

Words like these can be useful qualifiers in your reports. They indicate to the reader that what you have stated is often, but not always, the case. Make certain to place these adverb modifiers as close as possible to the words they modify. In the first example here, it would be inaccurate to write "were typically sampled," in that the adverb modifies the entire verb phrase "were sampled." EXAMPLES:

- "Cohesionless soils *typically* were sampled by driving a 2-in. diameter, split-barrel sampler." (Active-voice alternative: "*Typically,* we sampled cohesionless soils by driving a 2-in. diameter, split-barrel sampler.")
- "For projects like the one you propose, the technician *usually* will clean the equipment before returning to the office."
- "It is *generally* known that sites for dumping waste should be equipped with appropriate liners."

Good/Well

Though similar in meaning, "good" is used as an adjective and "well" is used as an adverb. A common usage error occurs when writers use the adjective when the adverb is required. EXAMPLES:

- "It is *good* practice to submit three-year plans on time."
- "He did *well* to complete the three-year plan on time, considering the many reports he had to finish that same week."

Imply/Infer

Remember that the person doing the speaking or writing implies, whereas the person hearing or reading the words infers. In other words, the word "imply" requires an active role; the word "infer" requires a passive role. When you imply a point, your words suggest rather than state a point. When you infer a point, you form a conclusion or deduce meaning from someone else's words or actions. EXAMPLES:

- "The contracts officer *implied* that there would be stiff competition for that $20 million waste-treatment project."
- "We *inferred* from her remarks that any firm hoping to secure the work must have completed similar projects recently."

Its/It's

These words are often confused. You can avoid error by remembering that "it's" with the apostrophe is used *only* as a contraction for "it is" or "it has." The other form—"its"—is a possessive pronoun. EXAMPLES:

- "Because of the rain, *it's* [or *it is*] going to be difficult to move the equipment to the site."
- "*It's* [or *it has*] been a long time since we submitted the proposal."
- "The company completed *its* part of the agreement on time."

Loose/Lose

"Loose," which rhymes with "goose," is an adjective that means "unfastened, flexible, or unconfined." "Lose," which rhymes with "ooze," is a verb that means "to misplace." EXAMPLES:

- "The power failure was linked to a *loose* connection at the switchbox."
- "Because of poor service, the photocopy machine company may *lose* its contract with McDuff's San Francisco office."

Modifiers: Dangling and Misplaced

This section includes guidelines for avoiding the most common modification errors—dangling modifiers and misplaced modifiers. But first we need to define the term "modifier." Words, phrases, and even dependent clauses can serve as modifiers. They serve to qualify, or add meaning to, other elements in the sentence. For our purposes here, the most important point is that modifiers need to be clearly connected to what they modify.

Modification errors occur most often with verbal phrases. A phrase is a group of words that lacks either a subject or predicate. The term "verbal" refers to (1) gerunds (*-ing* form of verbs used as nouns, such as "He likes *skiing*"), (2) participles (*-ing* form of verbs used as adjectives, such as "Skiing down the hill, he lost a glove"), or (3) infinitives (the word "to" plus the verb root, such as "To attend the opera was his favorite pastime"). Now let's look at the two main modification errors.

Dangling Modifiers. When a verbal phrase "dangles," the sentence in which it is used contains no specific word for the phrase to modify. As a result, the meaning of the sentence can be confusing to the reader. For example, "In designing the foundation, several alternatives were discussed." It is not at all clear exactly who is doing the "designing." The phrase dangles because it does not modify a specific word. The modifier does not dangle in this version of the sentence: "In designing the foundation, we discussed several alternatives."

Misplaced Modifiers. When a verbal phrase is misplaced, it may appear to refer to a word that it, in fact, does not modify. EXAMPLE: "Floating peacefully near the oil rig, we saw two humpback whales." Obviously, the whales are doing the floating, and the rig workers are doing the seeing here. Yet because the verbal phrase is placed at the beginning of the sentence, rather than at the end immediately after the word it modifies, the sentence presents some momentary confusion.

Misplaced modifiers can lead to confusion about the agent of action in technical tasks. EXAMPLE: "Before beginning to dig the observation trenches, we recommend that the contractors submit their proposed excavation program for our review." On quick reading, the reader is not certain about who will be "beginning to dig"—the contractors or the "we" in the sentence. The answer is the contractors. Thus a correct placement of the modifier would be "We recommend the following: Before the contractors begin digging observation trenches, they should submit their proposed excavation for our review."

Solving Modifier Problems. At best, dangling and misplaced modifiers produce a momentary misreading by the audience. At worst, they can lead to confusion that results in disgruntled readers, lost customers, or liability problems. To prevent modification problems, place all verbal phrases—indeed, all modifiers—as close as possible to the word they modify. If you spot a modification error while you are editing, correct it in one of two ways:

1. Leave the modifier as it is and rework the rest of the sentence. Thus you would change "Using an angle of friction of 20 degrees and a vertical weight of 300 tons, the sliding resistance would be . . ." to the following: "Using an angle of friction of 20 degrees and a vertical weight of 300 tons, we computed a sliding resistance of. . . ."

2. Rephrase the modifier as a complete clause. Thus you would change the previous original sentence to "If the angle of friction is 20 degrees and the vertical weight is 300 tons, the sliding resistance should be. . . ."

In either case, your goal is to link the modifier clearly and smoothly with the word or phrase it modifies.

Numbers

Like rules for abbreviations, those for numbers vary from profession to profession and even from company to company. Most technical writing subscribes to the approach that numbers are best expressed in figures (45) rather than words (forty-five). Note that this style may differ from that used in other types of writing, such as this textbook. Unless the preferences of a particular reader suggest that you do otherwise, follow these common rules for use of numbers in writing your technical documents:

■ *Rule 1: Follow the 10-or-Over Rule*

In general, use figures for numbers of 10 or more, words for numbers below 10. EXAMPLES: three technicians at the site/15 reports submitted last month/one rig contracted for the job.

■ *Rule 2: Do Not Start Sentences With Figures*

Begin sentences with the word form of numbers, not with figures. EXAMPLE: "Forty-five containers were shipped back to the lab."

■ *Rule 3: Use Figures as Modifiers*

Whether above or below 10, numbers are usually expressed as figures when used as modifiers with units of measurement, time, and money, especially when these units are abbreviated. EXAMPLES: 4 in., 7 hr, 17 ft, $5 per hr. Exceptions can be made when the unit is not abbreviated. EXAMPLE: five years.

■ *Rule 4: Use Figures in a Group of Mixed Numbers*

Use only figures when the numbers grouped together in a passage (usually *one* sentence) are both above and below 10. EXAMPLE: "For that project they assembled 15 samplers, 4 rigs, and 25 containers." In other words, this rule argues for consistency within a writing unit.

■ *Rule 5: Use the Figure Form in Illustration Titles*

Use the numeric form when labeling specific tables and figures in your reports. EXAMPLES: Figure 3, Table 14–B.

■ *Rule 6: Be Careful With Fractions*

Express fractions as words when they stand alone, but as figures when they are used as a modifier or are joined to whole numbers. EXAMPLE: "We have completed two-thirds of the project using the 2 1/2 in. pipe."

■ *Rule 7: Use Figures and Words With Numbers in Succession*

When two numbers appear in succession in the same unit, write the first as a word and the second as a figure. EXAMPLE: "We found fifteen 2-ft pieces of pipe in the machinery."

■ *Rule 8: Only Rarely Use Numbers in Parentheses*

Except in legal documents, avoid the practice of placing figures in parentheses after their word equivalents. EXAMPLE: "The second party will send the first party forty-five (45) barrels on or before the first of each month." Note that the parenthetical amount is placed immediately after the figure, not after the unit of measurement.

■ Rule 9: Use Figures With Dollars

Use figures with all dollar amounts, with the exception of the context noted in Rule 8. Avoid cents columns unless exactness to the penny is necessary.

■ Rule 10: Use Commas in Five-Digit Figures

To prevent possible misreading, use commas in figures of five digits or more. Include the comma with four-digit numbers only when they are grouped together with five-digit numbers, in the text or in tables. EXAMPLES: 15,000; 1,247; 6,003.

■ Rule 11: Use Words for Ordinals

Usually spell out the ordinal form of numbers. EXAMPLE: "The government informed all parties of the first, second, and third [not 1st, 2nd, and 3rd] choices in the design competition." A notable exception is tables and figures, where space limitations could argue for the abbreviated form.

Oral/Verbal

"Oral" refers to words that are spoken, as in "oral presentation." The term "verbal" refers to spoken or written language. To prevent confusion, avoid the word "verbal" and instead specify your meaning with the words "oral" and "written." EXAMPLES:

- "In its international operations, McDuff, Inc., has learned that some countries still rely upon *oral* [not *verbal*] contracts."
- "Their *oral* agreement last month was followed by a *written* [not *verbal*] contract this month."

Parts of Speech

This term refers to the eight main groups of words in English grammar. A word's placement in one of these groups is based upon its function within the sentence.

Noun. Words in this group name persons, places, objects, or ideas. The two major categories are (1) proper nouns and (2) common nouns. Proper nouns name specific persons, places, objects, or ideas, and they are capitalized. EXAMPLES: Cleveland; Mississippi River; McDuff, Inc.; Student Government Association; Susan Jones; Existentialism. Common nouns name general groups of persons, places, objects, and ideas, and they are not capitalized. EXAMPLES: trucks, farmers, engineers, assembly lines, philosophy.

Verb. A verb expresses action or state of being. Verbs give movement to sentences and form the core of meaning in your writing. EXAMPLES: explore, grasp, write, develop, is, has.

Pronoun.   A pronoun is a substitute for a noun. Some sample pronoun categories include (1) personal pronouns (I, we, you, she, he), (2) relative pronouns (who, whom, that, which), (3) reflexive and intensive pronouns (myself, yourself, itself), (4) demonstrative pronouns (this, that, these, those), and (5) indefinite pronouns (all, any, each, anyone).

Adjective.   An adjective modifies a noun. EXAMPLES: horizontal, stationary, green, large, simple.

Adverb.   An adverb modifies a verb, an adjective, another adverb, or a whole statement. EXAMPLES: soon, generally, well, very, too, greatly.

Preposition.   A preposition shows the relationship between a noun or pronoun (the object of a preposition) and another element of the sentence. Forming a prepositional phrase, the preposition and its object can reveal relationships such as location (''They went *over the hill*''), time (''He left *after the meeting*''), and direction (''She walked toward the office'').

Conjunction.   A conjunction is a connecting word that links words, phrases, or clauses. EXAMPLES: and, but, for, nor, although, after, because, since.

Interjection.   As an expression of emotion, an interjection can stand alone (''Look out!'') or can be inserted into another sentence.

Per Cent/Percent/Percentage

''Per cent'' and ''percent'' have basically the same usage and are used with exact numbers. The one word ''percent'' is preferred. Even more common in technical writing, however, is the use of the percent sign (%) after numbers. The word ''percentage'' is only used to express general amounts, not exact numbers. EXAMPLES:

- ''After completing a marketing survey, McDuff, Inc., discovered that 83 *percent* [or 83%] of its current clients have hired McDuff for previous projects.''
- ''A large *percentage* of the defects can be linked to the loss of two experienced quality-control inspectors.''

Principal/Principle

When these two words are misused, the careful reader will notice. Keep them straight by remembering this simple distinction: ''Principle'' is always a noun that means ''basic truth, belief, or theorem.'' EXAMPLE: ''He believed in the principle of free speech.'' ''Principal'' can be either a noun or an adjective and has three basic uses:

- **As a noun meaning "head official" or "person who plays a major role."** EXAMPLE: ''We asked that a *principal* in the firm sign the contract.''

- **As a noun meaning "the main portion of a financial account upon which interest is paid."** EXAMPLE: "If we deposit $5,000 in *principal,* we will earn 9 percent interest."
- **As an adjective meaning "main or primary."** EXAMPLE: "We believe that the *principal* reason for contamination at the site is the leaky underground storage tank."

Pronouns: Agreement and Reference

A pronoun is a word that replaces a noun, which is called the "antecedent" of the pronoun. EXAMPLES: this, it, he, she, they. Pronouns, as such, provide you with a useful strategy for varying your style by avoiding repetition of nouns. Here are some rules to prevent pronoun errors:

■ Rule 1: Make Pronouns Agree With Antecedents

Check every pronoun to make certain it agrees with its antecedent in number. That is, both noun and pronoun must be singular, or both must be plural. Of special concern are the pronouns "it" and "they." EXAMPLES:

- Change "McDuff, Inc., plans to complete their Argentina project next month" to this sentence: "McDuff, Inc., plans to complete its Argentina project next month."
- Change "The committee released their recommendations to all departments" to this sentence: "The committee released its recommendations to all departments."

■ Rule 2: Be Clear About the Antecedent of Every Pronoun

There must be no question about what noun a pronoun replaces. Any confusion about the antecedent of a pronoun can change the entire meaning of a sentence. To avoid such reference problems, you may need to rewrite a sentence or even use a noun rather than a pronoun. Do whatever is necessary to prevent misunderstanding by your reader. EXAMPLE: Change "The gas filters for these tanks are so dirty that they should not be used" to this sentence: "These filters are so dirty that they should not be used."

■ Rule 3: Avoid Using "This" as the Subject Unless a Noun Follows It

A common stylistic error is the vague use of "this," especially as the subject of a sentence. Sometimes the reference is not clear at all; sometimes the reference may be clear after several readings. In almost all cases, however, the use of "this" as a pronoun reflects poor technical style and tends to make the reader want to ask, "This what?" Instead, make the subject of your sentences concrete, either by adding a noun after the "this" or by recasting the sentence. EXAMPLE: Change "He talked constantly about the project to be completed at the Olympics. This made his office-mates irritable" to the following: "His constant talk about the Olympics project irritated his office-mates."

Punctuation: General

Commas.   Most writers struggle with commas, so you are not alone. The problem is basically threefold. First, the teaching of punctuation has been approached in different, and sometimes quite contradictory, ways. Second, comma rules themselves are subject to various interpretations. And third, problems with comma placement often mask more fundamental problems with the structure of a sentence itself.

You need to start by knowing the basic rules of comma use. The rules that follow are fairly simple. If you learn them now, you will save yourself a good deal of time later in that you will not be constantly questioning usage. In other words, the main benefit of learning the basics of comma use is increased confidence in your own ability to handle the mechanics of editing. (If you do not understand some of the grammatical terms that follow, such as "compound sentence," refer to the section on sentence structure.)

■ Rule 1: Commas in a Series

Use commas to separate words, phrases, and short clauses written in a series of three or more items. EXAMPLE: "The samples contained gray sand, sandy clay, and silty sand." According to current usage, a comma always comes before the "and" in a series.

■ Rule 2: Commas in Compound Sentences

Use a comma before the conjunction that joins main clauses in a compound sentence. EXAMPLE: "We completed the drilling at the Smith Industries location, and then we grouted the holes with Sakrete." The comma is needed here because it separates two complete clauses, each with its own subject and verb ("we completed" and "we grouted"). If the second "we" had been deleted, there would be only one clause containing one subject and two verbs ("we completed and grouted"). Thus no comma would be needed. Of course, it may be that a sentence following this comma rule is far too long; do not use the rule to string together intolerably long sentences.

■ Rule 3: Commas With Nonessential Modifiers

Set off nonessential modifiers with commas—either at the beginning, middle, or end of sentences. Nonessential modifiers are usually phrases that add more information to a sentence, rather than greatly changing its meaning. When you speak, often there is a pause between this kind of modifier and the main part of the sentence, giving you a clue that a comma break is needed. EXAMPLE: "The report, which we submitted three weeks ago, indicated that the company would not be responsible for transporting hazardous wastes." But—"The report that we submitted three weeks ago indicated that the company would not be responsible for transporting hazardous wastes." The first example includes a nonessential modifier, would be spoken with pauses, and therefore uses separating commas. The second example includes an *essential* modifier, would be spoken *without* pauses, and therefore includes *no* separating commas.

■ *Rule 4: Commas With Adjectives in a Series*

Use a comma to separate two or more adjectives that modify the same noun. To help you decide if adjectives modify the same noun, use this test: If you can reverse their positions and still retain the same meaning, then the adjectives modify the same word and should be separated by a comma. EXAMPLE: "Jason opened the two containers in a clean, well-lighted place."

■ *Rule 5: Commas With Introductory Elements*

Use a comma after introductory phrases or clauses of about five words or more. EXAMPLE: "After completing the topographic survey of the area, the crew returned to headquarters for its weekly project meeting." Commas like the one after "area" help readers separate secondary or modifying points from your main idea, which of course should be in the main clause. Without these commas, there may be difficulty reading such sentences properly.

■ *Rule 6: Commas in Dates, Titles, Etc.*

Abide by the conventions of comma usage in punctuating dates, titles, geographic place names, and addresses. EXAMPLES:

- "May 3, 1996, is the projected date of completion." (But note the change in the "military" form of dates: "We will complete the project on 3 May 1996.")
- "John F. Dunwoody, Ph.D., has been hired to assist on the project."
- "McDuff, Inc., has been selected for the project."
- "He listed Dayton, Ohio, as his permanent residence."

Note the need for commas after the year "1996," the title "Ph.D.," the designation "Inc.," and the state name "Ohio." Also note that if the day had not been in the first example, there would be *no* comma between the month and year and no comma after the year.

Semicolons.   The semicolon is easy to use if you remember that it, like a period, indicates the end of a complete thought. Its most frequent use is in situations where grammar rules would allow you to use a period but where your stylistic preference is for a less abrupt connector. EXAMPLE: "Five engineers left the convention hotel after dinner; only two returned by midnight."

One of the most common punctuation errors, the comma splice, occurs when a comma is used instead of a semicolon or period in compound sentences connected by words such as "however," "therefore," "thus," and "then." When you see that these connectors separate two main clauses, make sure either to use a semicolon or to start a new sentence. EXAMPLE: "We made it to the project site by the agreed-upon time; however, [or ". . . time. However, . . .] the rain forced us to stay in our trucks for two hours."

As noted in the "Lists" entry, there is another instance in which you might use semicolons. Place them after the items in a list when you are treating the list like a sentence and when any one of the items contains internal commas.

Colons.   As mentioned in the "Lists" entry, you should place a colon immediately after the last word in the lead-in before a formal list of bulleted or numbered items. EXAMPLE: "Our field study involved these three steps:" or "In our field study we were asked to:" The colon may come after a complete clause, as in the first example, or it may split a grammatical construction, as in the second example. However, it is preferable to use a complete clause before a formal list.

The colon can also be used in sentences in which you want a formal break before a point of clarification or elaboration. EXAMPLE: "They were interested in just one result: quality construction." In addition, use the colon in sentences in which you want a formal break before a series that is not part of a listing. EXAMPLE: "They agreed to perform all on-site work required in these four cities: Houston, Austin, Laredo, and Abilene." But note that there is no colon before a sentence series without a break in thought. EXAMPLE: "They agreed to perform all the on-site work required in Houston, Austin, Laredo, and Abilene."

Apostrophes.   The apostrophe can be used for contractions, for some plurals, and for possessives. Only the latter two uses cause confusion. Use an apostrophe to indicate the plural form of a word as a word. EXAMPLE: "That redundant paragraph contained seven *area's* and three *factor's* in only five sentences." Although some writers also use apostrophes to form the plurals of numbers and full-cap abbreviations, the current tendency is to include only the "s." EXAMPLES: 7s, ABCs, PCBs, P.E.s.

As for possessives, you probably already know that the grammar rules seem to vary, depending on the reference book you are reading. Here are some simple guidelines:

■ *Possessive Rule 1*

Form the possessive of multisyllabic nouns that end in "s" by adding just an apostrophe, whether the nouns are singular or plural. EXAMPLES: actress' costume, genius' test score, the three technicians' samples, Jesus' parables, the companies' joint project.

■ *Possessive Rule 2*

Form the possessive of one-syllable, singular nouns ending in "s" or an "s" sound by adding an apostrophe plus "s." EXAMPLES: Hoss's horse, Tex's song, the boss's progress report.

■ *Possessive Rule 3*

Form the possessive of all plural nouns ending in ''s'' or an ''s'' sound by adding just an apostrophe. EXAMPLES: the cars' engines, the ducks' flight path, the trees' roots.

■ *Possessive Rule 4*

Form the possessive of all singular and plural nouns not ending in ''s'' by adding an apostrophe plus ''s.'' EXAMPLES: the man's hat, the men's team, the company's policy.

■ *Possessive Rule 5*

Form the possessive of paired nouns by first determining whether there is joint ownership or individual ownership. For joint ownership, make only the last noun possessive. For individual ownership, make both nouns possessive. EXAMPLE: ''Susan and Terry's project was entered in the science fair; but Tom's and Scott's projects were not.''

Quotation Marks.   In technical writing, you may want to use this form of punctuation to draw attention to particular words, to indicate passages taken directly from another source, or to enclose the titles of short documents such as reports or book chapters. The rule to remember is this: Periods and commas go inside quotation marks; semicolons and colons go outside quotation marks.

Parentheses.   Use parentheses carefully, since long parenthetical expressions can cause the reader to lose the train of thought. This form of punctuation can be used when you (1) place an abbreviation after a complete term, (2) add a brief explanation within the text, or (3) include reference citations within the document text (as explained in chapter 13). The period goes after the close parenthesis when the parenthetical information is part of the sentence, as in the previous sentence. (However, it goes inside the close parenthesis when the parenthetical information forms its own sentence, as in the sentence you are reading.)

Brackets.   Use a pair of brackets for these purposes: (1) to set off parenthetical material already contained within another parenthetical statement and (2) to draw attention to a comment you are making within a quoted passage. EXAMPLE: ''Two McDuff studies have shown that the Colony Dam is up to safety standards. (See Figure 4–3 [Dam Safety Record] for a complete record of our findings.) In addition, the county engineer has a letter on file that will give further assurance to prospective homeowners on the lake. His letter notes that 'After finishing my three-month study [he completed the study in July 1993], I conclude that the Colony Dam meets all safety standards set by the county and state governments.' ''

Hyphens.　　The hyphen is used to form certain word compounds in English. Although the rules for its use sometimes seem to change from handbook to handbook, those that follow are the most common.

■ *Hyphen Rule 1*

Use hyphens with compound numerals. EXAMPLE: twenty-one through ninety-nine.

■ *Hyphen Rule 2*

Use hyphens with most compounds that begin with "self." EXAMPLES: self-defense, self-image, self-pity. Other "self" compounds, like "selfhood" and "selfsame," are written as unhyphenated words.

■ *Hyphen Rule 3*

Use hyphens with group modifiers when they precede the noun but not when they follow the noun. EXAMPLES: a well-organized paper, a paper that was well organized, twentieth-century geotechnical technology, bluish-gray shale, fire-tested material, thin-bedded limestone.

However, remember that when the first word of the modifier is an adverb ending in "-ly," place no hyphen between the words. EXAMPLES: carefully drawn plate, frightfully ignorant teacher.

■ *Hyphen Rule 4*

Place hyphens between prefixes and root words in the following cases: (a) between a prefix and a proper name (ex-Republican, pre-Sputnik); (b) between some prefixes that end with a vowel and root words beginning with a vowel, particularly if the use of a hyphen would prevent an odd spelling (semi-independent, re-enter, re-elect); and (c) between a prefix and a root when the hyphen helps to prevent confusion (re-sent, not resent; re-form, not reform; re-cover, not recover).

Punctuation: Lists

As noted in chapter 4 ("Page Design"), listings draw attention to parallel pieces of information whose importance would be harder to grasp in paragraph format. In other words, employ lists as an attention-getting strategy. Following are some general pointers for punctuating lists. (See pp. 94–96 in chapter 4 for other rules for lists.)

You have three main options for punctuating a listing. The common denominators for all three are that you (1) always place a colon after the last word of the lead-in and (2) always capitalize the first letter of the first word of each listed item.

Option A: Place no punctuation after listed items. This style is appropriate when the list includes only short phrases. More and more writers are choosing this option, as opposed to Option B. EXAMPLE:

"In this study, we will develop recommendations that address these six concerns in your project:

- Site preparation
- Foundation design
- Sanitary-sewer design
- Storm-sewer design
- Geologic surface faulting
- Projections for regional land subsidence"

Option B: Treat the list like a sentence series. In this case, you place commas or semicolons between items and a period at the end of the series. Whether you choose Option A or B largely depends on your own style or that of your employer. EXAMPLE:

"In this study, we developed recommendations that dealt with four topics:

- Site preparation,
- Foundation design,
- Sewer construction, and
- Geologic faulting."

Note that this option requires you to place an "and" after the comma that appears before the last item. Another variation of Option B occurs when you have internal commas within one or more of the items. In this case, you need to change the commas that follow the listed items into semicolons. Yet you still keep the "and" before the last item. EXAMPLE:

"Last month we completed environmental assessments at three locations:

- A gas refinery in Dallas, Texas;
- The site of a former chemical plant in Little Rock, Arkansas; and
- A waste pit outside of Baton Rouge, Louisiana."

Option C: Treat each item like a separate sentence. When items in a list are complete sentences, you may want to punctuate each one like a separate sentence, placing a period at the end of each. You *must* choose this option when one or more of your listed items contain more than one sentence. EXAMPLE:

"The main conclusions of our preliminary assessment are summarized here:

- At five of the six borehole locations, petroleum hydrocarbons were detected at concentrations greater than a background concentration of 10 mg/kg.
- No PCB concentrations were detected in the subsurface soils we analyzed. We will continue the testing, as discussed in our proposal.
- Sampling and testing should be restarted three weeks from the date of this report."

Sic

Latin for "thus," this word is most often used when a quoted passage contains an error or other point that might be questioned by the reader. Inserted within brackets, "sic" shows the reader that the error was included in the original passage—and that it was not introduced by you. EXAMPLE: "The customer's letter to our sales department claimed that 'there are too [sic] or three main flaws in the product.' "

Spelling

All writers find at least some words difficult to spell, and some writers have major problems with spelling. Automatic spell-checking software helps solve the problem. Yet you still need to remain vigilant during the proofreading stage. One or more misspelled words in an otherwise well-written document may cause readers to question professionalism in other areas.

This entry includes a list of commonly misspelled words. However, you should keep your own list of words you most frequently have trouble spelling. Like most writers, you probably have a relatively short list of words that give you repeated difficulty.

absence	column	exaggerate	interfered
accessible	commitment	existence	interference
accommodate	committee	experience	interrupt
accumulate	compatible	familiar	irrelevant
accustomed	compelled	favorite	judgment
achievement	conscience	February	knowledge
acknowledgment	conscientious	foreign	later
acquaintance	conscious	foresee	latter
admittance	controlled	forfeit	liable
advisable	convenient	forty	liaison
aisle	definitely	fourth	library
allotting	dependable	genius	lightning
analysis	descend	government	likely
analyze	dilemma	guarantee	loneliness
arctic	disappear	guidance	maintenance
athlete	disappoint	handicapped	manageable
athletic	disaster	harass	maneuver
awful	disastrous	height	mathematics
basically	efficient	illogical	medieval
believable	eligible	incidentally	mileage
benefited	embarrass	independence	miscellaneous
bulletin	endurance	indispensable	misspelled
calendar	environment	ingenious	mortgage
career	equipment	initially	movable
changeable	equipped	initiative	necessary
channel	essential	insistence	noticeable

nuisance	preference	safety	undoubtedly
numerous	preferred	similar	unmistakably
occasionally	privilege	sincerely	until
occurred	profession	specifically	useful
occurrence	professor	subtle	usually
omission	pronunciation	temperament	valuable
pamphlet	publicly	temperature	various
parallel	quantity	thorough	vehicle
pastime	questionnaire	tolerance	wholly
peculiar	recession	transferred	writing
possess	reference	truly	written
practically			

Subject-Verb Agreement

Subject-verb agreement errors are quite common in technical writing. They occur when writers fail to make the subject of a clause agree in number with the verb. EXAMPLE: "The nature of the diverse geological deposits are explained in the report." (The verb should be "is," since the singular subject is "nature.")

Writers who tend to make these errors should devote special attention to them. Specifically, isolate the subjects and verbs of all the clauses in a document and make certain that they agree. Here are seven specific rules for making subjects agree with verbs:

■ Rule 1: Subjects Connected by "And" Take Plural Verbs

This rule applies to two or more words or phrases that, together, form one subject phrase. EXAMPLE: "The site preparation section and the foundation design portion of the report are to be written by the same person."

■ Rule 2: Verbs After "Either/Or" Agree With the Nearest Subject

Subject words connected by "either" or "or" confuse many writers, but the rule is very clear. Your verb choice depends on the subject nearest the verb. EXAMPLE: "He told his group that neither the three reports nor the proposal was to be sent to the client that week."

■ Rule 3: Verbs Agree With the Subject, Not With the Subjective Complement

Sometimes called a predicate noun or adjective, a subjective complement renames the subject and occurs after verbs such as "is," "was," "are," and "were." EXAMPLE:

"The theme of our proposal is our successful projects in that region of the state." But the same rule would permit this usage: "Successful projects in that part of the state are the theme we intend to emphasize in the proposal."

■ Rule 4: Prepositional Phrases Do Not Affect Matters of Agreement

"As long as," "in addition to," "as well as," and "along with" are prepositions, not conjunctions. A verb agrees with its subject, not with the object of a prepositional phrase. EXAMPLE: "The manager of human resources, along with the personnel director, is supposed to meet with the three applicants."

■ Rule 5: Collective Nouns Usually Take Singular Verbs

Collective nouns have singular form but usually refer to a group of persons or things (for example, "team," "committee," or "crew"). When a collective noun refers to a group as a whole, use a singular verb. EXAMPLE: "The project crew was ready to complete the assignment." Occasionally, a collective noun refers to the members of the group acting in their separate capacities. In this case, either use a plural verb or, to avoid awkwardness, reword the sentence. EXAMPLE: "The crew were not in agreement about the site locations." Or, "Members of the crew were not in agreement about the site locations."

■ Rule 6: Foreign Plurals Usually Take Plural Verbs

Although usage is gradually changing, most careful writers still use plural verbs with "data," "strata," "phenomena," "media," and other irregular plurals. EXAMPLE: "The data he asked for in the request for proposal are incorporated into the three tables."

■ Rule 7: Indefinite Pronouns Like "Each" and "Anyone" Take Singular Verbs

Writers often fail to follow this rule when they make the verb agree with the object of a prepositional phrase, instead of with the subject. EXAMPLE: "Each of the committee members are ready to adjourn" (incorrect). "Each of the committee members is ready to adjourn" (correct).

To/Too/Two

"To" is part of the infinitive verb form *or* is a preposition in a prepositional phrase. "Too" is an adverb that suggests an excessive amount *or* that means "also." "Two" is a noun or an adjective that stands for the numeral "2." EXAMPLES:

- "He volunteered *to* go [infinitive verb] *to* Alaska [prepositional phrase] *to* work [another infinitive verb form] on the project."
- "Stephanie explained that the proposed hazardous-waste dump would pose *too* many risks *to* the water supply. Scott made this point, *too.*"

Utilize/Use

"Utilize" is simply a long form for the preferred verb "use." Although some verbs that end in "-ize" are useful words, most are simply wordy substitutes for shorter forms. As some writing teachers say, "Why use 'utilize' when you can use 'use.' "

Who/Whom

These two words give writers (and speakers) fits, but the importance of their correct use probably has been exaggerated. If you want to be one who uses them properly, remember this basic point: "Who" is a subjective form that can only be used in the subject slot of a clause; "whom" is an objective form that can only be used as a direct object or other nonsubject noun form of a sentence. EXAMPLES:

- "The man *who* you said called me yesterday is a good customer of the firm." (The clause "who . . . called me yesterday" modifies "man." Within this clause, "who" is the subject of the verb "called." Note that the subject role of "who" is not affected by the two words "you said," which interrupt the clause.)
- "They could not remember the name of the person *whom* they interviewed." (The clause "whom they interviewed" modifies "person." Within this clause, "whom" is the direct object of the verb "interviewed.")

EXERCISE 1: GRAMMAR AND MECHANICS

The following passages contain a variety of grammatical and mechanical errors covered in the handbook. The major focus is punctuation. Rewrite each passage.

1. Some concerns regarding plumbing design are mentioned in our report, however, no unusual design problems are expected.
2. An estimate of the total charges for an audit and for three site visits are based on our standard fee schedules.
3. The drill bit was efficient cheap and available.
4. The plan unless we have completely misjudged it, will increase sales markedly.
5. Our proposal contains design information for these two parts of the project; Phase 1 (evaluating the 3 computers) and Phase 2 (installing the computer selected).
6. If conditions require the use of all-terrain equipment to reach the construction locations, this will increase the cost of the project slightly.
7. An asbestos survey was beyond the scope of this project, if you want one, we would be happy to submit a proposal.
8. Jones-Simon Company, the owners of the new building, were informed of the problem with the foundation.
9. Also provided is the number and type of tests to be given at the office.
10. Calculating the standard usages by the current purchase order prices result in a downward adjustment of $.065.

11. Data showing the standard uses of the steel, including allowances for scrap, waste and end pieces of the tube rolls, are included for your convenience at the end of this report in Table 7.
12. This equipment has not been in operation for 3 months, and therefore, its condition could not be determined by a quick visual inspection.
13. Arthur Jones Manager of the Atlanta branch wrote that three proposals had been accepted.
14. The generator that broke yesterday has been shipped to Tampa already by Harry Thompson.
15. The first computer lasted eight years the second two years.
16. He wants one thing out of their work speed.
17. On 25 September 1993 the papers were signed.
18. On March 23 1993 the proposal was accepted.
19. The meeting was held in Columbus the Capital of Ohio.
20. McDuff, Inc. completed its Indonesia project in record time.
21. He decided to write for the brochure then he changed his mind.
22. Interest by the Kettering Hospital staff in the development of a masterplan for the new building wings have been expressed.
23. However much he wants to work for Gasion engineering he will turn the job down if he has to move to another state.
24. 35 computer scientists attended the convention, but only eleven of them were from private industry.
25. Working at a high salary gives him some satisfaction still he would like more emotional satisfaction from his job.
26. His handwriting is almost unreadable therefore his secretary asked him to dictate letters.
27. Any major city especially one that is as large as Chicago is bound to have problems with mass transit.
28. He ended his speech by citing the company motto; "Quality first, last, and always".
29. Houston situated on the Gulf of Mexico is an important international port.
30. The word "effect" is in that student's opinion a difficult one to use.
31. All persons who showed up for the retirement party, told stories about their association with Charlie over the years.
32. The data that was included in the study seems inconclusive.
33. My colleague John handled the presentation for me.
34. Before he arrived failure seemed certain.
35. While evaluating the quality of her job performance a study was made of her writing skills by her supervisor.
36. I shall contribute to the fund for I feel that the cause is worthwhile.
37. James visited the site however he found little work finished.
38. There are three stages cutting grinding and polishing.
39. The three stages are cutting grinding and polishing.
40. Writers occasionally create awkward verbs "prioritize" and "terminate" for example.

41. Either the project engineers or the consulting chemist are planning to visit with the client next week.
42. Besides Gerry Dave worked on the Peru project.
43. The corporation made a large unexpected gift to the university.
44. The reason for his early retirement are the financial incentives given by his employer.
45. Profit, safety and innovation are the factors that affect the design of many foundations.
46. No later than May 1995 the building will be finished.
47. Each of the committee members complete a review of the file submitted by the applicant.
48. The team completed their collaborative writing project on schedule.
49. Both the personnel officers and the one member of the quality team is going to attend the conference in Fargo.
50. He presented a well organized presentation but unfortunately the other speakers on the panel were not well-prepared.

EXERCISE 2: USAGE

For each of the following passages, select the correct word or phrase from the choices within the parentheses. Be ready to explain the rationale for your choice.

1. John (implied, inferred) in his report that TransAm Oil should reject the bid.
2. Before leaving on vacation, the company president left instructions for the manner in which responsibilities should be split (among, between) the three vice presidents.
3. Harold became (uninterested, disinterested) in the accounting problem after working on it for 18 straight hours.
4. A large (percent, percentage) of the tellers is dissatisfied with the revised work schedule.
5. The typist responded that he would make (less, fewer) errors if the partner would spell words correctly in the draft.
6. From her reading of the annual report, Ms. Jones (inferred, implied) that the company might expand its operations.
7. The president's decision concerning flextime will be (effected, affected) by the many conversations he is having with employees about scheduling difficulties.
8. His (principal, principle) concern was that the loan's interest and (principle, principal) remain under $500.
9. Throughout the day, his concentration was interrupted (continuously, continually) by phone calls.
10. He jogged (continuously, continually) for 20 minutes.
11. Five thousand books (compose, comprise) his personal library.
12. The clients (who, whom) he considered most important received Christmas gifts from the company.

13. The company decided to expand (its, it's) operations in the hope that (its, it's) the right time to do so.
14. The (nonflammable, flammable, inflammable) liquids were kept in a separate room, because of their danger.
15. They waited for (awhile, a while) before calling the subcontractor.
16. Caution should be taken to (ensure, insure) that the alarm system will not go off accidentally.
17. The new floors (are comprised of, are composed of, comprise) a thick concrete mixture.
18. He (expects, anticipates) that 15 new employees will be hired this year.
19. The main office offered to (augment, supplement) the annual operating budget of the Boston office with an additional $100,000 in funds.
20. It was (all together, altogether) too late to make changes in the proposal.
21. The arbitrator made sure that both parties (agreed to, agreed with) the terms and conditions of the contract before it was submitted to the board.
22. Option 1 calls for complete removal of the asbestos. (Alternately, Alternatively), Option 2 would only require that the asbestos material be thoroughly covered.
23. They had not considered the (amount, number) of cement blocks needed for the new addition.
24. (Due to, Because of) the change in weather, they had to reschedule the trip to the project site.
25. The health inspector found (too, to) many violations in that room (to, too).
26. They claimed that the old equipment (used, utilized) too much fuel.
27. Gone are the days when a major construction job gets started with a handshake and (a verbal, an oral) agreement.
28. The complex project has 18 (discreet, discrete) phases; each part deals with confidential information that must be handled (discretely, discreetly).
29. He was (definitive, definite) about the fact that he would not be able to complete the proposal by next Tuesday.
30. He usually received (complementary, complimentary) samples from his main suppliers.
31. To (lose, loose) a client for whom they had worked so hard was devastating.
32. It was (fortunate, fortuitous) he was there at the exact moment the customer needed to order a year's worth of supplies, for the sales commission was huge.
33. Among all the information on the graph, he located the one (data, datum) that shows the price of tuna on the Seattle market at 5 P.M. on August 7.
34. Each (principle, principal) of the corporation was required to buy stock.
35. He returned to the office to (assure, ensure) that the safe was locked.

Index

DATE DUE

PLANNING FORM

NAME: _____   ASSIGNMENT: _____

I. Purpose: Answer each question in one or two sentences.

   A. Why are you writing this document? _____
_____

   B. What response do you want from readers? _____
_____

II. Reader Matrix: Fill in names and positions of people who may read the document.

	Decision-Makers	Advisors	Receivers
Managers	_____ _____ _____	_____ _____ _____	_____ _____ _____
Experts	_____ _____ _____	_____ _____ _____	_____ _____ _____
Operators	_____ _____ _____	_____ _____ _____	_____ _____ _____
General Readers	_____ _____ _____	_____ _____ _____	_____ _____ _____

III. Information on Individual Readers: Answer these questions about selected members of your audience. Attach additional sheets as is necessary.

   1. What is this reader's technical or educational background?
_____
_____

   2. What main question does this person need answered?
_____
_____

   3. What main action do you want this person to take?
_____
_____

   4. What features of this person's personality might affect his or her reading?
_____
_____

   5. What features does this person prefer in

      Format? _____

      Style? _____

      Organization? _____

IV. Outline: Attach an outline (topic) to use in drafting the BODY of this document.

970113